ASTRONOMY AND
ASTROPHYSICS LIBRARY

W0232125

Springer
New York
Berlin
Heidelberg
Hong Kong
London
Milan
Paris
Tokyo

ASTRONOMY AND
ASTROPHYSICS LIBRARY

Continued after Index

A.K. Raychaudhuri   S. Banerji
A. Banerjee

General Relativity, Astrophysics, and Cosmology

With 44 Illustrations

Springer

A.K. Raychaudhuri
Physics Department
Presidency College
Calcutta 700073
India

S. Banerji
Burdwan University
Burdwan
West Bengal
India

A. Banerjee
Physics Department
Jadavpur University
Calcutta 32
India

Series Editors

Immo Appenzeller
Landessternwarte, Königstuhl
D-69117 Heidelberg
Germany

Gerhard Börner
MPI für Physik und Astrophysik
Institut für Astrophysik
Karl-Schwarzschild-Str. 1
D-85748 Garching
Germany

Michael A. Dopita
The Australian National University
    Institute of Advanced Studies
Research School of Astronomy
    and Astrophysics
Cotter Road, Weston Creek
Mount Stromlo Observatory
Canberra, ACT 2611
Australia

Martin Harwit
Department of Astronomy
Space Sciences Building
Cornell University
Ithaca, NY 14853-6801
USA

Rudolf Kippenhahn
Rautenbreite 2
D-37077 Göttingen
Germany

James Lequeux
Observatoire de Paris
61, Avenue de l'Observatoire
75014 Paris
France

André Maeder
Observatoire de Genève
CH-1290 Sauverny
Switzerland

Peter A. Strittmatter
Steward Observatory
The University of Arizona
Tuscon, AZ 85721
USA

Virginia Trimble
Astronomy Program
University of Maryland
College Park, MD 20742
and Department of Physics
University of California
Irvine, CA 92717
USA

Library of Congress Cataloging-in-Publication Data
Raychaudhuri, A.K.
    General relativity, astrophysics, and cosmology/
A.K. Raychaudhuri, S. Banerji, A. Banerjee.
        p.   cm.—(Astronomy and astrophysics library)
    Includes bibliographical references and index.
    1. Astrophysics. 2. General relavitiy (Physics) 3. Cosmology.
I. Banerji, Sriranjan, 1938– . II. Banerjee, Asit, 1940–
III. Title. IV. Series.
QB461.R28 1992
523.01—dc20

92-2780
CIP

ISBN-13:978-0-387-40628-2   e-ISBN-13:978-1-4612-2754-0
DOI: 10.1007/978-1-4612-2754-0

First softcover printing, 2003.

9 8 7 6 5 4 3 2 1   SPIN 10947661

www.springer-ny.com

Springer-Verlag   New York Berlin Heidelberg
A member of BertelsmannSpringer Science+Business Media GmbH

Preface

For about half a century the general theory of relativity attracted little attention from physicists. However, the discovery of compact objects such as quasars and pulsars, as well as candidates for black holes on the one hand, and the microwave background radiation on the other hand completely changed the picture. In addition, developments in elementary particle physics, such as predictions of the behavior of matter at the ultrahigh energies that might have prevailed in the early stages of the big bang, have greatly enhanced the interest in general relativity.

These developments created a large body of readers interested in general relativity, and its applications in astrophysics and cosmology. Having neither the time nor the inclination to delve deeply into the technical literature, such readers need a general introduction to the subject before exploring applications. It is for these readers that the present volume is intended. Keeping in mind the broad range of interests and wanting to avoid mathematical complications as much as possible, we have ventured to combine all three topics—relativity, astrophysics, and cosmology—in a single volume. Naturally, we had to make a careful selection of topics to be discussed in order to keep the book to a manageable length.

However, we expect that even the prospective researchers in these fields will have something to gain from this book. They will get an introduction to the subject that will enable them to consult more detailed treatises as well as the current literature. To facilitate these we have added an appendix on differential forms which, although presented at an elementary level, will enable the reader to use this language without much difficulty. The bibliography and the references, though not exhaustive, will also be of help for further studies.

We have borrowed liberally from published books and literature—we claim very little of this book to be strictly original. We are grateful to various authors and publishers for giving permission to reproduce some figures, and our indebtedness is duly acknowledged at the places where the figures appear.

We have omitted any discussion of quantum gravity and quantum cosmology. Our feeling in this matter is that these disciplines have not yet reached the stage where they can find a place in an elementary book such as this.

Our thanks are due to the referees of Springer-Verlag for their very constructive criticisms and suggestions. Thanks are due to the Indian National Science Academy and the Jadavpur University for financial support. Mention should be made of the cooperation and help from the S.N. Bose Centre of Basic Sciences and especially to the director Professor C.K. Majumdar.

Contents

Part I
The General Theory
of Relativity

1. Introduction

1.1. The Case for Nonflat Space–Time

Newtonian mechanics singles out a set of frames in uniform relative motion with a specially simple property, namely, in these frames Newton's first law of motion holds good. These frames are called inertial frames. The relation between the coordinates of an event in one inertial frame to those in another inertial frame were taken to be given by the Galilean transformation formulas; these are (with rectangular Cartesian coordinates)

$$x' = x - vt, \qquad y' = y, \qquad z' = z, \qquad t' = t. \tag{1.1}$$

Here v stands for the relative velocity between the two frames, assumed to be in the direction of the x-axis. These formulas are based on the ideas of an absolute time and an invariance of spatial distances. Going over to infinitesimals, we may write

$$dr^2 \equiv dx^2 + dy^2 + dz^2 = dx'^2 + dy'^2 + dz'^2 \equiv dr'^2,$$

$$dt = dt'. \tag{1.2}$$

In special relativity, the corresponding transformation is given by the Lorentz formulas

$$x' = \frac{x - vt}{\sqrt{1 - v^2/c^2}}, \qquad y' = y, \qquad z' = z, \tag{1.3}$$

$$t' = \frac{t - vx/c^2}{1 - v^2/c^2}.$$

This leads to the invariance of the "space–time interval" ds given by

$$ds^2 = c^2 \, dt^2 - dx^2 - dy^2 - dz^2. \tag{1.4}$$

There is a major departure in (1.4) from the physics of Galileo–Newton in that dt and dr are no longer separately invariants, as in (1.2). The relation (1.2) is an assertion of the existence of a three-dimensional space, which is Euclidean in character, and an invariant universal time. There is nothing to distinguish between 3-space directions indicating spatial isotropy. However, (1.3) and (1.4) indicate the following changes:

(i) there is no longer the independence between time t and the space coordinates; and

(ii) the geometry in (1.4) may be called pseudo-Euclidean.

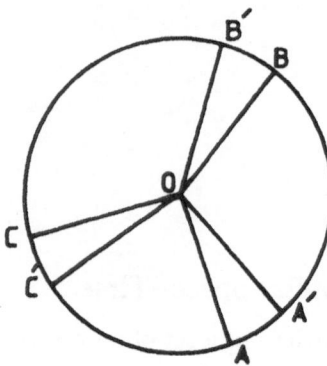

Figure 1.1

It is Euclidean because the right-hand side of (1.4) contains only terms involving squares of coordinate differentials with constant coefficients—what in three dimensions is a consequence of the Pythagoras theorem (a basis of Euclidean geometry). The qualifying prefix pseudo is to take note of the fact that all coefficients are not of the same sign. All this is, however, about inertial frames. Would the invariant (1.4) retain its form or, in other words, would Euclidean geometry continue to hold when we go over to accelerated frames? The answer is easily seen to be negative if we consider the case of a uniformly rotating disk.

A circular disk ABC is rotating in its own plane about an axis passing through its center θ (Fig 1.1). Any element of a radius is perpendicular to the instantaneous direction of the velocity of the element, hence its length will remain unaltered by a Lorentz transformation. On the other hand, each element of the circumference has a velocity parallel to its length, and looked at from the frame at rest appears to be contracted due to Lorentz contraction. Thus, the familiar result of Euclidean geometry, namely, the ratio

$$\frac{\text{the length of the circumference of the circle}}{\text{the length of the radius of the circle}} = 2\pi,$$

would no longer be valid. In other words, the line element (1.4) would no longer be valid in such cases. Similar results may be shown to hold in cases of other types of acceleration.

1.2. The Principle of Equivalence

The second difficulty with (1.4) can be understood by the consideration of what is known as the principle of equivalence. Galileo discovered a fundamental feature of gravity, namely, that in a gravitational field all bodies experience the same acceleration. The equation of motion in a gravitational field is

$$m_i \frac{d^2 \mathbf{r}}{dt^2} = -m_g \nabla \psi, \tag{1.5}$$

where ψ is the gravitational potential, m_i is the inertial mass of the body, and m_g its gravitational mass; m_i and m_g are, in fact, the measure of a body's "resistance" to the action of the force and its capability to respond to a gravitational field, respectively. So

$$\frac{d^2 \mathbf{r}}{dt^2} = -\frac{m_g}{m_i} \nabla \psi. \tag{1.6}$$

It is found observationally that the acceleration of all particles in the same gravitational field is the same, which demands that m_g is proportional to m_i. The proportionality constant may be reduced to unity by a suitable choice of units. Thus, we have $m_g = m_i$. This equality of gravitational and inertial mass is referred to as the Eötvös law. Eötvös et al. (1922) experimentally concluded that the equality is true to one part in 10^9. Dicke and his collaborators (Dicke, 1964) repeated the Eötvös experiment with greatly refined aparatus and could reduce the uncertainty to less than one part in 10^{11}. A further reduction in the uncertainty, to less than 10^{-12}, is due to Braginskii and Panov (1972).

It is obvious that in view of the above situation, there exists in any static homogeneous gravitational field a frame of reference in which the gravitational field is absent. Indeed, the equation of motion of a particle is now, in view of the equality of the inertial mass and the gravitational mass,

$$\frac{d^2 \mathbf{r}}{dt^2} = \mathbf{g}, \tag{1.7}$$

where $-\nabla \psi \equiv \mathbf{g}$ is independent of space and time coordinates, reflecting the homogeneous and static character of the gravitational field. If we make the transformation

$$\mathbf{r}' = \mathbf{r} - \tfrac{1}{2}\mathbf{g}t^2,$$

the equation of motion reduces to

$$-\frac{d^2 \mathbf{r}'}{dt^2} = 0, \tag{1.8}$$

The gravitational field is seen to vanish in the primed coordinates. The primed frame is actually a frame moving with the same acceleration $\mathbf{g}$ with respect to the "unprimed frame." As viewed by an observer falling with the same acceleration as the body and in the same direction, the body appears to be at rest or moving with a uniform velocity. Thus the action of a gravitational field on a body is indistinguishable from the effect of an acceleration with respect to an inertial frame. Due to this equality, the dynamical observations as made:

(a) in the presence of a uniform gravitational field; and
(b) in a uniformly accelerated frame;

are of a similar nature, in both cases all freely falling particles are found to move with the same uniform acceleration. Einstein raised this fact to the status of a principle; namely, that all physical phenomena (not only mechanical) would appear the same in a uniform gravitational field and in an accelerated frame. This was named the equivalence principle.

An Illustration: Einstein's Elevator

To illustrate the principle of equivalence Einstein gives the example of a man in an elevator. As long as the elevator is at rest the man can determine the strength of the gravitational field on the surface of the Earth by noting the acceleration of a freely falling particle. If he has no means of getting information from outside he might as well argue that his elevator is accelerating upwards with respect to the body at rest till it meets the floor. Imagine now that the elevator is broken from the cable and falls freely in the gravitational field of the Earth. During the fall the bodies inside the elevator undergo the same acceleration as the elevator itself and, as a result, they remain unaccelerated relative to the elevator. The observer inside the elevator might interpret that the bodies are in an inertial frame and that there is no gravitational field.

How can we then distinguish between an accelerated frame and a true gravitational field? The gravitational force, however, in a realistic situation is not a fictitious force and cannot be eliminated at all places by simply choosing a noninertial frame. This is because of the nonuniformity of any real gravitational field, and it is then impossible to introduce a frame of reference which cancels the effect of gravitation at all points. Moreover, for fields due to bounded distributions, the gravitational field vanishes at a large distance from the source, which, however, is not the case with accelerated frames. Thus, we can introduce a Lorentzian coordinate system locally in a small region of space only.

1.3. Conflict Between the Equivalence Principle and the Pseudo-Euclidean Metric: Gravitational Redshift

Consider the source B at the bottom and receiver A at the top of a system at rest in a uniform gravitational field **g**. To an observer falling freely with acceleration g, both of them will apparently accelerate upwards. From the principle of equivalence, the gravitational field may be replaced by the accelerated frame. We can now calculate the redshift of the signal sent by the

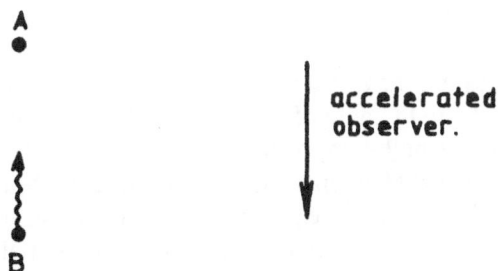

Figure 1.2

source B and received by A (Fig. 1.2). Let their separation be h and a light signal is sent in the direction of acceleration at $t = 0$ from B. This signal reaches A at time t, so that

$$ct = h + \tfrac{1}{2}gt^2 \tag{1.9}$$

because, by the time the signal reaches A, A has shifted from its original location by $\tfrac{1}{2}gt^2$. Another signal emitted at a later time Δt_B will reach A at time $(t + \Delta t_A)$. Therefore

$$c(t + \Delta t_A - \Delta t_B) = h + \tfrac{1}{2}g(t + \Delta t_A)^2 - \tfrac{1}{2}g(\Delta t_B)^2$$

$$= h + \tfrac{1}{2}g(t^2 + 2t \cdot \Delta t_A), \tag{1.10}$$

where we have neglected squares of Δt_A and Δt_B being small quantities of a higher order. From (1.9) and (1.10) we have

$$c(\Delta t_A - \Delta t_B) = gt \cdot \Delta t_A,$$

or,

$$\Delta t_A = \Delta t_B \frac{c}{c - gt} = \frac{\Delta t_B}{(1 - gh/c^2)}. \tag{1.11}$$

The final expression in (1.11) has been obtained by using (1.9) and considering $gt \ll c$. If λ_B and λ_A are the wavelengths at emission and reception, respectively, we get, using the invariance of the velocity of light,

$$\lambda_A = \frac{\lambda_B}{1 - gh/c^2}. \tag{1.12}$$

As $\lambda_A > \lambda_B$ there is a redshift according to the accelerated observer and from the equivalence principle there should be such a shift in the gravitational field. However, if we assume that even in a gravitational field the Minkowski metric holds, then the time for light to go from B to A will always be the same, as the conditions are static, and hence there would not be any spectral shift. Thus, either the equivalence principle enunciated by Einstein is wrong or the Minkowski metric needs changing in the presence of a gravitational field.

1.4. A Fifth Force

In recent years, doubts have been expressed about the correctness of the conclusion drawn from the data of Eötvös et al. It has been claimed that the Eötvös data show a small composition-dependent medium range force besides the inverse square Newtonian gravitational force (Stacey et al., 1987). This force may depend on B/μ, i.e., the baryon number per unit atomic mass. Thus Fischbach et al. (1986) suggest an interaction potential of the form

$$V(r) = -G_\infty \frac{mm'}{r}(1 + \alpha e^{-r/\lambda}), \tag{1.13}$$

where the constants α and λ specify the intensity and range of the fifth force. The constant α, supposed to depend on B/μ, is negative and has a value of the order 10^{-2} or 10^{-3}, while the range λ may be a few hundred meters. Observationally, such a potential will mimic a distance-dependent gravitational constant. Some support for such a hypothesis comes from recent geophysical determinations of the gravitational constant (see, e.g., Stacey and Tuck, 1981; Holding et al., 1986), which give a value of G a little higher than the value found from laboratory experiments. It may be noted that the results of Dicke et al. or Braginskii and Panov do not rule out a medium range force, as they studied the gravitational acceleration due to the Sun whose distance far exceeds the suggested range of the force. While some earlier experiments led to conflicting conclusions about this fifth force, more recent experiments apparently rule out the existence of such a force (Nelson et al., 1990; Kuroda and Mio, 1990).

2. Tensor Calculus and Riemannian Geometry

2.1. Riemannian Geometry and the Metric Tensor

The fundamental new idea introduced by Einstein in the general theory of relativity is that in the presence of gravitational fields, although (1.4) is no longer an invariant, there exists an invariant which is a homogeneous quadratic expression in the four coordinate differentials. We write this in the form

$$ds^2 = g_{\mu\nu} \, dx^\mu \, dx^\nu, \tag{2.1}$$

where a summation over the repeated indices μ and ν is to be understood. This convention of summation over any repeated index is known after Einstein and will be used throughout. An important thing to note is that the particular index which is repeated (called the dummy index) may be arbitrarily changed, i.e., $ds^2 = g_{\mu\nu} \, dx^\mu \, dx^\nu = g_{\alpha\beta} \, dx^\alpha \, dx^\beta$ or $A^\alpha B_\alpha = A^\sigma B_\sigma$, etc. The coefficients $g_{\mu\nu}$ are called metric tensor components and were assumed to represent the gravitational field. But before proceeding further, we discuss a little about tensors and Riemannian geometry; in this discussion we shall not restrict ourselves to four dimensions.

In an n-dimensional space, let x^α ($\alpha = 1, 2, \ldots, n$) represent a coordinate system. We may transform to another coordinate system x'^α (say) where we shall have functional relations of the form

$$x'^\mu = f^\mu(x^\alpha). \tag{2.2}$$

The condition that x'^μ's are independent demands that the Jacobian of the transformation does not vanish, i.e.,

$$J \equiv \begin{vmatrix} \dfrac{\partial x'^1}{\partial x^1} & \dfrac{\partial x'^1}{\partial x^2} & \cdots & \dfrac{\partial x'^1}{\partial x^n} \\ \cdots & \cdots & \cdots & \cdots \\ \dfrac{\partial x'^n}{\partial x^1} & \dfrac{\partial x'^n}{\partial x^2} & \cdots & \dfrac{\partial x'^n}{\partial x^n} \end{vmatrix} \neq 0. \tag{2.3}$$

The coordinate differentials transform in the following manner:

$$dx^\alpha = \frac{\partial x^\alpha}{\partial x'^\beta} \, dx'^\beta. \tag{2.4}$$

(Note that, although we have introduced the possibility of general nonlinear transformations, the coordinate differentials transform linearly.)

2.2. Vectors and Tensors

We begin by discussing the transformation property of some common physical entities. For simplicity, we shall consider only a transformation of special coordinates. The simplest objects are the scalars which do not undergo any change in a transformation of coordinates. Familiar examples are the mass of a body, volume, temperature, and pressure. The magnitude of the velocity is also a scalar but if we take into account the direction of the velocity, then the velocity is defined by its three components dx/dt, dy/dt, and dz/dt. After a transformation to another system of rectangular Cartesian coordinates, the new velocity components are

$$\frac{d\bar{x}}{dt} = \frac{\partial \bar{x}}{\partial x} \cdot \frac{dx}{dt} + \frac{\partial \bar{x}}{\partial y} \cdot \frac{dy}{dt} + \frac{\partial \bar{x}}{\partial z} \cdot \frac{dz}{dt},$$

$$\frac{d\bar{y}}{dt} = \frac{\partial \bar{y}}{\partial x} \cdot \frac{dx}{dt} + \frac{\partial \bar{y}}{\partial y} \cdot \frac{dy}{dt} + \frac{\partial \bar{y}}{\partial z} \cdot \frac{dz}{dt},$$

$$\frac{d\bar{z}}{dt} = \frac{\partial \bar{z}}{\partial x} \cdot \frac{dx}{dt} + \frac{\partial \bar{z}}{\partial y} \cdot \frac{dy}{dt} + \frac{\partial \bar{z}}{\partial z} \cdot \frac{dz}{dt}.$$

Or, in index notation, using the summation convention we have

$$\frac{d\bar{x}^i}{dt} = \frac{\partial \bar{x}^i}{\partial x^k} \cdot \frac{dx^k}{dt},$$

equivalently,

$$v^i = \frac{\partial x^i}{\partial x'^k} \cdot v'^k. \tag{2.5}$$

Obviously, (2.4) and (2.5) are of the same form. Now consider the gradient of a scalar ϕ. This, too, has three components and the transformation relation is

$$\frac{\partial \phi}{\partial x'^i} = \frac{\partial \phi}{\partial x^k} \cdot \frac{\partial x^k}{\partial x'^i}. \tag{2.6}$$

For transformations from one rectangular Cartesian system to another $\partial x'^i/\partial x^k = \partial x^k/\partial x'^i$ (both being simply the cosine of the angle between the ith primed axis and the kth unprimed axis), and hence the three transformations (2.4), (2.5), and (2.6) are of identical form. However, if we consider a transformation to a spherical polar coordinate system, then (2.5) is replaced by

$$v^r = \frac{\partial r}{\partial x} \cdot \frac{dx}{dt} + \frac{\partial r}{\partial y} \cdot \frac{dy}{dt} + \frac{\partial r}{\partial z} \cdot \frac{dz}{dt},$$

$$v^\theta = \frac{\partial \theta}{\partial x} \cdot \frac{dx}{dt} + \frac{\partial \theta}{\partial y} \cdot \frac{dy}{dt} + \frac{\partial \theta}{\partial z} \cdot \frac{dz}{dt}, \tag{2.5'}$$

$$v^\phi = \frac{\partial \phi}{\partial x} \cdot \frac{dx}{dt} + \frac{\partial \phi}{\partial y} \cdot \frac{dy}{dt} + \frac{\partial \phi}{\partial z} \cdot \frac{dz}{dt},$$

$$dr = \frac{\partial r}{\partial x} dx + \frac{\partial r}{\partial y} dy + \frac{\partial r}{\partial z} dz, \quad \text{etc.} \tag{2.4'}$$

and

$$\frac{\partial \phi}{\partial r} = \frac{\partial \phi}{\partial x} \cdot \frac{\partial x}{\partial r} + \frac{\partial \phi}{\partial y} \cdot \frac{\partial y}{\partial r} + \frac{\partial \phi}{\partial z} \cdot \frac{\partial z}{\partial r}. \tag{2.6'}$$

Obviously, while (2.5′) and (2.4′) have the same transformation coefficients, it is not so with (2.6′) for

$$\left(\frac{\partial r}{\partial x}\right)_{y,z} \neq \left(\frac{\partial x}{\partial r}\right)_{\theta,\phi}, \qquad \left(\frac{\partial \theta}{\partial x}\right)_{y,z} \neq \left(\frac{\partial x}{\partial \theta}\right)_{r,\phi}, \qquad \text{etc.}$$

To distinguish between these two types of objects we introduce the following definitions; considering the general case of n-dimensional space:

(a) a set of n quantities A^α ($\alpha = 1, 2, 3 \ldots, n$) which transforms like the coordinate differentials are said to constitute a contravariant vector

$$\bar{A}^\beta = \frac{\partial \bar{x}^\beta}{\partial x^\alpha} A^\alpha; \tag{2.7}$$

(b) a set of n quantities A_α ($\alpha = 1, 2, 3 \ldots, n$) which transforms like the gradients of a scalar are said to constitute a covariant vector

$$\bar{A}_\beta = A_\alpha \frac{\partial x^\alpha}{\partial \bar{x}^\beta}. \tag{2.8}$$

Note that the equation of motion

$$m\ddot{\mathbf{r}} = -\nabla V$$

of classical mechanics is valid in rectangular Cartesian coordinates, but not so in general. The reason is that while the left-hand side is a contravariant vector, the right-hand side is a covariant vector. However, in Lagrange's equation,

$$\frac{d}{dt}\left(\frac{\partial L}{\partial \dot{q}_k}\right) - \frac{\partial L}{\partial q_k} = 0,$$

both the terms are covariant vectors in the q space, and the equation holds quite generally for any transformation of the type $Q_i = f_i(q_k)$. (That the equation holds good for the more general transformation $Q_i = f_i(q_k, t)$ requires some additional consideration.)

Vector Algebra

The sum of two vectors A^α and B^β is defined to be C^γ where

$$C^\gamma = A^\gamma + B^\gamma, \tag{2.8a}$$

and is itself a vector. Similarly, D^γ, which is the difference of the two vectors, is defined as

$$D^\gamma = A^\gamma - B^\gamma. \tag{2.8b}$$

Obviously, D^γ is also a vector.

A product of the components of two vectors would give rise to n^2 elements. To see their transformation property, note that

$$A^\alpha B^\beta = A'^\mu \frac{\partial x^\alpha}{\partial x'^\mu}, \qquad B'^\nu \frac{\partial x^\beta}{\partial x'^\nu} = A'^\mu B'^\nu \frac{\partial x^\alpha}{\partial x'^\mu} \frac{\partial x^\beta}{\partial x'^\nu}, \qquad (2.9)$$

writing $A^\alpha B^\beta = C^{\alpha\beta}$ we have

$$C^{\alpha\beta} = C'^{\mu\nu} \frac{\partial x^\alpha}{\partial x'^\mu} \frac{\partial x^\beta}{\partial x'^\nu}. \qquad (2.10)$$

Definition. A set of n^2 quantities, which in a coordinate transformation transforms according to formula (2.10), constitutes a contravariant tensor of rank 2. In general, a contravariant tensor of rank p in n-dimensional space is an object with n^p components, the transformation law for such a tensor being

$$S^{\alpha_1\alpha_2\cdots\alpha_p} = S'^{\mu_1\mu_2\cdots\mu_p} \frac{\partial x^{\alpha_1}}{\partial x'^{\mu_1}} \cdots \frac{\partial x^{\alpha_p}}{\partial x'^{\mu_p}}. \qquad (2.11)$$

In a similar manner the covariant tensor of rank 2, $S_{\alpha\beta}$, obeys the transformation rule

$$S_{\alpha\beta} = S'_{\mu\nu} \frac{\partial x'^\mu}{\partial x^\alpha} \frac{\partial x'^\nu}{\partial x^\beta}, \qquad (2.12)$$

while for a mixed tensor S^α_β,

$$S^\alpha_\beta = S'^\mu_\nu \frac{\partial x^\alpha}{\partial x'^\mu} \frac{\partial x'^\nu}{\partial x^\beta}, \qquad (2.13)$$

covariant and mixed tensors of higher rank can be defined in a similar manner. Thus a vector is just a tensor of rank 1 with a single index, and a scalar is a tensor of rank 0 with no index at all.

Contraction and Reduction of the Rank of Tensors

A tensor of rank 2 (contravariant, covariant, or mixed) can be written out in matrix form, e.g., for the tensor C^α_β,

$$\begin{vmatrix} C_1^1 & C_2^1 & \cdots & C_n^1 \\ C_1^2 & C_2^2 & \cdots & C_n^2 \\ \cdots\cdots\cdots\cdots\cdots\cdots \\ C_1^n & C_2^n & \cdots & C_n^n \end{vmatrix}.$$

The reader may recall some objects which obey the tensor transformation law. Thus in the dynamics of rotating rigid bodies, we get a set of six quantities, I_{ik}, defined by

$$I^{ik} = \int \rho(x^i x^k - r^2 \delta^{ik}) \, dv,$$

where the x^i's are the rectangular Cartesian coordinates of the element of volume dv, ρ is the mass density, and $r^2 = x^2 + y^2 + z^2$. In an orthogonal transformation of the space coordinates, I^{ik} will then transform to I'^{lm} such that

$$I^{ik} = I'^{lm} \frac{\partial x^i}{\partial x'^l} \frac{\partial x^k}{\partial x'^m}$$

as ρ and dv are scalars. (Note that the Kronecker δ^{ik} obeys the transformation law (2.10); we need not distinguish between covariant and contravariant indices in the present case as we are restricting ourselves to rectangular Cartesian coordinates.) Thus the I^{ik}'s form a symmetric tensor of rank 2. Another familiar tensor of rank 2 is the stress tensor appearing in the theory of elasticity. The reader may easily check their tensor character.

If we consider the trace of the above matrix, we find that after a transformation

$$\sum_\alpha C'^\alpha_\alpha = \sum_\alpha C^\beta_\gamma \frac{\partial x'^\alpha}{\partial x^\beta} \frac{\partial x^\gamma}{\partial x'^\alpha} = \sum C^\beta_\gamma \delta^\gamma_\beta = \sum_\beta C^\beta_\beta.$$

Thus the trace is a scalar. This process of putting a contravariant and co-variant index identical in a tensor expression, and adding over all values of that index is called contraction. A contraction, as we shall see, lowers the rank of a tensor by two. In particular, if the tensor C^α_β is the product of two vectors A^α and B_β, then the trace $C^\alpha_\alpha = A^\alpha B_\alpha$, and this scalar is called the scalar product of the two vectors A^α and B_α. The positive square root of $|A^\alpha A_\alpha|$ is called the norm (or magnitude) of the vector A^α.

Two vectors are said to be orthogonal if their scalar product vanishes and, in general, the angle θ between the two vectors A^α and B_β is defined as

$$\cos \theta = \frac{A^\alpha B_\alpha}{|A^\mu A_\mu|^{1/2} |B^\nu B_\nu|^{1/2}}.$$

We shall call a vector A^α timelike, spacelike, or null according to whether $A^\alpha A_\alpha$ is greater than, less than, or equal to zero. A null vector is also some-times called lightlike. Note that the definition of angle fails if both the vectors are not spacelike. Nevertheless, any two vectors are said to be orthogonal if $A^\alpha B_\alpha = 0$; in particular, we say that a null vector is self-orthogonal.

Symmetric and Antisymmetric Tensors

A tensor $C_{\alpha\beta}$ is said to be symmetric if $C_{\alpha\beta} = C_{\beta\alpha}$, and anti- (or skew-) symmetric if $C_{\alpha\beta} = -C_{\beta\alpha}$. However, the idea is not limited to second rank tensors only. A tensor of arbitrary rank $C_{\alpha\beta\gamma\delta\varepsilon}$ will be called symmetric (or antisymmetric) with respect to, say, the first and fourth indices if

$$C_{\alpha\beta\gamma\delta\varepsilon} = \pm C_{\delta\beta\gamma\alpha\varepsilon},$$

the upper (lower) sign corresponding to the symmetric (antisymmetric) case.

It is important to notice that it is a coordinate independent property. For

$$C'_{\alpha\beta} = C_{\sigma\mu} \frac{\partial x^\sigma}{\partial x'^\alpha} \frac{\partial x^\mu}{\partial x'^\beta}$$

$$= \pm C_{\mu\sigma} \frac{\partial x^\sigma}{\partial x'^\alpha} \frac{\partial x^\mu}{\partial x'^\beta} \qquad (\text{+ for symmetric})$$

$$= \pm C'_{\beta\alpha}. \qquad (\text{– for antisymmetric}).$$

Similar definitions hold for contravariant tensors as well. In general, any tensor of rank 2 can be broken up into its symmetric and antisymmetric parts, which are themselves tensors. Thus writing

$$C_{(\alpha\beta)} \equiv \tfrac{1}{2}(C_{\alpha\beta} + C_{\beta\alpha})$$

and

$$C_{[\alpha\beta]} \equiv \tfrac{1}{2}(C_{\alpha\beta} - C_{\beta\alpha}),$$

we see that $C_{(\alpha\beta)}$ and $C_{[\alpha\beta]}$ are both tensors with characteristic symmetry properties.

Raising and Lowering of Indices

That $g_{\mu\nu} dx^\mu dx^\nu$ is an invariant determines the transformation rule for $g_{\mu\nu}$:

$$g_{\mu\nu} dx^\mu dx^\nu = g'_{\alpha\beta} dx'^\alpha dx'^\beta,$$

or

$$g_{\mu\nu} = g'_{\alpha\beta} \frac{\partial x'^\alpha}{\partial x^\mu} \frac{\partial x'^\beta}{\partial x^\nu}. \qquad (2.14)$$

Thus it is clear that $g_{\mu\nu}$ is a covariant tensor of rank 2. Its reciprocal $g^{\mu\nu}$ is defined by

$$g^{\mu\nu} g_{\nu\sigma} = \delta^\mu_\sigma, \qquad (2.15)$$

where

$$\delta^\mu_\sigma = \begin{cases} 0 & \text{if } \mu \neq \sigma, \\ 1 & \text{if } \mu = \sigma, \end{cases}$$

where δ^μ_σ is called the Krönecker delta. Note that $g^{\mu\nu}$ may alternatively be defined as the minor of $g_{\mu\nu}$ divided by the determinant $\|g_{\mu\nu}\|$. We now proceed to show that the quantities $g^{\mu\nu}$ transform as a contravariant tensor of rank 2. We note

$$g'_{\alpha\beta} g^{\mu\nu} \frac{\partial x'^\alpha}{\partial x^\mu} \frac{\partial x'^\lambda}{\partial x^\nu} = g_{\rho\sigma} \frac{\partial x^\rho}{\partial x'^\alpha} \frac{\partial x^\sigma}{\partial x'^\beta} g^{\mu\nu} \frac{\partial x'^\alpha}{\partial x^\mu} \frac{\partial x'^\lambda}{\partial x^\nu}$$

$$= g_{\rho\sigma} g^{\mu\nu} \delta^\rho_\mu \frac{\partial x^\sigma}{\partial x'^\beta} \frac{\partial x'^\lambda}{\partial x^\nu}$$

$$= g_{\mu\sigma} g^{\mu\nu} \frac{\partial x^\sigma}{\partial x'^\beta} \frac{\partial x'^\lambda}{\partial x^\nu} = \delta^\lambda_\beta, \qquad (2.16)$$

where we have used (2.14) and (2.15).

Now contracting (2.16) with $g'^{\beta\sigma}$ we have

$$g'_{\alpha\beta} g'^{\beta\sigma} g^{\mu\nu} \frac{\partial x'^{\alpha}}{\partial x^{\mu}} \frac{\partial x'^{\lambda}}{\partial x^{\nu}} = \delta^{\lambda}_{\beta} g'^{\beta\sigma},$$

or

$$g^{\mu\nu} \frac{\partial x'^{\sigma}}{\partial x^{\mu}} \frac{\partial x'^{\lambda}}{\partial x^{\nu}} = g'^{\lambda\sigma}, \tag{2.17}$$

which shows the tensor character of $g^{\mu\nu}$. The Kronecker δ^{μ}_{ν} is a mixed tensor because

$$\delta^{\mu}_{\nu} \frac{\partial x'^{\rho}}{\partial x^{\mu}} \frac{\partial x^{\nu}}{\partial x'^{\sigma}} = \frac{\partial x'^{\rho}}{\partial x^{\mu}} \frac{\partial x^{\mu}}{\partial x'^{\sigma}} = \delta'^{\rho}_{\sigma}. \tag{2.18}$$

The metric tensor and its reciprocal are used to lower or raise tensor and vector indices. Thus, we say that A^{α} and A_{α} represent the same vector in contra- and covariant form if

$$A^{\alpha} = g^{\alpha\beta} A_{\beta} \qquad \text{or} \qquad A_{\beta} = g_{\alpha\beta} A^{\alpha}. \tag{2.19}$$

The above procedure of raising or lowering indices can be extended to a tensor of arbitrary rank so that

$$S^{\mu\nu} = S_{\alpha\beta} g^{\mu\alpha} g^{\nu\beta}; \qquad S^{\alpha}_{\beta\gamma} = g^{\alpha\sigma} S_{\sigma\beta\gamma}. \tag{2.20}$$

Tensor Densities

Reading the transformation relation as a determinantal equation

$$g_{\mu\nu} = g'_{\alpha\beta} \frac{\partial x'^{\alpha}}{\partial x^{\mu}} \frac{\partial x'^{\beta}}{\partial x^{\nu}},$$

we get

$$g \equiv \det |g_{\mu\nu}| = \det \left| g'_{\alpha\beta} \frac{\partial x'^{\alpha}}{\partial x^{\mu}} \frac{\partial x'^{\beta}}{\partial x^{\nu}} \right| = J^2 \det|g'_{\alpha\beta}| = J^2 g',$$

that is,

$$g' = J^{-2} g = J'^2 g, \tag{2.21}$$

where J is the Jacobian of the transformation $x \to x'$ and J' is that for the reciprocal transformation, i.e.,

$$J' = \det \left| \frac{\partial x^{i}}{\partial x'^{k}} \right|. \tag{2.22}$$

A quantity like g, which transforms as a scalar except for some extra factors of the Jacobian, is called a scalar density and, similarly, a quantity that transforms like a tensor except for the factor of the Jacobian determinant is called the tensor density. The power (exponent) of the Jacobian J' involved is called the weight of the density. Thus, g is a scalar density of weight $+2$.

The Levi-Civita Tensor

Consider, in a four-dimensional space, the four index object defined by

$$\varepsilon^{\mu\nu\rho\sigma} = \begin{cases} 0 & \text{if any two of the indices are equal;} \\ +1 & \text{if } \mu\nu\rho\sigma \text{ can be brought to the order 0 1 2 3} \\ & \quad \text{by an even number of interchanges;} \\ -1 & \text{if } \mu\nu\rho\sigma \text{ is brought to the order 0 1 2 3} \\ & \quad \text{by an odd number of interchanges.} \end{cases} \tag{2.23}$$

Obviously, this definition implies a complete antisymmetry of $\varepsilon^{\mu\nu\rho\sigma}$, the interchange of any pair of indices changes the sign without changing the magnitude. Further, although a four index symbol possesses 256 elements, in the present case $\varepsilon^{\mu\nu\rho\sigma}$ has only one independent element.

What is the transformation property of $\varepsilon^{\mu\nu\rho\sigma}$? We recall the rule for the evaluation of determinant and apply it to the case of the Jacobian of transformation $x \rightarrow x'$. We have

$$\varepsilon^{\mu\nu\rho\sigma} \frac{\partial x'^{\alpha}}{\partial x^{\mu}} \frac{\partial x'^{\beta}}{\partial x^{\nu}} \frac{\partial x'^{\gamma}}{\partial x^{\rho}} \frac{\partial x'^{\delta}}{\partial x^{\sigma}} = \varepsilon^{\alpha\beta\gamma\delta} J.$$

However, we have already noted that $J = \sqrt{|g|}/\sqrt{|g'|}$. Hence

$$\sqrt{|g|}\,\varepsilon^{\mu\nu\rho\sigma} \frac{\partial x'^{\alpha}}{\partial x^{\mu}} \frac{\partial x'^{\beta}}{\partial x^{\nu}} \frac{\partial x'^{\gamma}}{\partial x^{\rho}} \frac{\partial x'^{\delta}}{\partial x^{\sigma}} = \sqrt{|g'|}\,\varepsilon^{\alpha\beta\gamma\delta}. \tag{2.24}$$

Thus $\varepsilon^{\mu\nu\rho\sigma}|g|^{-1/2}$ is a tensor of rank 4. This tensor is referred to as the Levi-Civita tensor and is be written $\eta^{\mu\nu\rho\sigma}$, and $\varepsilon^{\mu\nu\rho\sigma}$ is the corresponding tensor density of weight $+1$. The covariant tensor is $(|g|)^{1/2}\varepsilon_{\mu\nu\rho\sigma}$. A generalization to n dimensions is obvious.

2.3. Invariant Volume and Volume Integral

Consider the determinant formed by four infinitesimal vectors $A^{\alpha}_{(1)}$, $A^{\alpha}_{(2)}$, $A^{\alpha}_{(3)}$, and $A^{\alpha}_{(4)}$:

$$\begin{vmatrix} A^{0}_{(1)} & A^{1}_{(1)} & A^{2}_{(1)} & A^{3}_{(1)} \\ A^{0}_{(2)} & A^{1}_{(2)} & A^{2}_{(2)} & A^{3}_{(2)} \\ A^{0}_{(3)} & A^{1}_{(3)} & A^{2}_{(3)} & A^{3}_{(3)} \\ A^{0}_{(4)} & A^{1}_{(4)} & A^{2}_{(4)} & A^{3}_{(4)} \end{vmatrix}. \tag{2.25}$$

Calling dV as the value of the above determinant we have, in view of the transformation formula,

$$A^{\alpha} = A'^{\beta} \frac{\partial x^{\alpha}}{\partial x'^{\beta}}$$

and, using (2.21),

$$dV\sqrt{|g|} = dV'\sqrt{|g'|}. \tag{2.26}$$

Recalling the expression for volume in three-dimensional vector algebra, we may call dV the volume encompassed by the infinitesimal vectors and the above expression shows that $dV\sqrt{|g|}$ rather than dV is an invariant. We may build up a finite four-dimensional volume, by adding together such infinitesimal volumes encompassed by the infinitesimal displacement vectors, to obtain

$$\int \sqrt{|g|}\, dV \qquad \text{as an invariant,} \tag{2.27}$$

provided the boundary of the domain is invariantly specified. With this idea of invariant volume, we may go on to divide the domain into meshes of arbitrary form. It is usual to consider meshes formed by the coordinate differentials, i.e., $A_{(4)} = (dx^0, 0, 0, 0)$; $A_{(1)} = (0, dx^1, 0, 0)$; $A_{(2)} = (0, 0, dx^2, 0)$; $A_{(3)} = (0, 0, 0, dx^3)$. Then in the determinant only the diagonal elements survive and we get $dV = dx^0\, dx^1\, dx^2\, dx^3$. So that

$$\int \sqrt{|g|} \cdot dx^0\, dx^1\, dx^2\, dx^3 \qquad \text{is an invariant,}$$

provided the domain of integration is the same. It should, however, be noted that, in general,

$$\sqrt{|g|}\, dx^0\, dx^1\, dx^2\, dx^3 \neq \sqrt{|g'|}\, dx^0\, dx'^1\, dx'^2\, dx'^3.$$

Also, if ϕ is any scalar, then

$$\int \phi \sqrt{|g|}\, dx^0\, dx^1\, dx^2\, dx^3 \qquad \text{is an invariant,}$$

if the domain of integration is absolutely defined.

Note. As an example, note that in Euclidean 3-space, if we consider the rectangular Cartesian coordinates x, y, z and the spherical polar coordinates r, θ, ϕ; then $dx\, dy\, dz \neq r^2 \sin \theta\, dr\, d\theta\, d\phi$, but the volume integral

$$\int dx\, dy\, dz = \int r^2 \sin \theta\, dr\, d\theta\, d\phi$$

provided the boundary of the volume of integration is the same on either side.

2.4. Affine Connection—Parallel Transport

The idea of the parallel transport of a vector is in a sense more primitive than the idea of a metric, for even without a metric being specified (or independent of the metric tensor components), parallel displacement may be defined. Let us first of all consider a Euclidean space and a rectangular Cartesian coordinate system. Then if the vector A^α at the point x^μ is transported parallelly to $x^\mu + dx^\mu$ the change in the vector component dA^μ vanishes. However, this

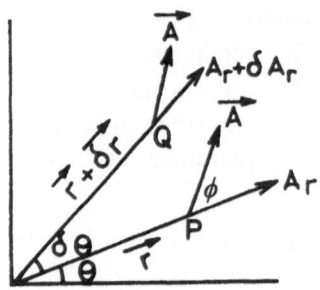

Figure 2.1

simple result does not hold good, even in Euclidean space, if we change to more general coordinate systems. We may, for example, consider the spherical polar coordinate system. In this case, obviously, for parallel displacement, dA^r, dA^θ, and dA^ϕ do not vanish.

Figure 2.1 illustrates that in a two flat, in a parallel displacement with polar coordinates, dA^r and dA^θ do not vanish, $A^r = |A| \cos \phi$, $A^r + \delta A^r = |A| \cos(\phi - \delta\theta)$. Another way of indicating this difficulty is to state that dA^μ, in general, is not a vector; had it been so, then the vanishing of dA^μ in any coordinate system would have ensured its vanishing in any other. Indeed, differentiating

$$A^\mu = A'^\nu \frac{\partial x^\mu}{\partial x'^\nu},$$

$$dA^\mu = dA'^\nu \frac{\partial x^\mu}{\partial x'^\nu} + A'^\nu \frac{\partial^2 x^\mu}{\partial x'^\alpha \partial x'^\nu} dx'^\alpha.$$

(2.28)

The second term on the right vanishes only if the transformation is linear in which case, of course, dA^μ behaves as a vector. We are thus faced with the problem of finding a definition of parallel displacement. We are guided by the following considerations:

(i) if the displacement dx^μ is infinitesimal, dA^α should involve dx^μ linearly; and

(ii) dA^α should be linear in the vector components A^μ as, otherwise, two vectors linearly related will not remain so after parallel displacement (i.e., say $B^\mu = aA^\mu$ at x^μ, where a is a constant number, then after parallel displacement this relation between $B^\mu + dB^\mu$ and $A^\mu + dA^\mu$ will persist only if the linear relation obtained).

The above two conditions lead to

$$dA^\mu = -\Gamma^\mu_{\alpha\beta} A^\alpha \, dx^\beta,$$

(2.29)

where the negative sign is conventional and the Γ's are independent of the vector A^μ and the displacement dx^β. We have already pointed out that dA^μ is not a vector; hence the Γ's, called the affinities, or affine connections, are also not tensors. How do the Γ's transform in a coordinate transformation? Using

(2.28) in (2.29) we get

$$dA'^\nu \frac{\partial x^\mu}{\partial x'^\nu} + A'^\nu \frac{\partial x^\mu}{\partial x'^\alpha \partial x'^\nu} dx'^\alpha = -\Gamma^\mu_{\alpha\beta} A'^\nu \frac{\partial x^\alpha}{\partial x'^\nu} \frac{\partial x^\beta}{\partial x'^\sigma} dx'^\sigma,$$

or

$$dA'^\nu \frac{\partial x^\mu}{\partial x'^\nu} = -A'^\nu dx'^\alpha \left[\frac{\partial^2 x^\mu}{\partial X'^\alpha \partial x'^\nu} + \Gamma^\mu_{\sigma\beta} \frac{\partial x^\sigma}{\partial x'^\nu} \frac{\partial x^\beta}{\partial x'^\alpha} \right]. \qquad (2.30)$$

Again multiplying (2.30) by $\partial x'^\rho / \partial x^\mu$ and using the definition of Γ', i.e.,

$$dA'^\rho = -\Gamma'^\rho_{\nu\alpha} A'^\nu dx'^\alpha,$$

we get

$$\Gamma'^\rho_{\nu\alpha} = \Gamma^\mu_{\sigma\beta} \frac{\partial x^\sigma}{\partial x'^\nu} \frac{\partial x^\beta}{\partial x'^\alpha} \frac{\partial x'^\rho}{\partial x^\mu} + \frac{\partial^2 x^\mu}{\partial x'^\alpha \partial x'^\nu} \frac{\partial x'^\rho}{\partial x^\mu}, \qquad (2.31)$$

which gives the desired transformation rule. Again we note that the Γ's, although not a tensor in general, behave as a tensor for the restricted group of linear transformations. The above transformation formula also indicates that if we have two affinities defined with respect to the same coordinates their difference is a tensor.

Note. We make clear the idea of two affinities with the same coordinates with an example. Consider a spherical surface. We may project the points of the surface onto a plane. Then the same coordinates will map both the spherical surface and the plane, and the affinities in the two cases will naturally be different.

As the Γ's are not tensor quantities, the question arises whether they may be made to vanish locally (i.e., at a certain point in space). We shall return to this question later. Here we note that $\Gamma^\alpha_{\beta\gamma} - \Gamma^\alpha_{\gamma\beta}$ (i.e., the part of affine connection antisymmetric with respect to the lower two indices) is a tensor in view of (2.31). Hence we have the theorem that if the Γ's are not symmetric in the lower two indices, then they cannot be made to vanish even at a single point. The antisymmetric part is referred to as torsion.

In Riemannian geometry, which is the basis of Einstein's theory, it is assumed that the Γ's are symmetric ($\Gamma^\alpha_{\beta\gamma} = \Gamma^\alpha_{\gamma\beta}$), while a geometry involving torsion has been considered by Cartan as an extension of the general theory of relativity. Henceforth, we shall restrict ourselves to Riemannian geometry unless specifically stated. We may introduce a set of affinities by considering the parallel displacement of a covariant vector

$$dA_\alpha = \overline{\Gamma}^\mu_{\alpha\beta} A_\mu dx^\beta. \qquad (2.32)$$

If we now demand that scalars (in particular, the scalar product of two vectors) remain unchanged in a parallel displacement we get

$$0 = d(A^\alpha B_\alpha) = A^\alpha dB_\alpha + dA^\alpha B_\alpha = A^\alpha B_\beta (\overline{\Gamma}^\beta_{\alpha\mu} - \Gamma^\beta_{\alpha\mu}) dx^\mu,$$

as A^α and B_β are arbitrary vectors, we get

$$\overline{\Gamma}^\beta_{\alpha\mu} = \Gamma^\beta_{\alpha\mu}. \qquad (2.33)$$

2.5. Covariant Differentiation

Let us suppose that a vector field A^μ is given, A^μ being differentiable functions of the coordinates. We may now construct two vectors at the point $x^\nu + dx^\nu$—one that of the vector field which is $A^\mu + \partial A^\mu/\partial x^\nu \, dx^\nu$ and the other by parallel displacement $A^\mu - \Gamma^\mu_{\alpha\nu} A^\alpha \, dx^\nu$. Both being vectors at the same point, their difference will also be a vector at $x^\nu + dx^\nu$, i.e.,

$$\left(\frac{\partial A^\mu}{\partial x^\nu} + \Gamma^\mu_{\alpha\nu} A^\alpha \right) dx^\nu$$

is a vector for arbitrary dx^ν. Using now the theorem that if $A^\alpha_\beta B^\beta$ is a vector for an arbitrary vector B^β, then A^α_β is a tensor of second rank, and we see that the expression within the parentheses above is a tensor. This tensor is called the covariant derivative of A^μ with respect to x^ν and will be written as $A^\mu_{;\nu}$ so that

$$A^\mu_{;\nu} \equiv A^\mu_{,\nu} + \Gamma^\mu_{\alpha\nu} A^\alpha, \tag{2.34}$$

where we have indicated an ordinary partial derivative by a comma.

Similarly, if we start with a covariant vector, the covariant derivative is given by

$$A_{\mu;\nu} = A_{\mu,\nu} - \Gamma^\alpha_{\mu\nu} A_\alpha. \tag{2.35}$$

We can extend the ideas of parallel transport and covariant differentiation to tensors of higher rank. Thus, under parallel transport

$$d(A^\mu B^\nu) = A^\mu \, dB^\nu + B^\nu \, dA^\mu$$

$$= -\Gamma^\nu_{\alpha\beta} B^\alpha A^\mu \, dx^\beta - \Gamma^\mu_{\alpha\beta} A^\alpha B^\nu \, dx^\beta$$

$$= -(\Gamma^\nu_{\alpha\beta} A^\mu B^\alpha + \Gamma^\mu_{\alpha\beta} A^\alpha B^\nu) \, dx^\beta. \tag{2.36}$$

Writing $A^\mu B^\nu = T^{\mu\nu}$ we have from (2.36)

$$dT^{\mu\nu} = -(\Gamma^\nu_{\alpha\beta} T^{\mu\alpha} + \Gamma^\mu_{\alpha\beta} T^{\alpha\nu}) \, dx^\beta,$$

and

$$DT^\mu \equiv \left(\frac{\partial T^\mu}{\partial x^\beta} + \Gamma^\nu_{\alpha\beta} T^{\mu\alpha} + \Gamma^\mu_{\alpha\beta} T^{\alpha\nu} \right) dx^\beta$$

$$\equiv T^{\mu\nu}_{;\beta} \, dx^\beta. \tag{2.37}$$

Thus, the covariant derivative of a contravariant tensor of rank 2 is given by

$$T^{\mu\nu}_{;\beta} = T^{\mu\nu}_{,\beta} + \Gamma^\nu_{\alpha\beta} T^{\mu\alpha} + \Gamma^\mu_{\alpha\beta} T^{\alpha\nu}. \tag{2.38}$$

Similarly,

$$T_{\mu\nu;\beta} = T_{\mu\nu,\beta} - \Gamma^\sigma_{\alpha\beta} T_{\sigma\nu} - \Gamma^\sigma_{\nu\beta} T_{\mu\sigma}.$$

As a scalar is unaffected by parallel transport there is no difference between the covariant derivative and the ordinary derivative; indeed, we have already seen that such a derivative forms a covariant vector. We now show that, with the assumption that the scalars remain unchanged in parallel displacement, we may deduce a relation between the Γ's and $g_{\mu\nu}$ and their first derivatives.

We have, in view of the above assumption, $d(g_{\alpha\beta}A^\alpha A^\beta) = 0$ where d indicates a parallel displacement from x^μ to $x^\mu + dx^\mu$. Thus

$$0 = d(g_{\alpha\beta}A^\alpha A^\beta) = d(g_{\alpha\beta})A^\alpha A^\beta + g_{\alpha\beta}(dA^\alpha)A^\beta + g_{\alpha\beta}A^\alpha\, d(A^\beta)$$

$$= g_{\alpha\beta,\mu}\, dx^\mu\, A^\alpha A^\beta - g_{\alpha\beta}\Gamma^\alpha_{\mu\nu}A^\nu\, dx^\mu\, A^\beta - g_{\alpha\beta}A^\alpha\Gamma^\beta_{\mu\nu}\, dx^\mu\, A^\nu$$

$$= [g_{\alpha\beta,\mu} - g_{\sigma\beta}\Gamma^\sigma_{\mu\alpha} - g_{\sigma\alpha}\Gamma^\sigma_{\mu\beta}]A^\alpha A^\beta\, dx^\mu.$$

Owing to the arbitrary character of the vector A^α and the displacement dx^μ we get

$$g_{\alpha\beta,\mu} - g_{\sigma\beta}\Gamma^\sigma_{\mu\alpha} - g_{\sigma\alpha}\Gamma^\sigma_{\mu\beta} = 0. \tag{2.39}$$

A cyclic change of the indices α, β, and μ gives two additional relations

$$g_{\beta\mu,\alpha} - g_{\sigma\mu}\Gamma^\sigma_{\alpha\beta} - g_{\sigma\beta}\Gamma^\sigma_{\alpha\mu} = 0, \tag{2.40}$$

$$g_{\mu\alpha,\beta} - g_{\sigma\alpha}\Gamma^\sigma_{\beta\mu} - g_{\sigma\mu}\Gamma^\sigma_{\beta\alpha} = 0. \tag{2.41}$$

We find from (2.39) + (2.40) − (2.41), using the symmetry of the Γ's,

$$g_{\alpha\beta,\mu} + g_{\beta\mu,\alpha} - g_{\mu\alpha,\beta} - 2g_{\sigma\beta}\Gamma^\sigma_{\mu\alpha} = 0,$$

or

$$\Gamma^\sigma_{\mu\alpha} = \tfrac{1}{2}g^{\sigma\beta}[g_{\mu\beta,\alpha} + g_{\beta\alpha,\mu} - g_{\mu\alpha,\beta}], \tag{2.42}$$

which is the desired relation.

With this form for the Γ's we obtain, for the covariant derivative of $g_{\mu\nu}$, using (2.38)

$$g_{\mu\nu;\alpha} = g_{\mu\nu,\alpha} - \Gamma^\sigma_{\mu\alpha}g_{\sigma\nu} - \Gamma^\sigma_{\nu\alpha}g_{\mu\sigma}$$

$$= g_{\mu\nu,\alpha} - \tfrac{1}{2}[g_{\mu\nu,\alpha} + g_{\nu\alpha,\mu} - g_{\mu\alpha,\nu}]$$

$$- \tfrac{1}{2}[g_{\mu\nu,\alpha} + g_{\mu\alpha,\nu} - g_{\nu\alpha,\mu}] = 0. \tag{2.43}$$

This result may not be true if the Γ's are nonsymmetric, i.e., in spaces with torsion.

The relation between the Γ's and the $g_{\mu\nu}$'s are sometimes obtained by assuming this result, namely, that the covariant derivative of $g_{\mu\nu}$ vanishes. Apparent justifications for such an assumption are the following. The Γ's involve only the first derivatives of $g_{\mu\nu}$ besides the $g_{\mu\nu}$'s and so they will vanish in the locally Lorentz frame. However, this idea is wrong if torsion exists, i.e., $\Gamma^l_{ik} \neq \Gamma^l_{ki}$; in that case the Γ^l_{ik}'s are tensor quantities and they cannot be reduced to zero merely by a coordinate transformation. Summing up, we may say that a Riemannian geometry may be specified by any of these equivalent assumptions; that the Γ's are symmetric and:

(a) the scalars remain unchanged in parallel displacement;
(b) the covariant derivatives of $g_{\mu\nu}$ vanish; and
(c) the Γ's involve only $g_{\mu\nu}$ and their first derivatives.

Although, in general, $g_{\mu\nu}$'s cannot be diagonalized (much less reduced to constant values) over the entire space, standard theorems of quadratic forms show that they can be reduced to constant values (in particular, Minkowski values) at any particular point. One way of seeing this is to consider the transformation formula for this to be the case

$$\eta_{\alpha\beta} = g_{\mu\nu} \frac{\partial x^\mu}{\partial x'^\alpha} \frac{\partial x^\nu}{\partial x'^\beta}.$$

This would give ten algebraic equations for the sixteen $(\partial x^\mu / \partial x'^\alpha)$ and hence can be satisfied. We can go further, consider the first derivatives of $g_{\mu\nu}$

$$\frac{\partial g'_{\mu\nu}}{\partial x'^\alpha} = \frac{\partial}{\partial x'^\alpha} \left(g_{\sigma\rho} \frac{\partial x^\sigma}{\partial x'^\mu} \frac{\partial x^\rho}{\partial x'^\nu} \right).$$

If these vanish, we would get 40 relations between the derivatives $\partial x^\mu / \partial x'^\alpha$ (16 in number) and $(\partial^2 x^\mu / \partial x'^\alpha \, \partial x'^\beta)$ (40 in number). Considering the 10 relations already introduced for reducing $g_{\mu\nu}$'s to $\eta_{\alpha\beta}$ we get $10 + 40 = 50$ constraints on $40 + 16 = 56$ variables. Hence they can be satisfied. It now follows from (2.42) that the Γ's can also be made to vanish at any particular point by a suitable choice of coordinates. Such coordinates are usually referred to as locally Lorentz (or geodesic) coordinates. Again, from (2.42), we have

$$\Gamma^\beta_{\alpha\beta} = \tfrac{1}{2} g^{\beta\sigma} [g_{\sigma\alpha,\beta} + g_{\sigma\beta,\alpha} - g_{\alpha\beta,\sigma}].$$

Now $g^{\beta\sigma} g_{\sigma\alpha,\beta} = g^{\beta\sigma} g_{\beta\alpha,\sigma}$ because of the dummy nature of the indices β, σ and the symmetry of $g^{\beta\sigma}$. So

$$\Gamma^\beta_{\alpha\beta} = \tfrac{1}{2} g^{\beta\sigma} g_{\beta\sigma,\alpha} = \frac{1}{2|g|} |g|_{,\alpha} = (\ln \sqrt{|g|})_{,\alpha}, \qquad (2.44)$$

where we have used the rule for the differentiation of determinants.*

We present an expression for the covariant divergence of a four vector

$$A^\mu_{;\mu} = A^\mu_{,\mu} + \Gamma^\mu_{\alpha\mu} A^\alpha$$

$$= A^\mu_{,\mu} + \tfrac{1}{2}(\ln \sqrt{|g|})_{,\alpha} A^\alpha = \frac{1}{\sqrt{|g|}} (A^\alpha \sqrt{|g|})_{,\alpha}. \qquad (2.45)$$

For the divergence of a tensor $A^{\mu\nu}$, we have

$$A^{\mu\nu}_{;\nu} = A^{\mu\nu}_{,\nu} + \Gamma^\nu_{\alpha\nu} A^{\mu\alpha} + \Gamma^\mu_{\alpha\nu} A^{\alpha\nu}.$$

In the case where the tensor $A^{\mu\nu}$ is antisymmetric, the last term on the right vanishes because of the symmetry of $\Gamma^\alpha_{\mu\nu}$ with respect to μ and ν, and we get

$$A^{\mu\nu}_{;\nu} = A^{\mu\nu}_{,\nu} + \frac{1}{2} \frac{(\sqrt{|g|})_{,\alpha}}{\sqrt{|g|}} A^{\mu\alpha}$$

$$= \frac{1}{\sqrt{|g|}} (A^{\mu\nu} \sqrt{|g|})_{,\nu}. \qquad (2.46)$$

---

* $\partial g / \partial x^\alpha = g g^{\mu\nu} g_{\mu\nu,\alpha}$. (The differential coefficient of a determinant is the contracted product of the differential coefficient of an element with its cofactor in the determinant.)

2.6. The Differential Equation of a Geodesic

A straight line in flat space may be defined by any of its two properties:

(1) it is a line such that the tangent vector maintains its direction unchanged; and
(2) it is the shortest distance between any two points.

It turns out that the geodesics in Riemannian space have two exactly analogous properties. In view of (1) we may demand that the tangent vector must be displaced parallelly along the geodesic. This requires that if v^μ is the unit tangent vector

$$v^\mu_{;\beta} v^\beta = (v^\mu_{,\beta} + \Gamma^\mu_{\alpha\beta} v^\alpha) v^\beta = 0. \tag{2.47}$$

Thus the above equation becomes, using the arc length ds,

$$\frac{dv^\mu}{ds} + \Gamma^\mu_{\alpha\beta} v^\alpha v^\beta = 0. \tag{2.48}$$

Adopting the analogy of the second property we may demand an extremal property of the geodesic, i.e., given any two points, the geodesic is the line joining them of extremal length. Thus, the geodesic is given by

$$\delta \int ds = 0,$$

or

$$\delta \int \frac{ds}{d\lambda} \, d\lambda = 0,$$

which is written as

$$\delta \int \left(g_{\mu\nu} \frac{dx^\mu}{d\lambda} \frac{dx^\nu}{d\lambda} \right)^{1/2} d\lambda = 0, \tag{2.49}$$

the parameter λ is not to be varied. This is exactly analogous to the Hamilton variational principle in mechanics

$$\delta \int L \, dt = 0 \tag{2.50}$$

with λ taking the role of t and x^μ that of the generalized coordinates. The Lagrangian differential equations

$$\frac{d}{dt} \left(\frac{\partial L}{\partial q_k} \right) - \frac{\partial L}{\partial q_k} = 0 \tag{2.51}$$

now become

$$\frac{d}{d\lambda} \left[\left(g_{\mu\alpha} \frac{dx^\mu}{d\lambda} \right) \Big/ \left(\frac{ds}{d\lambda} \right) \right] - \frac{1}{2} \frac{g_{\mu\nu,\alpha} (dx^\mu/d\lambda)(dx^\nu/d\lambda)}{(ds/d\lambda)} = 0,$$

or

$$g_{\mu\alpha}\frac{d^2x^\mu}{d\lambda^2} + g_{\mu\alpha,\nu}\frac{dx^\mu}{d\lambda}\frac{dx^\nu}{d\lambda} - \frac{g_{\mu\alpha}(dx^\mu/d\lambda)(d^2s/d\lambda^2)}{(ds/d\lambda)}$$

$$-\tfrac{1}{2}g_{\mu\nu,\alpha}\frac{dx^\mu}{d\lambda}\frac{dx^\nu}{d\lambda} = 0.$$

or

$$g_{\mu\alpha}\frac{d^2x^\mu}{d\lambda^2} + \tfrac{1}{2}(g_{\mu\alpha,\nu} + g_{\nu\alpha,\mu} - g_{\mu\nu,\alpha})\frac{dx^\mu}{d\lambda}\frac{dx^\nu}{d\lambda} - \frac{g_{\mu\nu}(dx^\mu/d\lambda)(d^2s/d\lambda^2)}{(ds/d\lambda)} = 0.$$
(2.52)

In the case of a null line $ds = 0$ and in the nonnull case it is usual to identify the parameter λ with the arc length s up to a constant multiplier, so that in either case the last term falls off and we get

$$\frac{d^2x^\mu}{d\lambda^2} + \Gamma^\mu_{\alpha\beta}\frac{dx^\alpha}{d\lambda}\frac{dx^\beta}{d\lambda} = 0$$
(2.53)

which is identical to (2.48) as $v^\mu = dx^\mu/ds = $ constant $dx^\mu/d\lambda$. In general, the geodesic equation is (2.53).

We have taken the Lagrangian in (2.49) as $(g_{\mu\nu}\,dx^\mu/d\lambda\,dx^\nu/d\lambda)^{1/2}$. The same equation (2.53) follows if we take the Lagrangian as $(g_{\mu\nu}\,dx^\mu/d\lambda\,dx^\nu/d\lambda)$, i.e., free of the radical sign. This may be seen as follows. For the variational principle $\delta \int f(L)\,dt = 0$, the Euler–Lagrange equations are (considering $f(L)$ to be a differentiable function of L)

$$\frac{d}{d\lambda}\left[f'(L)\frac{\partial L}{\partial \dot{q}_k}\right] - f'(L)\frac{\partial L}{\partial q_k} = 0,$$
(2.54)

where $f'(L) = df(L)/dL$. Now if L, and consequently $f(L)$, is a constant of motion, the above equation reduces to the Euler–Lagrange equation with L as the Lagrangian. In our case, L is indeed a constant of motion ($L = 0$ for null lines and $L = 1$ for timelike lines with $d\lambda = dS$).

However, we may look upon the constancy of L in a more interesting manner. The Lagrangian L does not involve the parameter λ explicitly, hence the "Hamiltonian" $H \equiv \partial L/\partial \dot{q}_k \cdot \dot{q}_k - L$ is a constant of motion. But H in this case is simply L, thus L itself is a constant of motion. If we are allowed to borrow the language of mechanics, the Lagrangian does not involve "time" and is also free of any "potential energy" term—hence it is a conserved quantity. Thus, very often, we shall take the geodesic equation to follow from the variational principle

$$\delta \int \left(g_{\mu\nu}\frac{dx^\mu}{d\lambda}\frac{dx^\nu}{d\lambda}\right) d\lambda = 0.$$
(2.55)

Problems

1. Prove the following relations:

   (i) $g^{\mu\sigma}_{,\beta} = -g^{\alpha\sigma}\Gamma^{\mu}_{\alpha\beta} - g^{\mu\alpha}\Gamma^{\sigma}_{\alpha\beta}$; and

   (ii) $g^{\mu\nu}\Gamma^{\alpha}_{\mu\nu} = -\dfrac{1}{\sqrt{|g|}}(\sqrt{|g|}g^{\alpha\beta})_{,\beta}$.

2. Prove the following results: If v^{μ} and u^{μ} are two unit timelike vectors, then $(v^{\mu}u_{\mu})^2 \geq 1$. When does the sign of equality hold? (Note that the definition of $\cos\theta$ does not always lead to real θ in the case of an indefinite metric.)

3. Consider the surface of a sphere with the metric

$$dS^2 = d\theta^2 + \sin^2\theta\, d\phi^2.$$

A vector A^{α}, which has the components $A^{\theta} = A^{\phi} = 1$ at a point on the equator $(\theta = \phi = \pi/2)$, is parallely transported to the pole $(\theta = 0)$ along two distinct routes:
   (i) the great circle $\phi = \pi/2$; and
   (ii) the equator $\theta = \pi/2$ from $\phi = \pi/2$ to $\phi = 0$ and then along the great circle $\phi = 0$.
Find the two values of the vector.

4. Show that the metric

$$dS^2 = dt^2 - dr^2 - dz^2 - \left(r^2 - \frac{r^4}{a^2}\right)d\phi^2 + \frac{2r^2}{a}d\phi\, dt,$$

where a is a constant and satisfies the signature condition everywhere. Comment on the spacelike or timelike nature of the coordinate ϕ. Find the condition that a line of constant r and Z can be a geodesic.

5. If l^{μ} is an eigenvector of an antisymmetric tensor $F_{\alpha\beta}$ with the eigenvalue λ, show that either $\lambda = 0$ or that l^{μ} is a null vector. (Hence the theorem for spaces with a definite metric, an antisymmetric tensor, does not have any nonzero eigenvalue.)

2.7. The Integrability of Parallel Displacement

An affine connection $\Gamma^{\alpha}_{\mu\nu}$ enables us to transport a vector along a curve uniquely from one point to the other. When the space is Euclidean or pseudo-Euclidean, we can introduce a rectangular Cartesian coordinate system, in which the components of the metric tensor are constants. In these cases, the Γ's vanish and the parallel vectors have the same components at every point, independent of the path followed in the displacement process. There is thus a unique parallel at every point and we say the parallel displacement is integrable. Again, the integrability is by its definition an invariant property independent of the choice of the coordinate system.

In a parallel displacement of two vectors the angle between them clearly remains unchanged. Also, if $x^{\mu} = x^{\mu}(S)$ is the parametric equation of a certain geodesic curve, the unit tangent vector $v^{\mu} = dx^{\mu}/dS$ satisfies the equation $Dv^{\mu} = 0$ along the curve. In other words, v^{μ} at x^{μ}, being parallelly transported to another point $(x^{\mu} + dx^{\mu})$ on the same geodesic curve, coincides with the vector $(v^\mu + dv^\mu)$ tangent to the curve at the point $(x^\mu + dx^\mu)$. Thus the tangent vectors at the points to the geodesic are autoparallel. Thus a vector parallelly transported along a geodesic curve will make the same angle with the tangent vector at every point. We now show that in a curved space the parallel displacement of a vector from one given point to another, in general, yields different results if it is carried along different paths. Particularly, by a parallel displacement along a closed contour, we do not in general recover the original vector. Consider, for example, a simple two-dimensional spherical surface, which is a nonflat two-dimensional space. What happens to a vector subject to a parallel transport along a closed contour made up of three geodesic curves and returning to the original starting point? In moving along OP the vector A^μ retains its angle fixed with the curve and goes over to B^μ, which again moves in a similar way along PQ and goes over to C^μ. Finally, C^μ moves along QO being parallelly transported and attains the value A'^μ at the point O. The vector does not go over to the original A^μ after completing the closed circuit (see Fig. 2.2). Similar considerations are also valid in an n-dimensional space–time. The total change in a vector when it undergoes parallel displacement along a certain closed circuit is

$$\Delta A^\alpha = -\oint \Gamma^\alpha_{\mu\nu} A^\mu \, dx^\nu. \tag{2.56}$$

The condition of integrability of a parallel displacement equation

$$dA^\mu = -\Gamma^\mu_{\alpha\beta} A^\alpha \, dx^\beta \tag{2.57}$$

is

$$\frac{\partial}{\partial x^\gamma} (\Gamma^\mu_{\alpha\beta} A^\alpha) = \frac{\partial}{\partial x^\beta} (\Gamma^\mu_{\alpha\gamma} A^\alpha), \tag{2.58}$$

which gives, using (2.57),

$$(\Gamma^\mu_{\alpha\beta,\gamma} - \Gamma^\mu_{\sigma\beta}\Gamma^\sigma_{\gamma\alpha} - \Gamma^\mu_{\beta\alpha,\gamma} + \Gamma^\mu_{\sigma\gamma}\Gamma^\sigma_{\beta\alpha}) A^\alpha = 0. \tag{2.59}$$

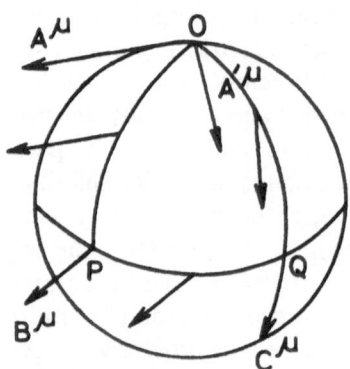

Figure 2.2

If this holds for any arbitrary vector A^α, we must have

$$R^\mu_{\beta\alpha\gamma} \equiv \Gamma^\mu_{\alpha\beta,\gamma} - \Gamma^\mu_{\delta\beta}\Gamma^\sigma_{\gamma\alpha} - \Gamma^\mu_{\beta\alpha,\gamma} + \Gamma^\mu_{\sigma\gamma}\Gamma^\sigma_{\beta\alpha} = 0. \qquad (2.60)$$

(If the reader is not familiar with Pfaffians, we may cite an example from mechanics with which he is presumably familiar. The elementary work done is $dw = F_i\, dx^i$; the condition for this to be integrable, i.e., that the work done be path independent, is that $F_{i,k} = F_{k,i}$. We are using exactly the same relation, only dw is being replaced by dA^μ.)

We now show that (2.60) is the necessary and sufficient condition for the flatness of the space. That the condition is necessary is readily seen; in flat space it is possible to introduce rectangular Cartesian coordinates, then all Γ's vanish and consequently (2.60) is satisfied.

To see that the condition is also sufficient, at any point P take n ortho-normal vectors $A_{(1)}, A_{(2)}, \ldots, A_{(n)}$ (orthonormal means that the vectors are each a unit vector and they are mutually orthogonal). Now, displace these vectors parallelly. Because of the vanishing of $R^\mu_{\beta\alpha\gamma}$, these parallel vectors are unique, irrespective of the path of displacement. Thus, we have a field of uniquely defined parallel vectors through all space. Because of the parallelism

$$A^\mu_{(i);\nu} = 0,$$

or

$$A^{\mu;\nu}_{(i)} - A^{\nu;\mu}_{(i)} = 0,$$

which leads to the relation

$$A_{(i)\mu,\nu} - A_{(i)\nu,\mu} = 0. \qquad (2.61)$$

This means that the $A_{(i)\mu}$'s are gradient vectors, i.e., there exist scalars $\phi^{(i)}$ such that

$$A_{(i)\mu} = \frac{\partial\phi^{(i)}}{\partial x^\mu}. \qquad (2.62)$$

The n scalars $\phi^{(i)}$ may now be chosen to be a new system of coordinates. This is permissible since the Jacobian $\|\partial\phi^{(i)}/\partial x^\mu\|$ is simply the determinant formed by the vector components $A_{(i)\mu}$ and hence does not vanish. In the transformed system, the contravariant metric tensor $\bar{g}^{ik}$ is given by

$$\bar{g}^{ik} = g^{\mu\nu}\frac{\partial\phi^{(i)}}{\partial x^\mu}\frac{\partial\phi^{(k)}}{\partial x^\nu} = g^{\mu\nu}A_{(i)\mu}A_{(k)\nu}, \qquad (2.63)$$

which, owing to the orthonormal property of the $A_{(i)}$ vectors, vanishes if $i \neq k$ and is ±1 when $i = k$ (the two signs occur as $A_{(i)}$ may be a timelike or spacelike vector). Thus, in the new coordinates, $\bar{g}^{ik}$'s are characteristics of the flat space form.

2.8. The Riemann–Christoffel Tensor

It is easy to show that $R^{\alpha}_{\beta\gamma\delta}$, as defined by (2.60), is a mixed tensor of rank 4. We consider the second covariant differentiation of a vector A^{α}.

$$(A^{\alpha}_{;\beta})_{;\gamma} = (A^{\alpha}_{;\beta})_{,\gamma} + \Gamma^{\alpha}_{\gamma\delta}A^{\delta}_{;\beta} - \Gamma^{\delta}_{\beta\gamma}A^{\alpha}_{;\delta}$$

$$= A^{\alpha}_{,\beta\gamma} + (\Gamma^{\alpha}_{\sigma\beta}A^{\sigma})_{,\gamma} + \Gamma^{\alpha}_{\gamma\delta}(A^{\delta}_{,\beta} + \Gamma^{\delta}_{\sigma\beta}A^{\sigma}) - \Gamma^{\delta}_{\beta\gamma}(A^{\alpha}_{,\delta} + \Gamma^{\alpha}_{\sigma\delta}A^{\sigma}). \quad (2.64)$$

Interchanging β and γ and taking the difference we get, after a little adjustment of the dummy indices,

$$A^{\alpha}_{;\beta;\gamma} - A^{\alpha}_{;\gamma;\beta} = [\Gamma^{\alpha}_{\lambda\beta,\gamma} - \Gamma^{\alpha}_{\lambda\gamma,\beta} + \Gamma^{\alpha}_{\lambda\beta}\Gamma^{\alpha}_{\mu\gamma} - \Gamma^{\mu}_{\lambda\gamma}\Gamma^{\alpha}_{\mu\beta}]A^{\lambda}$$

$$\equiv R^{\alpha}_{\lambda\gamma\beta}A^{\lambda}. \quad (2.65)$$

The left-hand side of (2.65) is the commutator of the second covariant derivative and is obviously a tensor. The right-hand side must therefore be a tensor. Since A^{λ} is an arbitrary vector, $R^{\alpha}_{\lambda\gamma\beta}$ is itself a tensor and is called the Riemann–Christoffel tensor or simply the curvature tensor. In flat space, the curvature tensor identically vanishes.

Symmetry Properties of the Curvature Tensor

The symmetry properties of the curvature tensor become apparent when it is expressed in a fully covariant form by lowering the upper index

$$R_{\mu\lambda\gamma\beta} = g_{\mu\alpha}R^{\alpha}_{\lambda\gamma\beta}$$

$$= g_{\mu\alpha}[\Gamma^{\alpha}_{\lambda\beta,\gamma} - \Gamma^{\alpha}_{\lambda\gamma,\beta} + \Gamma^{\alpha}_{\lambda\beta}\Gamma^{\alpha}_{\mu\gamma} - \Gamma^{\alpha}_{\lambda\gamma}\Gamma^{\alpha}_{\mu\beta}]. \quad (2.66)$$

Introducing a locally Lorentz coordinate system at a point, we have, at that point,

$$R_{\mu\lambda\nu\beta} = \tfrac{1}{2}(g_{\mu\beta,\nu\lambda} + g_{\lambda\nu,\mu\beta} - g_{\mu\nu,\beta\lambda} - g_{\lambda\beta,\mu\nu}). \quad (2.67)$$

Since the symmetry properties of a tensor do not depend on coordinates, we can use (2.67) to note the following properties:

$$R_{\mu\lambda\nu\beta} = R_{\nu\beta\mu\lambda}, \qquad R_{\mu\lambda\nu\beta} = -R_{\lambda\mu\nu\beta} = -R_{\mu\lambda\beta\nu} = R_{\lambda\mu\beta\nu}. \quad (2.68)$$

From (2.68) we observe that the curvature tensor $R_{\mu\lambda\nu\beta}$ is:

(i) antisymmetric in each of the index pairs (μ, λ) and (ν, β); and also
(ii) symmetric under interchange of the first pair with the second pair.

It can be further verified that

$$R_{\mu\lambda\nu\beta} + R_{\mu\nu\beta\lambda} + R_{\mu\beta\lambda\nu} = 0 \quad (2.69)$$

for any μ, λ, ν, and β.

Number of Algebraically Independent Components of the Riemann–Christoffel Curvature Tensor

Let us here consider for simplicity the case of four dimensions. We treat the first index pair as one index capable of assuming only six values in view of the antisymmetry property such as (1.2), (1.3), (1.4), (2.3), (2.4), (3.4), and similarly for the second pair. Thus $R_{\mu\lambda\nu\beta}$ is expressed as objects S_{ik} where i stands for the pair $(\mu\lambda)$ and k for the pair $(\nu\beta)$, and because of the symmetry property already studied $S_{ik} = S_{ki}$. In consequence, there are 21 independent components. One more nontrivial relation between the components of $R_{\mu\lambda\nu\beta}$ with $\mu\lambda\nu\beta$ all different still remains to be accounted for. It is the relation following (2.69)

$$R_{0123} + R_{0231} + R_{0312} = 0. \tag{2.70}$$

The above relation finally reduces the total number of independent components to 20. For an n-dimensional space the number of independent components is

$$\tfrac{1}{12}n^2(n^2 - 1).$$

2.9. The Bianchi Identity

In addition to the algebraic identities (2.68) and (2.69), the curvature tensor satisfies an important differential identity known as the Bianchi identity. To derive this we adopt a locally Lorentz frame. In this frame the tensor reduces to the form (2.67) and the covariant derivative is identical to the ordinary derivative, so that we get

$$R_{\alpha\lambda\nu\beta;\sigma} = R_{\alpha\lambda\nu\beta,\sigma} = \tfrac{1}{2}(g_{\mu\beta,\nu\lambda\sigma} + g_{\lambda\nu,\mu\beta\sigma} - g_{\mu\nu,\beta\lambda\sigma} - g_{\lambda\beta,\mu\nu\sigma}). \tag{2.71}$$

From the above, by cyclic permutation, we arrive at the Bianchi identity, which written in covariant form is

$$R_{\alpha\lambda\nu\beta;\sigma} + R_{\alpha\lambda\beta\sigma;\nu} + R_{\alpha\lambda\sigma\nu;\beta} = 0. \tag{2.72}$$

The above relation is a tensor relation and is manifestly covariant. Since it holds in a locally Lorentz frame it holds in any arbitrary frame. This relation is sometimes symbolically written as

$$R_{\alpha\lambda[\gamma\beta;\sigma]} = 0. \tag{2.73}$$

2.10. The Ricci Tensor and the Einstein Tensor

The Ricci tensor is defined as the contraction of the Riemann tensor in the following manner:

$$R_{\lambda\nu} = g^{\alpha\beta}R_{\alpha\lambda\nu\beta} = R^{\beta}_{\lambda\nu\beta},$$

that is,

$$R_{\lambda\nu} = \Gamma^{\alpha}_{\lambda\beta}\Gamma^{\beta}_{\mu\nu} - \Gamma^{\mu}_{\lambda\nu}\Gamma^{\beta}_{\mu\beta} + \Gamma^{\mu}_{\lambda\mu,\nu} - \Gamma^{\mu}_{\lambda\nu,\mu}. \tag{2.74}$$

We see that one independent scalar alone is available on contraction of the Riemann curvature tensor. It is the Ricci scalar R defined as

$$R = g^{\lambda\nu}g^{\alpha\beta}R_{\alpha\lambda\nu\beta}, \tag{2.75}$$

R is sometimes called the scalar curvature. Another possibility of constructing a scalar from the curvature tensor is by contracting it with the Levi-Civita antisymmetric tensor and this scalar is

$$\eta^{\alpha\lambda\nu\beta} R_{\alpha\lambda\nu\beta} = 0,$$

which is a consequence of the cyclic relation (2.69). A nontrivial scalar $R_{\alpha\beta\mu\nu}R^{\alpha\beta\mu\nu}$ is called the curvature scalar.

We introduce at this stage the Einstein tensor, which plays a significant role in the general theory of relativity. It is defined as

$$G_{\mu\nu} = (R_{\mu\nu} - \tfrac{1}{2} R g_{\mu\nu}). \tag{2.76}$$

A familiar form of the Bianchi identity is $G^{\mu\nu}_{;\mu} = 0$, which can be deduced from (2.72) by contracting. We obtain, on contraction,

$$R_{\lambda\nu;\sigma} - R_{\lambda\sigma;\nu} + (g^{\alpha\beta} R_{\alpha\lambda\sigma\nu})_{;\beta} = 0.$$

Contracting, again with $g^{\lambda\nu}$, we get

$$R_{,\sigma} - R^{\nu}_{\sigma;\nu} - R^{\beta}_{\sigma;\beta} = 0,$$

that is,

$$R^{\nu}_{\sigma;\nu} - \tfrac{1}{2}(R\delta^{\nu}_{\sigma})_{,\nu} = 0,$$

which can be written in a more compact form

$$(R^{\nu}_{\sigma} - \tfrac{1}{2} R\delta^{\nu}_{\sigma})_{;\nu} = 0,$$

which is the same as

$$G^{\nu}_{\sigma;\nu} = 0. \tag{2.77}$$

2.11. The Weyl Tensor

The tensor

$$C_{\mu\nu\alpha\beta} \equiv R_{\mu\nu\alpha\beta} + \frac{1}{(n-2)}(g_{\mu\alpha}R_{\nu\beta} - g_{\mu\beta}R_{\nu\alpha} + g_{\nu\beta}R_{\mu\alpha} - g_{\nu\alpha}R_{\mu\beta})$$

$$+ \frac{R}{(n-1)(n-2)}(g_{\mu\beta}g_{\nu\alpha} - g_{\mu\alpha}g_{\nu\beta}) \tag{2.78}$$

is called the Weyl tensor defined in n-dimensional space. It possesses all the symmetry properties of the Riemann–Christoffel tensor and is traceless, i.e.,

$$g^{\nu\alpha}C_{\mu\nu\alpha\beta} = g^{\mu\beta}C_{\mu\nu\alpha\beta} = 0,$$

also

$$C^{\mu}_{\nu\alpha\beta;\mu} = R^{\mu}_{\nu\alpha\beta;\mu} + \frac{1}{(n-2)}(R_{\nu\beta;\alpha} - R_{\nu\alpha;\beta} + g_{\nu\beta}R^{\mu}_{\alpha;\mu} - g_{\gamma\alpha}R^{\mu}_{\beta;\mu})$$

$$+ \frac{R, \mu}{(n-1)(n-2)}(\delta^{\mu}_{\beta}\delta_{\nu\alpha} - \delta^{\mu}_{\alpha}g_{\nu\beta}). \tag{2.79}$$

A very important property of the Weyl tensor is its conformal invariance; for this reason it is sometimes referred to as the conformal curvature tensor. A conformal change is one in which $g_{\mu\nu}$ changes to $\bar{g}_{\mu\nu} = \phi^2 g_{\mu\nu}$ where ϕ may be any function of the coordinates. A tedious but straightforward calculation shows that $C_{\mu\nu\alpha\beta}$ (Eisenhart, 1925; Carmelie, 1982) remains the same after this change from $g_{\mu\nu}$ to $\bar{g}_{\mu\nu}$. As a result we may say that the necessary and sufficient condition for a space to be conformally flat (i.e., for the metric to be transformable to the form $dS^2 = \phi^2 \sum_\alpha \pm (dx^\alpha)^2)$ is that the tensor $C_{\mu\nu\alpha\beta}$ must vanish.

The Weyl tensor has played a crucial role in the Petrov classification. This is based on the degeneracy of the eigenvalues of the Weyl tensor. For details the reader may consult Petrov (1969).

Problems

1. If $A_\beta^\alpha B^\beta$ is a contravariant vector for arbitrary vector B^β, then A_β^α is a mixed tensor of rank 2. Prove this.

2. Show that

$$\eta^{\alpha\beta\gamma\delta}\eta_{\alpha\beta\gamma\delta} = -4!,$$

$$\eta^{\alpha\beta\gamma\mu}\eta_{\alpha\beta\gamma\rho} = -6\delta_\rho^\mu,$$

$$\eta^{\alpha\beta\gamma\mu}\eta_{\alpha\beta\mu\rho} = -2\delta_{\mu\rho}^{\gamma\sigma},$$

$$\eta^{\alpha\beta\gamma\sigma}\eta_{\alpha\mu\rho\nu} = -\delta_{\mu\rho\nu}^{\beta\gamma\sigma}.$$

3. Prove that:
   (i) a vector orthogonal to a timelike vector is necessarily spacelike;
   (ii) a vector orthogonal to a null vector is either null or spacelike; and
   (iii) a vector orthogonal to a spacelike vector may be of any type.
   (The vectors are to be assumed real.)

4. Show that the geodesics on a two-dimensional spherical surface are the great circles.

5. Show that $g^{\mu\nu}\phi_{;\mu;\nu} = (1/\sqrt{|g|})(g^{\alpha\beta}\phi_{,\alpha}\sqrt{|g|})_{,\beta}$, and hence obtain the expression for the Laplacian operator in spherical polar coordinates.

6. Noting that if η^μ is a null vector, $\phi\eta^\mu$, where ϕ is a scalar, and represents the same vector, show that η^μ is geodetic if

$$\eta_{;\nu}^\mu \eta^\nu = \alpha\eta^\mu,$$

where α is any scalar.

7. Show that the necessary and sufficient condition for a 3-space to be flat is the vanishing of the Ricci tensor.

8. Show that the metric of a 3-space can always be diagonalized.

9. Give an example to show that with an indefinite metric, a geodesic can be a curve of maximal length.

10. Show that the metric

$$dS^2 = dt^2 - \frac{R^2}{[1 + kr^2/4]^2} [dr^2 + r^2 \, d\theta^2 + r^2 \sin^2 \theta \, d\phi^2],$$

where R is a function of t alone, and $k = 0, +1$, or -1 is conformally flat, and reduce it to the conformally flat form in all three cases.

11. Calculate the Ricci tensor components for the metric

$$dS^2 = e^\nu \, dt^2 - e^\lambda \, dr^2 - r^2(d\theta^2 + \sin^2 \theta \, d\phi^2),$$

where ν and λ are functions of r alone.

12. Show that the metric of Problem 11 may be transformed to the "isotropic" form

$$dS^2 = e^\nu \, dt^2 - e^\mu(dR^2 + R^2 \, d\theta^2 + R^2 \sin^2 \theta \, d\phi^2).$$

If, in Problem 11, $e^\nu = e^{-\lambda} = (1 - 2m/r)$ where m is a constant, investigate whether the entire domain of r from 0 to ∞ is mapped out in the R coordinate.

13. Show that if F_{ik} is an antisymmetric tensor,

$$F_{ik,l} + F_{kl,i} + F_{li,k}$$

is a tensor of rank 3 antisymmetric in all three indices.

14. Show that $A_{\mu\nu}B^{\mu\nu} = 0$ if one of the tensors is symmetric and the other antisymmetric.

2.12. Geodesic Deviation

In Newtonian mechanics and gravitation, if two test particles are in a non-uniform gravitational field, then their separation vector δx^μ changes according to the equation

$$\frac{d^2(\delta x^\mu)}{dt^2} = \frac{\partial g^\mu}{\partial x^\alpha}, \qquad \delta x^\alpha = \frac{\partial^2 \phi}{\partial x^\mu \, \partial x^\alpha} \delta x^\alpha, \tag{2.80}$$

where g^μ is the gravitational intensity, ϕ is the gravitational potential and the x^μ's are rectangular Cartesian coordinates. (Note with flat space and rectangular Cartesian coordinates covariant and contravariant indices have the same significance.) Thus (2.80) contains terms involving the second derivatives of the potential, these are called tidal force terms.

The corresponding problem in the general theory of relativity will be to study the separation of neighboring geodesics, as they represent the paths of free particles. Thus, we consider a *singly infinite* pencil of geodesics so that individual geodesics are specified by a parameter v which varies continuously as we go from one geodesic to another. Let u be an affine parameter running along each individual geodesic. Then the coordinates x^μ of a point on any member of this congruence of geodesics will be uniquely determined by the parameters u and v, i.e., $x^\mu \equiv x^\mu(u, v)$. Let A and A' represent two neighboring

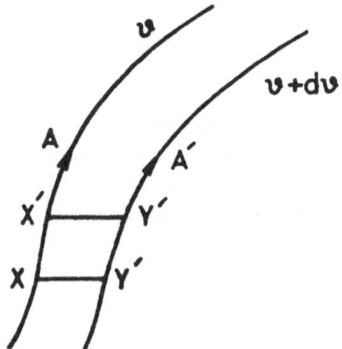

Figure 2.3

geodesics of the congruence and let XY be a vector normal to A directed from A to A' (Fig. 2.3). We are interested in the change of XY as X moves along A so that XY changes to $X'Y'$. Of course, $X'Y'$ is just the vector XY parallelly displaced along the geodesic A.

We define two vectors ξ^μ and η^μ,

$$\xi^\mu \equiv \frac{\partial x^\mu}{\partial u} \quad \text{and} \quad \eta^\mu \equiv \frac{\partial x^\mu}{\partial v},$$

so that the vector **XY** is $= \eta^\mu \, dv$ and ξ^μ is tangential to the geodesic.

Now

$$\xi^\mu_{;\alpha}\eta^\alpha - \eta^\mu_{;\alpha}\xi^\alpha = \xi^\mu_{,\alpha}\eta^\alpha - \eta^\mu_{,\alpha}\xi^\alpha = \frac{\partial^2 x^\mu}{\partial v \, \partial u} - \frac{\partial^2 x^\mu}{\partial u \, \partial v} = 0. \tag{2.81}$$

Thus the Lie derivative of η^μ with respect to ξ^μ vanishes. (The definition of the Lie derivative is given later.) Consider next the second derivative of η^μ along the geodesic A, i.e., $(\eta^\mu_{;\alpha}\xi^\alpha)_{;\beta}\xi^\beta$. We have, using (2.81),

$$\begin{aligned}
(\eta^\mu_{;\alpha}\xi^\alpha)_{;\beta}\xi^\beta &= (\xi^\mu_{;\alpha}\eta^\alpha)_{;\beta}\xi^\beta \\
&= \xi^\mu_{;\alpha;\beta}\eta^\alpha\xi^\beta + \xi^\mu_{;\alpha}\eta^\alpha_{;\beta}\xi^\beta \\
&= \xi^\mu_{;\beta;\alpha}\eta^\alpha\xi^\beta + R^\mu_{\gamma\alpha\beta}\eta^\alpha\xi^\beta\xi^\gamma + \xi^\mu_{;\alpha}\xi^\alpha_{;\beta}\eta^\beta \\
&= (\xi^\mu_{;\beta}\xi^\beta)_{;\alpha}\eta^\alpha + R^\mu_{\gamma\alpha\beta}\eta^\alpha\xi^\beta\xi^\gamma \\
&= R^\mu_{\gamma\alpha\beta}\eta^\alpha\xi^\beta\xi^\gamma. \tag{2.82}
\end{aligned}$$

The first term in the last but one step vanishing because ξ^μ is geodetic. Equation (2.82) may be written as

$$\frac{D^2\eta^\mu}{D\mu^2} = R^\mu_{\gamma\alpha\beta}\eta^\alpha\xi^\beta\xi^\gamma, \tag{2.83}$$

where in the case of nonnull geodesics u is identified with the arc length s. Equation (2.83) is known as the equation of geodesic deviation.

Problems

1. Study (2.83) for flat space, integrate it, and obtain the integrated result directly from elementary geometry.

2. Use the equation of geodetic deviation to show that, under the action of a plane wave, only the transverse separation of two neighboring particles undergoes a periodic acceleration, and the component of separation in the direction of propagation remains unaffected by the wave.

3. Einstein's Field Equations

3.1. Einstein's Formulation of the Field Equations

While Einstein was guided by a variety of empirical considerations, the same equations were obtained by Hilbert by a formal procedure which has since been generally applied in almost all branches of physics. The considerations of Einstein were along the following lines:

(1) The Minkowski metric appeared inconsistent with the gravitational shift of spectral lines. This suggests a change to Riemannian geometry—this was consistent with the idea of the principle of equivalence, as geometrically any infinitesimal region of Riemannian space–time would be identified with the tangent flat space where the Minkowski metric holds and, consequently, the special theory of relativity remains valid.
(2) The basic field variables are the metric tensor components, they are ten in number in the case of four dimensions. While they are independent, they are arbitrary to the extent of the fourfold transformation of coordinates, i.e., a coordinate transformation does not change the physics of the problem. This leads to the principle of covariance. The principle of covariance insists on the frame independence of physical laws. This would be automatically satisfied if the equations are tensor equations; hence, it was insisted upon that the basic field equations must have a tensor form.
(3) As the Newtonian laws of gravitation have been so successful in the description of planetary motion, Newton's laws of motion and gravitation should be obtainable from the new theory at some level of approximation. In particular, as Poisson's equation is a differential equation of second order, so should be the new field equations.
(4) In the special theory of relativity, the conservations of momentum and energy lead to the equation $T^{\mu\nu}_{,\nu} = 0$ where $T^{\mu\nu}$ is the energy–stress–momentum tensor. Hence, its generally covariant form $T^{\mu\nu}_{;\nu} = 0$ should be derivable from the new field equation.

These conditions lead uniquely to the field equations

$$R_{\mu\nu} - \tfrac{1}{2}Rg_{\mu\nu} = -kT_{\mu\nu}, \tag{3.1}$$

where k is the coupling constant to be determined by comparison with observation. Note that (3.1) does indeed satisfy the conditions set forth above.

3.2. Weak Field Approximation (Static Case)

As stated earlier, results of Newtonian physics should be obtainable from the general theory of relativity at some level of approximation. We show this in two steps: first, at the linearized level we shall show that the field equations in the static case reduce to the Poisson equation. In the second step we introduce the assumption that a test particle describes a geodesic and this leads to the Newtonian equation of motion. For a static gravitational field the metric can be written in the form

$$ds^2 = g_{00}\, dt^2 + g_{ik}\, dx^i\, dx^k, \tag{3.2}$$

where the g_{00}'s and g_{ik}'s are independent of the time coordinate t.

The word static implies the ideas:

(i) that the metric tensor components are independent of the timelike coordinate;
(ii) that the timelike coordinate so defined is hypersurface orthogonal, i.e., there exists a family of three-dimensional spaces in which all directions are orthogonal to these time lines, and
(iii) that the matter is at rest in this coordinate system so that the matter velocity is along the timelike direction.

The field will be called weak when the g_{ik}'s depart only slightly from the Minkowski metric $\eta_{\mu\nu}$, i.e.,

$$g_{\mu\nu} = \eta_{\mu\nu} + h_{\mu\nu}, \tag{3.3}$$

with $h_{\mu\nu} \ll 1$ and $\eta_{\mu\nu} = \text{diag}(1, -1, -1, -1)$ (we are choosing units such that $c = 1$). Of course, this implies a choice of coordinate system.

The Newtonian Approximation

For a static weak gravitational field, as already described, the products of the perturbation term $h_{\mu\nu}$'s and their higher powers may be neglected. The same conditions are valid for their derivatives also. Writing the subscripts 0, i for time and space coordinates, respectively, we have, to the order of approximation specified,

$$R_{00} = -\Gamma^{\alpha}_{00,\alpha} = \tfrac{1}{2}\eta^{\alpha\beta}h_{00,\alpha\beta}$$

$$= -\tfrac{1}{2}\nabla^2 h_{00}, \tag{3.4}$$

where ∇^2 is the ordinary Laplacian operator, i.e.,

$$\nabla^2 \equiv \frac{\partial^2}{\partial x^2} + \frac{\partial^2}{\partial y^2} + \frac{\partial^2}{\partial z^2}.$$

For the energy–momentum tensor we use the expression $T^{\mu\nu} = \rho v^\mu v^\nu$, neglecting the stress components compared to the energy density. This is justified for ordinary matter. As we are considering a static field, $(v^0)^2 = 1$ and the v^i's vanish. Then from Einstein's field equations (3.1)

$$R_{00} = -k(T_{00} - \tfrac{1}{2}g_{00} T) = -\frac{k}{2}\rho.$$ (3.5)

Equations (3.4) and (3.5) together give

$$\nabla^2 h_{00} = k\rho.$$ (3.6)

This is of the Poisson form.

Now, for a test particle moving along a geodesic in a weak gravitational field, we have from the geodesic equation (2.51)

$$\frac{d^2 x^i}{dt^2} = -\Gamma^i_{00} \left(\frac{dx^0}{dt}\right)^2 \qquad (x^0 \text{ stands for } t),$$

where we have assumed the velocity to be small compared to that of light so that $ds = dt$, and in view of the same approximation procedure the above equation can be written as

$$\frac{d^2 x^i}{dt^2} = -\tfrac{1}{2} h_{00,i}.$$ (3.7)

The corresponding Newtonian equation of motion in a gravitational field is given by

$$\frac{d^2 x^i}{dt^2} = -\phi_{,i},$$ (3.8)

where ϕ stands for the Newtonian potential of the gravitational field.

Comparing (3.7) and (3.8) leads us to assume a relation between the metric tensor component g_{00} and ϕ

$$\tfrac{1}{2} h_{00} = \phi,$$ (3.9)

or

$$g_{00} = 1 + 2\phi.$$ (3.10)

Again, in view of (3.9), (3.6) is just the Poisson equation

$$\nabla^2 \phi = 4\pi G\rho,$$ (3.11)

if

$$k \equiv 8\pi G.$$ (3.12)

In the above analysis the velocity of light c has been put at unity. In general, (3.12) will be replaced by

$$k = \frac{8\pi G}{c^2}.$$

Problems

1. Deduce the Poisson-type equation in case the source of a weak static field is black body radiation. Explain the result.

2. Show that $R_0^0 \sqrt{|g|} = [g^{ik} g^{00} \sqrt{|g|} g_{00,k}]_{,i}$ in the case of a general static field; hence, prove that a solution which is free of singularities cannot represent a mass particle.

3.3. Gravitational Waves in Weak Field Approximation

Maxwell's equations, expressed in terms of the electromagnetic potential, are second-order differential equations and lead to the wave equation if we use a suitable gauge condition. Einstein's field equations are also second-order differential equations involving both time and space coordinates and, as we shall presently show with the linearized form, they lead to a wave equation with a suitable choice of coordinates. Further, the equation shows that the waves propagate with the velocity of light. This latter result is not surprising, for any propagation with a smaller velocity may be transformed away, and hence a physical propagation of any gravitational phenomena may be expected to have the invariant velocity c. However, the question as to whether the waves carry energy is a complicated one.

We restrict ourselves to the linearized field equations; of course, the field can now no longer be regarded as static. However, as in the previous section, the $h_{\mu\nu}$ are small quantities and their products will be neglected. Hence, to this order of approximation,

$$R_{\mu\nu} = \Gamma^\alpha_{\mu\alpha,\nu} - \Gamma^\alpha_{\mu\nu,\alpha}. \tag{3.13}$$

A direct calculation with $g_{\mu\nu} = \eta_{\mu\nu} + h_{\mu\nu}$, to the order of approximation used, yields

$$R_{\mu\nu} = \tfrac{1}{2}\eta^{\alpha\beta}[h_{\mu\beta,\alpha\nu} + h_{\alpha\beta,\mu\nu} - h_{\mu\alpha,\beta\nu}] - \tfrac{1}{2}\eta^{\alpha\beta}[h_{\mu\beta,\nu\alpha} + h_{\nu\beta,\mu\alpha} - h_{\mu\nu,\beta\alpha}]$$
$$= \tfrac{1}{2}\square h_{\mu\nu} + \tfrac{1}{2}[-(h^\beta_\mu - \tfrac{1}{2}\delta^\beta_\mu h)_{,\beta\nu} - (h^\alpha_\nu - \tfrac{1}{2}\delta^\alpha_\nu h)_{,\mu\alpha}], \tag{3.14}$$

where $h^\alpha_\beta \equiv \eta^{\alpha\sigma} h_{\sigma\beta}$ and $h \equiv h^\alpha_\alpha$,

$$\square \equiv \eta^{\alpha\beta} \frac{\partial^2}{\partial x^\alpha \, \partial x^\beta}.$$

We now show that by a suitable coordinate transformation it is possible to simplify (3.14) to a considerable extent. Of course, the transformation has to be infinitesimal as otherwise it would disturb the smallness of $h_{\mu\nu}$. Consider the transformation

$$x'^\mu = x^\mu + \xi^\mu, \tag{3.15}$$

where the ξ^μ's are infinitesimal. Indicating the transformed quantities by primes, we have to the first order

$$g'_{\alpha\beta} = g_{\alpha\beta} - \xi_{\alpha,\beta} - \xi_{\beta,\alpha}, \tag{3.16}$$

and

$$h'_{\alpha\beta} = h_{\alpha\beta} - \xi_{\alpha,\beta} - \xi_{\beta,\alpha}, \qquad h' = h - 2\eta^{\alpha\beta}\xi_{\alpha,\beta},$$

so that

$$(h_\mu'^\beta - \tfrac{1}{2}\delta_\mu^\beta h')_{,\beta} = (h_\mu^\beta - \tfrac{1}{2}\delta_\mu^\beta h)_{,\beta} - \eta^{\mu\nu}\xi_{\beta,\mu\nu}. \tag{3.17}$$

Hence, if we choose ξ_β as the solution of the differential equation (such a choice is always possible)

$$\Box\xi_\alpha \equiv \eta^{\mu\nu}\xi_{\alpha,\mu\nu} = (h_\alpha^\beta - \tfrac{1}{2}\delta_\alpha^\beta h)_{,\beta}, \tag{3.18}$$

then in the transformed coordinates

$$(h_\mu'^\beta - \tfrac{1}{2}\delta_\mu^\beta h')_{,\beta} = 0. \tag{3.19}$$

This is usually referred to as a coordinate condition and (3.14) simplifies to

$$R_{\mu\nu} = \tfrac{1}{2}\Box h_{\mu\nu}, \tag{3.20}$$

where we have dropped the primes. The reader will recognize that the coordinate conditions are the analogues of the gauge conditions used in the solution of Maxwell's equations. Hence, for empty space, we have the standard wave equation propagating with velocity c

$$\Box h_{\mu\nu} = 0. \tag{3.21}$$

Plane Waves

It is interesting to consider the case of plane waves within the limitation of linearized approximation. Thus, all the $g_{\mu\nu}$'s are taken as functions of the single argument $(x^1 \pm Vx^0)$, representing a wave progressing in time x^0 along the x^1-axis with velocity V. To begin with we are not assuming that $V = c$. A straightforward calculation gives, for the Ricci tensor components,

$$R_{00} = -\frac{V^2}{2}(h''_{11} + h''_{22} + h''_{33}) \pm Vh''_{10} - \tfrac{1}{2}h''_{00},$$

$$R_{11} = \tfrac{1}{2}(h''_{00} - h''_{22} - h''_{33}) \mp Vh''_{01} + \frac{V^2}{2}h''_{11},$$

$$R_{01} = \mp\frac{V}{2}(h''_{22} + h''_{33}),$$

$$R_{02} = \pm\frac{V}{2}h''_{12} - \tfrac{1}{2}h''_{02},$$

$$R_{12} = \mp\tfrac{1}{2}h''_{02} - \frac{V^2}{2}h''_{12},$$

$$R_{23} = \tfrac{1}{2}h''_{23}(V^2 - 1),$$

$$R_{22} = \tfrac{1}{2}h''_{22}(V^2 - 1),$$

$$R_{33} = \tfrac{1}{2}h''_{33}(V^2 - 1), \tag{3.22}$$

where the primes indicate differentiation with respect to the argument $(x^1 \pm Vx^0)$. Solutions of the equation $R_{\mu\nu} = 0$ in (3.22) may be of three types:

(a) $V^2 = 1$, $h_{22} + h_{33} = 0$, h_{22}, h_{33}, $h_{23} \neq 0$, all other $h_{\alpha\beta}$ vanish. These are called completely transverse waves or TT waves.

(b) Only h_{02}, h_{03}, h_{12}, h_{13} nonvanishing and

$$h_{02} = \mp V h_{12}, \qquad \tfrac{1}{2} V^2 h_{11} \mp V h_{10} \pm \tfrac{1}{2} h_{00} = 0.$$

A direct calculation shows that the Riemann–Christoffel tensor vanishes in this case, so that the geometry is Euclidean. This trivial case has appeared only because we have not used the coordinate condition (3.19) so far.

(c) Only h_{00}, h_{11}, and h_{01} survive, here also the disturbance is trivial, as the Riemann–Christoffel tensor vanishes.

Thus the only nontrivial waves are of the TT type. The condition $h_{22} = -h_{33} \neq 0$ represents a stretching and shrinkage of equal magnitude in the perpendicular direction, which is equivalent to a shear in the $x^2 x^3$ plane. A varying h_{23} indicates a change in the angle between the x^2- and x^3-axes, which is also a shear. Thus the T–T waves are of the nature of a shear. The two modes correspond to two states of polarization for a particular direction of propagation. Consider a wave with $h_{22} = -h_{33} \neq 0, h_{23} \neq 0$, then if we make a rotation of axes in the 2, 3 plane

$$x^2 = \bar{x}^2 \cos\theta + \bar{x}^3 \sin\theta,$$

$$x^3 = -\bar{x}^2 \sin\theta + \bar{x}^3 \cos\theta,$$

$$\bar{g}_{22} = g_{22} \cos^2\theta + g_{33} \sin^2\theta + 2g_{23} \sin\theta \cos\theta,$$

or

$$\bar{h}_{22} = h_{22} \cos 2\theta + h_{23} \sin 2\theta.$$

Similarly,

$$\bar{h}_{33} = h_{33} \cos 2\theta + h_{23} \sin 2\theta,$$

$$\bar{h}_{23} = h_{22} \sin 2\theta + h_{23} \cos 2\theta.$$

The occurrence of 2θ is characteristic of the tensor character of the wave, a rotation by π leaves the metric unchanged. This indicates a helicity 2 for the gravitons. The gravitational field, like electromagnetic fields, propagates in vacuum with the velocity of light. This result, although here obtained by a consideration of weak fields, holds quite generally for fields of arbitrary strength.

3.4. Detection of Gravitational Waves

Suppose that two free test particles with the separation vector V^α are exposed to a gravitational wave. In the radiation field we have a time dependence of $R^\alpha_{\beta\nu\delta}$ which, in turn, causes V^α to change according to the equation of geodesic deviation (2.83). For a periodically changing Riemann tensor the separation between these points changes periodically. The pioneering experiment of Weber (1961), to detect the existence of gravitational waves, utilized two aluminum bars of cylindrical shape. They were of length 153 cm, diameter 66 cm, and weight 1.4×10^3 kg. Each bar was suspended by a wire in vacuum and was mechanically decoupled from its surroundings. Around the middle regions there were piezoelectric strain transducers, which couple into electronic circuits sensitive to the end-to-end oscillations of the bars. When the gravitational waves hit the bar broadside-on the relative acceleration between the two ends carried by the waves excited the fundamental modes of both the bars, one at the University of Maryland and the other at Argonne National Laboratory near Chicago. Weber observed a number of sudden simultaneous excitations in these two bars and these coincidences might have been caused by gravitational waves coming from outer space.

Estimates based on Weber's data and on conventional gravitational radiation theory suggested a total radiated power of 1–1000 $M_\odot$ per year. If the galactic enter is losing energy at such a rate the dynamical equilibrium of stars in its neighborhood would be seriously disrupted. Later experiments, however, yielded negative results.

In recent years, binary pulsar observations have brought indirect confirmation of the existence of gravitational waves. We shall describe that in a later chapter.

Problems

1. Show that the in the case of a plane wave, coordinate condition (3.19) may be used to make all the $h_{\alpha\beta}$'s, other than h_{22}, h_{33}, and h_{23}, vanish and $h_{22} = -h_{33}$.

2. Show that the two waves $(h_{22} = -h_{33} \neq 0, h_{23} = 0)$ and $(h_{23} \neq 0, h_{22} = -h_{33} = 0)$ are polarized in directions $\pi/4$ with one another (i.e., show that the wave of one type is transformed to the other type by a rotation of $\pi/4$).

3.5. Integration of the Linearized Equations for a Stationary Axially Symmetric Distribution

It is already shown that by a suitable choice of the coordinate system the perturbation term $h_{\mu\nu}$ may be made to satisfy the gauge condition $(h_\nu^\mu - \frac{1}{2}\delta_\nu^\mu h)_{,\mu} = 0$, so that Einstein's field equation in the linear approximation reduces to

$$\Box \gamma_{\mu\nu} = -\frac{16\pi G}{c^4} T_{\mu\nu} = -2k T_{\mu\nu}, \qquad (3.23)$$

with the substitution $k = 8\pi G/c^4$. In (3.23), $\gamma_{\mu\nu}$ is written for $(h_{\mu\nu} - \frac{1}{2}\eta_{\mu\nu}h)$, $\eta_{\mu\nu}$ and $h_{\mu\nu}$ having the usual meanings. We assume that although there is a rotation, the matter distribution and the gravitational field produced are time-independent. Such a situation can be obtained only if there is rotational symmetry of the matter distribution about the axis of rotation, which we call here the x^3-axis. The rotation is slow, i.e., in the coordinate system in which the field is stationary, the velocity of each constituent particle of the source is small compared to the velocity of light. Thus the energy–momentum tensor can be written to the first-order approximation (i.e., retaining only terms linear in V/c, V being the spatial velocity of the particles)

$$T_{00} = \rho(v_0)^2 = \rho, \tag{3.24a}$$

$$T_{01} = \rho v_0 v_1 = \rho \frac{v}{c} \sin \phi, \tag{3.24b}$$

$$T_{02} = \rho v_0 v_2 = -\rho \frac{v}{c} \cos \phi, \tag{3.24c}$$

where v_1 and v_2 are the components of velocity v along the x^1- and x^2-axes, respectively, and ϕ is the angle between the x^1-axis and OQ (see Fig. 3.1); Q being the location of a particle under consideration and O being the origin of the coordinate system. The other components T_{ij} of the energy–momentum tensors can be neglected because they involve higher powers of V/c. The stresses are also assumed sufficiently small. Because of axial symmetry, ρ and V are functions of $\bar{R}$ and x^3 only, where $\bar{R}^2 = (x^1)^2 + (x^2)^2$. Since the gravitational field is stationary, i.e., independent of time, the linearized field equation (3.23) reduces to the Poisson-type equation

$$\nabla^2 \gamma_{\mu\nu} = 2k T_{\mu\nu}, \tag{3.25}$$

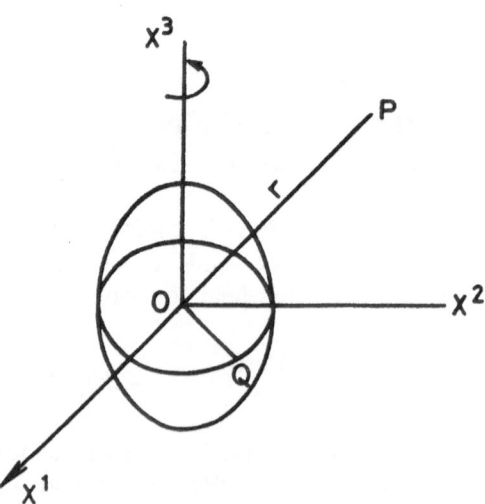

Figure 3.1

and the $\gamma_{\mu\nu}$'s are therefore given by

$$\gamma_{\mu\nu} = -\frac{k}{2\pi} \int \frac{T_{\mu\nu}(x'^i, t - R/c)}{R} d^3x', \tag{3.26}$$

where $R^2 = \sum_{i=1}^{3} (x^i - x'^i)^2$. (Note that we have taken only the retarded solution.) Here x^i and x'^i represent the field point and the source point, respectively. Remembering the definitions of r and R we have evidently the following relations:

$$r^2 = \sum_{i=1}^{3} (x^i)^2,$$

and

$$R^2 = r^2 + \sum_{i=1}^{3} (x'^i)^2 - 2 \sum_{i=1}^{3} x^i x'^i. \tag{3.27}$$

The relation (3.27) immediately leads to

$$\frac{1}{R} = \frac{1}{r} + \frac{\sum x^i x'^i}{r^3} - \frac{r'^2}{r^3} + \text{higher order terms in powers of } 1/r. \tag{3.28}$$

In what follows we shall assume that r is large compared to the dimensions of the source. The lowest order approximation (i.e., $1/R = 1/r$) yields for γ_{00}

$$\gamma_{00} = -\frac{k}{2\pi} \int \frac{T_{00}(x'^i)}{R} d^3x'. \tag{3.29}$$

In view of (3.24a), the solution (3.29) enables us to write

$$\gamma_{00} = -\frac{k}{2\pi} \frac{1}{r} \int \rho \, d^3x' = -\frac{4GM}{c^2r}, \tag{3.30}$$

where $M \equiv \int \rho \, d^3x'$. This is the monopole term. The next term, i.e., the $O(1/r^2)$ term in γ_{00}—the dipole term—vanishes if the origin O is also the center of mass of the source. So that

$$\int \rho x'^i \, d^3x' = 0$$

and

$$g_{00} = 1 - \frac{2GM}{C^2r} + O\left(\frac{1}{r^3}\right).$$

Expressions for the nonvanishing γ_{0i} are

$$\gamma_{01} = -\frac{k}{2\pi} \int \frac{T_{01}(x'^i) \, d^3x'}{R}, \tag{3.31}$$

and

$$\gamma_{02} = -\frac{k}{2\pi} \int \frac{T_{02}(x'^i)}{R} d^3x'. \tag{3.32}$$

Putting in the expression for T_{0i}, and using the expansion (3.28) of $1/R$, we see that the $1/r$ term vanishes on integration over the angle ϕ. Regarding the next term, only $\cos^2 \phi$ and $\sin^2 \phi$ survive on integration so that we get, for the $1/r^3$ term,

$$\gamma_{01} = -\frac{kx^2}{2\pi r^3} \int T_{01} x^{2\prime} \, d^3x',$$

$$\gamma_{02} = -\frac{kx^1}{2\pi r^3} \int T_{02} x^{1\prime} \, d^3x'. \tag{3.33}$$

In special relativity, the x^3 component of the angular momentum is given by (in the present units)

$$J = c \int (x^{1\prime} T_{02} - x^{2\prime} T_{01}) \, d^3x'. \tag{3.34}$$

In view of (3.24b, c) and axial symmetry we have

$$\int x^{1\prime} T_{02} \, d^3x' = -\int x^{2\prime} T_{01} \, d^3x' = \frac{T}{2c}, \tag{3.35}$$

so that we obtain

$$\gamma_{01} = -\frac{2Gx^2}{c^3 r^3} J, \qquad \gamma_{02} = \frac{2Gx^1}{c^3 r^3} J. \tag{3.36}$$

To our order of approximation $\gamma_{ij} = 0$ for $i, j = 1, 2, 3$. Hence, we have

$$\gamma = -h = -\frac{4GM}{c^2 r}. \tag{3.37}$$

Using (3.37) we find

$$h_{00} = -\frac{2GM}{c^2 r}, \tag{3.38}$$

$$h_{ij} = \eta_{ij} \frac{2GM}{c^2 r}. \tag{3.39}$$

Finally, using (3.36), (3.38), and (3.39) the line element is

$$ds^2 = \left(1 - \frac{2GM}{c^2 r}\right) c^2 \, dt^2 - \left(1 + \frac{2GM}{c^2 r}\right)(dx^{1^2} + dx^{2^2} + dx^{3^2})$$

$$- \frac{4GJ}{c^3 r^3}(x^2 \, dx^1 - x^1 \, dx^2)c \, dt. \tag{3.40}$$

Transforming into spherical polar coordinates r, θ, and ϕ

$$(x^2 \, dx^1 - x^1 \, dx^2) = -r^2 \sin^2 \theta \, d\phi,$$

$$x^3 = r \cos \theta,$$

$$x^2 = r \sin \theta \sin \phi,$$

$$x^1 = r \sin \theta \cos \phi,$$

so that the line element takes the form

$$ds^2 = \left(1 - \frac{2GM}{c^2 r}\right) c^2\, dt^2 - \left(1 + \frac{2GM}{c^2 r}\right)(dx^{1^2} + dx^{2^2} + dx^{3^2})$$

$$+ \frac{4GJ}{c^2 r} \sin^2 \theta\, d\phi\, dt. \tag{3.41}$$

Here

$$dx^{1^2} + dx^{2^2} + dx^{3^2} = dr^2 + r^2\, d\theta^2 + r^2 \sin^2 \theta\, d\phi^2.$$

The above metric (3.41) enables us to identify the mass and angular momentum of the source and, in particular, shows that the terms h_{0i} in the metric indicates the presence of angular momentum, its magnitude per unit mass being given by half the ratio of h_{0i} at the equator to h_{00}. More generally, considering an asymptotically flat space we may show that for $r \to \infty$ we may reduce the metric to the form, writing $GM = m$,

$$ds^2 = \left[1 - \frac{2m}{r} + \frac{2m^2}{r^2} + O\left(\frac{1}{r^3}\right)\right] dt^2$$

$$+ \left[4\varepsilon_{jkl} \frac{S^k x^l}{r^3} + O\left(\frac{1}{r^3}\right)\right] dt\, dx^j$$

$$- \left[\left(1 + \frac{2m}{r}\right)\delta_{ik} + O\left(\frac{1}{r^3}\right)\right] dx^i\, dx^k, \tag{3.42}$$

where S^k is the angular momentum vector of the source distribution.

3.6. The Action Principle and the Energy–Momentum Tensors

Hamilton's principle in classical mechanics starts with a scalar function $L = L(q_j, \dot{q}_j, t)$, built from the generalized coordinates q_i and the velocity $\dot{q}_i$. The action integral A is given by

$$A = \int_{t_i}^{t_f} L\, dt$$

from an initial state to a final state corresponding to times t_i and t_f, respectively. The actual motion is specified by that path $q_j = \bar{q}_j$ for which A is stationary, i.e., $\delta A = 0$ for the arbitrary infinitesimal variation of q_j, subject to the conditions that these variations vanish at the terminal points. The times t_i and t_f are not to be varied. Such a variation leads to the Lagrange equations. The procedure mentioned above can be extended to fields also. Suppose a field is described by a series of independent variables ϕ^A ($A = 1, 2, \ldots$) of space–time coordinates x^α. We can construct a scalar Lagrangian density from ϕ^A and its derivatives $\phi^A_{,\alpha}$. The corresponding action function is

$$A = \int L(\phi^A, \phi^A, \alpha)\sqrt{-g}\, d^4x, \tag{3.43}$$

integration extending over a domain of the space–time manifold, with the boundary denoted by $\sum$. The equations satisfied by ϕ^A can be derived by varying the action and putting $\delta A = 0$ for infinitesimal variations of ϕ^A. The variation $\delta\phi^A$, of course, must vanish at the boundary where is here a three-dimensional hypersurface. The procedure is illustrated in the following. In general,

$$\delta A = \delta \int L\sqrt{-g}\, d^4x = 0,$$

(3.44)

leads to

$$\int \sum_A \left(\frac{\partial L}{\partial \phi^A}\delta\phi^A + \frac{\partial L}{\partial \phi^A_{,\alpha}}\delta\phi^A_{,\alpha}\right)\sqrt{-g}\, d^4x = 0,$$

or

$$\int \sum_A \left[\frac{\partial L}{\partial \phi^A}\sqrt{-g} - \frac{\partial}{\partial x^\alpha}\left(\frac{\partial L}{\partial \phi^A_{,\alpha}}\sqrt{-g}\right)\right]\delta\phi^A\, d^4x = 0,$$

or

$$\frac{\partial L}{\partial \phi^A} - \frac{1}{\sqrt{-g}}\frac{\partial}{\partial x^\alpha}\left(\frac{\partial L}{\partial \phi^A_{,\alpha}}\sqrt{-g}\right) = 0.$$

(3.45)

For a scalar field ϕ the obvious scalar Lagrangian is

$$L = \phi_{,\beta}\phi^{,\beta} + m^2\phi$$

(if we are not to go beyond the first derivatives).

Then the above equation (3.45) becomes

$$m^2\phi - \frac{1}{\sqrt{-g}}\frac{\partial}{\partial x^\alpha}(\phi_{,\beta}g^{\beta\alpha}\sqrt{-g}) = 0,$$

which can also be written in the form

$$m^2\phi - \phi^{;\alpha}_{;\alpha} \equiv m^2\phi - \Box\phi = 0.$$

(3.46)

This is the Klein–Gordon equation for the massive scalar field with m as the mass term. In the second case, we consider the electromagnetic field. The appropriate Lagrangian for the source-free electromagnetic field is expressed as

$$L = -\frac{1}{16\pi}F^{\mu\nu}F_{\mu\nu} \quad \text{with} \quad F_{\mu\nu} \equiv A_{\mu,\nu} - A_{\nu;\mu}.$$

Here the field variables are the vector potential components A_μ. The variation of the Lagrangian, therefore, gives us

$$\delta L = -\frac{1}{8\pi}F^{\mu\nu}\delta F_{\mu\nu}$$

$$= -\frac{1}{8\pi}F^{\mu\nu}\delta(A_{\mu,\nu} - A_{\nu,\mu})$$

$$= -\frac{1}{4\pi}F^{\mu\nu}\delta A_{\mu,\nu}.$$

(3.47)

In deriving the expression (3.47) we have utilized the antisymmetric property of the field tensor $F^{\mu\nu}$. This relation immediately gives, for the derivative of the Lagrangian,

$$\frac{\partial L}{\partial A_{\mu,\nu}} = -\frac{1}{4\pi} F^{\mu\nu}, \tag{3.48}$$

which, when substituted in the general relation (3.45), gives

$$\frac{\partial}{\partial x^\nu}(F^{\mu\nu}\sqrt{-g}) = 0,$$

or, in other words,

$$F^{\mu\nu}_{;\nu} = 0. \tag{3.49}$$

This can immediately be recognized as one set of Maxwell's field equations for the source-free electromagnetic field. On the other hand, for an electromagnetic field in the presence of sources, the suitable Lagrangian is

$$L = -\frac{1}{16\pi} F^{\mu\nu} F_{\mu\nu} + A_\mu J^\mu. \tag{3.50}$$

Using this Lagrangian expression in the general relation (3.45) and putting $\phi^A = A^\mu$ we obtain

$$J^\mu - \frac{1}{\sqrt{-g}}\frac{\partial}{\partial x^\nu}\left(-\frac{1}{4\pi}F^{\mu\nu}\sqrt{-g}\right) = 0,$$

or

$$F^{\mu\nu}_{;\nu} = -4\pi J^\mu.$$

which can also be written, in view of the antisymmetric character of $F^{\mu\nu}$, in the form

$$F^{\nu\mu}_{;\nu} = 4\pi J^\mu. \tag{3.51}$$

The question (3.51) is the well-known Maxwell's equation in the presence of the sources.

3.7. The Energy–Stress Tensor

So far we have discussed the variation with respect to only the field variables. We now proceed to show that the variation with respect to the metric tensor yields a symmetric tensor whose covariant divergence vanishes. This tensor is identified with the energy–momentum complex.

An infinitesimal coordinate transformation $x'^\alpha = x^\alpha - \xi^\alpha$ changes the metric $g_{\mu\nu}$ to $g'_{\mu\nu}$:

$$g'_{\mu\nu} = g_{\alpha\beta}\frac{\partial x^\alpha}{\partial x'^\mu}\frac{\partial x^\beta}{\partial x'^\nu}$$

$$= g_{\alpha\beta}\left[\delta^\alpha_\mu + \frac{\partial \xi^\alpha}{\partial x'^\mu}\right]\left[\delta^\beta_\nu + \frac{\partial \xi^\beta}{\partial x'^\nu}\right]$$

$$= g_{\mu\nu} + g_{\alpha\nu}\frac{\partial\xi^\alpha}{\partial x'^\mu} + g_{\alpha\mu}\frac{\partial\xi^\alpha}{\partial x'^\nu}$$

$$= g_{\mu\nu} + \frac{\partial\xi_\nu}{\partial x'^\mu} + \frac{\partial\xi_\mu}{\partial x'^\nu} - \xi^\alpha\frac{\partial g_{\alpha\nu}}{\partial x'^\mu} - \xi^\alpha\frac{\partial g_{\alpha\mu}}{\partial x'^\nu}$$

$$= g_{\mu\nu} + \xi_{\nu;\mu} + \xi_{\mu;\nu} - \xi^\alpha\frac{\partial g_{\mu\nu}}{\partial x^\alpha}$$

$$= g_{\mu\nu}(x'^\alpha) + \xi_{\mu;\nu} + \xi_{\nu;\mu},$$

or

$$\delta g_{\mu\nu} = \xi_{\mu;\nu} + \xi_{\nu;\mu}. \tag{3.52}$$

Since the action A is a scalar, it does not change under a transformation of coordinates. Under the coordinate transformations, each of the field variables ϕ^A changes by $\delta\phi^A$. But the terms containing the change of the field variable vanish in view of the corresponding field equations. Thus, we need to consider only the terms containing $\delta g_{\mu\nu}$ and set these variations to zero at the integration limits. Thus

$$\delta A = \int\left[\frac{\partial(L\sqrt{-g})}{\partial g^{\mu\nu}}\delta g^{\mu\nu} + \frac{\partial(L\sqrt{-g})}{\partial g^{\mu\nu}_{,\alpha}}\delta(g^{\mu\nu}_{,\alpha})\right]d^4x$$

$$= \left[\frac{\partial(L\sqrt{-g})}{\partial g^{\mu\nu}}\delta g^{\mu\nu} + \frac{\partial}{\partial x^\alpha}\left(\frac{\partial(L\sqrt{-g}}{\partial g^{\mu\nu}_{,\alpha}}\delta g^{\mu\nu}\right) - \frac{\partial}{\partial x^\alpha}\left(\frac{\partial(L\sqrt{-g})}{\partial g^{\mu\nu}_{,\alpha}}\right)\delta g^{\mu\nu}\right]d^4x. \tag{3.53}$$

The contribution to δA from the second term of the integrand in (3.53) will be zero, as it can be reduced to a surface integral over the boundary where $\delta g^{\mu\nu} = 0$. Now defining a symmetric tensor $T_{\mu\nu}$ such that

$$\tfrac{1}{2}\sqrt{-g}\,T_{\mu\nu} \equiv \frac{\partial(L\sqrt{-g})}{\partial g^{\mu\nu}} - \frac{\partial}{\partial x^\alpha}\left(\frac{\partial(L\sqrt{-g})}{\partial g^{\mu\nu}_{,\alpha}}\right), \tag{3.54}$$

we have

$$\delta A = \tfrac{1}{2}\int T_{\mu\nu}\delta g^{\mu\nu}\sqrt{-g}\,d^4x$$

$$= -\tfrac{1}{2}\int T^{\mu\nu}\delta g_{\mu\nu}\sqrt{-g}\,d^4x. \tag{3.55}$$

In (3.55) we have utilized the relation $\delta g^{\mu\nu} = -g^{\mu\alpha}g^{\nu\beta}\delta g_{\alpha\beta}$. Using (3.52) in (3.55) we get

$$\delta A = -\tfrac{1}{2}\int T^{\mu\nu}(\xi_{\mu;\nu} + \xi_{\nu;\mu})\sqrt{-g}\,d^4x$$

$$= -\int T^{\mu\nu}\xi_{\mu;\nu}\sqrt{-g}\,d^4x$$

$$= -\int(T^{\mu\nu}\xi_\mu)_{;\nu}\sqrt{-g}\,d^4x + \int T^{\mu\nu}_{;\nu}\xi_\mu\sqrt{-g}\,d^4x. \tag{3.56}$$

The first term on the right-hand side can be converted into a surface integral where the ξ_μ's vanish. Further, due to arbitrariness of the ξ_μ, the $\delta A = 0$ leads us to $T^{\mu\nu}_{;\nu} = 0$. Comparing this with the conservation relation $T^{\mu\nu}_{;\nu} = 0$ in Lorentz space, it is natural to identify $T^{\mu\nu}$ with the energy–momentum tensor, at least to within a constant factor. From the expression (3.55), if L does not depend on $g^{\mu\nu}_{,\alpha}$ explicitly, we have

$$\tfrac{1}{2}\sqrt{-g}\,T_{\mu\nu} = \frac{\partial L}{\partial g^{\mu\nu}}\sqrt{-g} + L\frac{\partial(\sqrt{-g})}{\partial g^{\mu\nu}}, \tag{3.57}$$

or,

$$T_{\mu\nu} = 2\frac{\partial L}{\partial g^{\mu\nu}} - Lg_{\mu\nu}. \tag{3.58}$$

So once the suitable Lagrangian can be defined to yield the correct field equations from the variational procedure, the corresponding energy–momentum tensor can be obtained from the prescription (3.58). The above procedure is not only valid for sourceless fields but can also be used in the presence of sources along with the associated interaction terms.

We may illustrate the above procedure by calculating the energy–momentum tensor for a massive scalar field mentioned previously. It is a straightforward matter to calculate this from (3.58) utilizing the Lagrangian for such a field $L_s = m^2\phi^2 + \phi_{,\alpha}\phi^{,\alpha}$, and it is given by

$$
\begin{aligned}
(T_{\mu\nu})_s &= 2\frac{\partial}{\partial g^{\mu\nu}}(\phi_{,\alpha}\phi^{,\alpha} + m^2\phi^2) - (\phi_\alpha\phi^{,\alpha} + m^2\phi^2)g_{\mu\nu} \\
&= 2(\phi_\mu\phi_\nu - \tfrac{1}{2}g_{\mu\nu}\phi_{,\alpha}\phi^{,\alpha}) - m^2\phi^2 g_{\mu\nu}.
\end{aligned} \tag{3.59}
$$

We can proceed in a similar manner for an electromagnetic field using the appropriate Lagrangian

$$L_{em} = -\frac{1}{16\pi}F_{\mu\nu}F^{\mu\nu}$$

to obtain its energy–momentum tensor

$$(T_{\mu\nu})_{em} = \frac{1}{4\pi}\left[-F_{\mu\alpha}F^\alpha_\nu + \tfrac{1}{4}g_{\mu\nu}F_{\alpha\beta}F^{\alpha\beta}\right]. \tag{3.60}$$

3.8. The Einstein Equations from the Variational Principle

The Einstein field equation can be derived from the principle of least action $\delta(A_g + A_m) = 0$, where A_g and A_m are the actions of the gravitational field and matter, respectively. For the gravitational field we have to choose the Lagrangian suitably out of the many ways of constructing scalars from the metric tensors and their derivatives. The simplest of them is the Ricci scalar R, which is a function of the metric tensors and their derivatives. Unlike the usual expressions for the Lagrangian, the Ricci scalar contains second derivatives of $g_{\mu\nu}$, which, however, do not introduce any additional complications. This is because these terms build up a divergence expression which falls off in the variation process.

We consider the variation of the integral

$$A_g = \int R\sqrt{-g}\, d^4x,$$

defined on a space–time region V with a bounding 3-surface where the variation $\delta g_{\mu\nu}$ vanishes. Considering the variation in $g_{\mu\nu}$ we have

$$\delta A_g = \int [\delta R_{\mu\nu}g^{\mu\nu}\sqrt{-g} + R_{\mu\nu}\delta g^{\mu\nu}\sqrt{-g} + R_{\mu\nu}g^{\mu\nu}\delta\sqrt{-g}]\, d^4x$$

$$= \int (R_{\mu\nu} - \tfrac{1}{2}g_{\mu\nu}R)\delta g^{\mu\nu}\sqrt{-g}\, d^4x + \int g^{\mu\nu}\delta R_{\mu\nu}\sqrt{-g}\, d^4x. \quad (3.61)$$

The second integral on the right vanishes as may be seen in the following manner:

$$\delta R_{\mu\nu} = \Gamma^{\alpha}_{\mu\beta}(\delta\Gamma^{\beta}_{\alpha\nu}) + (\delta\Gamma^{\alpha}_{\mu\beta})\Gamma^{\beta}_{\alpha\nu} - \Gamma^{\alpha}_{\mu\nu}(\delta\Gamma^{\beta}_{\alpha\beta}) - (\delta\Gamma^{\alpha}_{\mu\nu})\Gamma^{\beta}_{\alpha\beta} + (\delta\Gamma^{\alpha}_{\mu\alpha})_{,\nu} - (\delta\Gamma^{\alpha}_{\mu\nu})_{,\alpha}.$$

$$(3.62)$$

Although the Γ's are not tensors, the $\delta\Gamma$'s being differentes of two Γ's are; hence, the above can be written as

$$\delta R_{\mu\nu} = (\delta\Gamma^{\alpha}_{\mu\alpha})_{;\nu} - (\delta\Gamma^{\alpha}_{\mu\nu})_{;\alpha}. \quad (3.63)$$

So, the integral becomes

$$\int g^{\mu\nu}[(\delta\Gamma^{\alpha}_{\mu\alpha})_{;\nu} - (\delta\Gamma^{\alpha}_{\mu\nu})_{;\alpha}]\sqrt{-g}\, d^4x$$

$$= \int [(g^{\mu\nu}\delta\Gamma^{\alpha}_{\mu\alpha})_{;\nu} - (g^{\mu\nu}\delta\Gamma^{\alpha}_{\mu\nu})_{;\alpha}]\sqrt{-g}\, d^4x$$

$$= \int C^{\nu}_{;\nu}\sqrt{-g}\, d^4x = \int (C^{\nu}\sqrt{-g})_{,\nu}\, d^4x, \quad (3.64)$$

where the vector

$$C^{\nu} \equiv g^{\mu\nu}\delta\Gamma^{\alpha}_{\mu\alpha} - g^{\mu\alpha}\delta\Gamma^{\nu}_{\mu\alpha}, \quad (3.65)$$

and by the usual argument the ordinary divergence integral vanishes. Thus (3.61) reduces to

$$\delta A_g = \int (R_{\mu\nu} - \tfrac{1}{2}g_{\mu\nu}R)\delta g^{\mu\nu}\sqrt{-g}\, d^4x. \quad (3.66)$$

Now putting the variation of the action function $\delta A_g = 0$ corresponding to the variations in the $g_{\mu\nu}$'s, we get field equations of gravitation in the absence of any matter

$$R_{\mu\nu} - \tfrac{1}{2}g_{\mu\nu}R = 0. \quad (3.67)$$

We can generalize these results in the presence of the matter field or any other field. The equations in this case are obtained from the principle of least action $\delta(A_g + kA_m) = 0$. The action function for the nongravitational field is

$$A_m = \int L_m \sqrt{-g}\, d^4x,$$

and its variation, according to the general relation (3.55), is given by

$$\delta A_m = \tfrac{1}{2} \int (T_{\mu\nu})_m \delta g^{\mu\nu} \sqrt{-g}\, d^4x. \tag{3.68}$$

Combining (3.66) with (3.68) we have, from $\delta(A_g + kA_m) = 0$, Einstein's field equations in the presence of a nongravitational field.

$$R_{\mu\nu} - \tfrac{1}{2} g_{\mu\nu} R + k T_{\mu\nu} = 0, \tag{3.69}$$

where k is the coupling constant, which from weak field limits can be identified with $8\pi G/C^4$.

Problems

1. Show that the variational principle $\delta \int R\sqrt{-g}\, d^4x = 0$ is identical to $\delta \int L\sqrt{-g}\, d^4x = 0$ where

$$L \equiv -g^{\mu\nu}(\Gamma^\beta_{\mu\alpha}\Gamma^\alpha_{\beta\nu} - \Gamma^\alpha_{\mu\nu}\Gamma^\beta_{\alpha\beta}).$$

(Note that L is not a scalar and unlike R does not involve second derivatives of $g_{\mu\nu}$.)

2. Show that if the equations of motion are obtained from an action principle $\delta \int L\sqrt{-g}\, d^4x = 0$, where L is a function of the field variables ϕ^α and their first derivatives $\phi^\alpha_{,\beta}$, then

$$T^\beta_\gamma \equiv \frac{\partial(L\sqrt{-g})}{\partial \phi^\alpha_{,\beta}} \phi^\alpha_{,\gamma} - L\sqrt{-g}\, \delta^\beta_\gamma$$

is ordinary divergence free, i.e., $T^\beta_{\gamma,\beta} = 0$.

3. Using the results of Problems 1 and 2, show that the ordinary (i.e., noncovariant) divergence of $T^\nu_\mu \sqrt{-g} + t^\nu_\mu$ vanishes where

$$t^\alpha_\beta = \frac{1}{16\pi} \left[-\sqrt{-g}\, g^{\mu\nu}_{,\beta} \frac{\partial(L\sqrt{-g})}{\partial(g^{\mu\nu}_{,\alpha}\sqrt{-g})} + \delta^\alpha_\beta L\sqrt{-g} \right],$$

L being defined as in Problem 1; t^α_β was identified by Einstein as the energy–momentum complex of the gravitational field but it is not a tensor.

4. The Schwarzschild Metric and Crucial Tests

4.1. The Schwarzschild Solution

The exact solution of Einstein's field equation in empty space, obtained by Schwarzschild in 1916, describes the geometry of space–time outside a spherically symmetric distribution of matter. Spherical symmetry means an invariance under any arbitrary rotation axes at a particular point called the center of symmetry. Using polar coordinates r, θ, and ϕ, and choosing the origin at the center of symmetry, we have the general form of the line element with spherical symmetry

$$ds^2 = A \, dt^2 + 2B \, dt \, dr - C \, dr^2 - D(d\theta^2 + \sin^2 \theta \, d\phi^2). \tag{4.1}$$

The invariants under the rotation mentioned above are the radial coordinates r, dr, and $(d\theta^2 + \sin^2 \theta \, d\phi^2)$. In the line element (4.1) A, B, C, and D are functions of the radial coordinate r and the time t, in general. In the present case, however, the field being static, we assume that they are functions of the radial coordinate alone. The coordinate transformation like $t' = t + \int BA^{-1} \, dr$ eliminates g_{01}, the nondiagonal element in the metric. Further, without any loss of generality, we make a transformation, so that D itself becomes the radial coordinate and (4.1) assumes the form

$$ds^2 = e^\nu \, dt^2 - e^\lambda \, dR^2 - R^2(d\theta^2 + \sin^2 \theta \, d\phi^2), \tag{4.2}$$

ν and λ being functions of R alone.

The coordinate R has the significance that the area of the surface of the sphere R = constant and is given by $4\pi R^2$. Now

$$g_{00} = e^\nu, \qquad g_{11} = -e^\lambda, \qquad g_{22} = -R^2, \qquad g_{33} = -R^2 \sin^2 \theta,$$

the coordinates t, R, θ, ϕ being numbered 0, 1, 2, 3, respectively, and the nonvanishing Christoffel symbols are

$$\Gamma_{10}^0 = \Gamma_{01}^0 = \frac{\nu'}{2}; \qquad \Gamma_{11}^1 = \frac{\lambda'}{2}; \qquad \Gamma_{22}^1 = -Re^{-\lambda}, \qquad \Gamma_{33}^1 = -R \sin^2 \theta e^{-\lambda},$$

$$\Gamma_{00}^1 = \frac{\nu'}{2} e^{\nu-\lambda}; \qquad \Gamma_{12}^2 = \Gamma_{21}^2 = \Gamma_{13}^3 = \Gamma_{31}^3 = \frac{1}{R}, \tag{4.3}$$

$$\Gamma_{33}^2 = -\sin \theta \cos \theta; \qquad \Gamma_{23}^3 = \Gamma_{32}^3 = \cot \theta,$$

where the primes indicate differentiation with respect to R.

A straightforward calculation of $R_{\mu\nu}$ from (4.2) and (4.3) leads to the following nontrivial equations:

$$G_1^1 \equiv R_1^1 - \tfrac{1}{2}R = e^{-\lambda}\left(\frac{v'}{R} + \frac{1}{R^2}\right) - \frac{1}{R^2} = 0, \tag{4.4}$$

$$G_3^3 = G_2^2 \equiv R_2^2 - \tfrac{1}{2}R = \tfrac{1}{2}e^{-\lambda}\left(v'' + \frac{v'^2}{2} + \frac{v' - \lambda'}{R} - \frac{v'\lambda'}{2}\right) = 0, \tag{4.5}$$

$$G_0^0 = R_0^0 - \tfrac{1}{2}R = e^{-\lambda}\left(\frac{1}{R^2} - \frac{\lambda'}{R}\right) - \frac{1}{R^2} = 0. \tag{4.6}$$

Here G_ν^μ is the Einstein tensor, i.e., $G_\nu^\mu \equiv R_\nu^\mu - \tfrac{1}{2}R\delta_\nu^\mu$. Adding (4.4) and (4.6) we have

$$\lambda' + v' = 0. \tag{4.7}$$

Equation (4.6) can be written as

$$\frac{d}{dR}(e^{-\lambda}R) = 1, \tag{4.8}$$

which on integration yields

$$e^{-\lambda} = \left(1 - \frac{2m}{R}\right), \tag{4.9}$$

where m is, so far, an ordinary constant of integration.

In view of (4.7) the solution for g_{00} can be easily obtained

$$e^{v} = e^{-\lambda} = \left(1 - \frac{2m}{R}\right). \tag{4.10}$$

Here we have used the boundary condition that at large distances from the source as $R \to \infty$, the space–time tends to be flat, i.e., $e^{v} \to 1$ and $e^{\lambda} \to 1$. To understand the significance of the integration constant m, we recall (3.10) for weak fields which shows that for large R the gravitational potential $\phi = m/R$. However, the Newtonian potential ϕ is $-GM/R$ where M is the mass of the gravitating body. Hence $m = GM$. All this has been obtained by taking the velocity of light c to be unity. If we retain c, we get $m = GM/c^2$. Hence the Schwarzschild metric is

$$ds^2 = \left(1 - \frac{2m}{R}\right)dt^2 - \left(1 - \frac{2m}{R}\right)^{-1}dR^2 - R^2(d\theta^2 + \sin^2\theta\, d\phi^2)$$

$$= \left(1 - \frac{2GM}{c^2R}\right)dt^2 - \left(1 - \frac{2GM}{c^2R}\right)^{-1}dR^2 - R^2(d\theta^2 + \sin^2\theta\, d\phi^2). \tag{4.11}$$

(Note that for the Sun, $M_\odot \approx 2 \times 10^{33}$ g and $m_\odot \approx 1.5$ km. For the Earth, the corresponding values are $M_\oplus \approx 6 \times 10^{27}$ g and $m_\oplus \approx 0.44$ cm.)

4.2. Birkhoff's Theorem

Birkhoff's theorem states that the spherically symmetric vacuum space–time has a Schwarzschild solution even if the metric is not explicity assumed to be static.* To see this, we write out the field equations taking e^ν and e^λ to depend on t as well

$$G_0^0 = e^{-\lambda}\left(\frac{1}{R^2} - \frac{\lambda'}{R}\right) + \frac{1}{R^2} = 0, \tag{4.12}$$

$$G_1^1 = e^{-\lambda}\left(\frac{\nu'}{R} + \frac{1}{R^2}\right) - \frac{1}{R^2} = 0, \tag{4.13}$$

$$G_2^2 = G_3^3 = -\tfrac{1}{2}e^{-\lambda}\left(\nu'' + \frac{\nu'^2}{2} + \frac{\nu' - \lambda'}{R} - \frac{\nu'\lambda}{2}\right) + \tfrac{1}{2}e^{-\nu}\left(\ddot\lambda + \frac{\dot\lambda^2}{2} - \frac{\dot\lambda\dot\nu}{2}\right), \tag{4.14}$$

$$G_0^1 = -\tfrac{1}{2}e^{-\lambda}\frac{\dot\lambda}{R} = 0. \tag{4.15}$$

The field equation (4.15) shows that λ is a function of the radial coordinate R alone; (4.12) and (4.13) together yield the same relation (4.7), which when integrated gives us $e^\nu = e^{f(t)}e^\lambda$. Now the transformation of the time coordinate $t' = \int e^{f(t)/2}\, dt$ leaves the spatial coordinates unaffected, so that we again have $g_{00} = |g_{11}|^{-1} = (1 - 2m/R)$ as in (4.10). We thus arrive at the conclusion that the spherically symmetric gravitational field in vacuum is necessarily static and of the Schwarzschild form. The consequence of Birkhoff's theorem is that a radially pulsating distribution of matter can emit no gravitational waves.

We now introduce the so-called isotropic form of the Schwarzschild metric. The relevant coordinate transformation is

$$R = r\left(1 + \frac{m}{2r}\right)^2. \tag{4.16}$$

A straightforward calculation then shows that, in the new coordinate system, the Schwarzschild line element becomes

$$ds^2 = \left(\frac{1 - m/2r}{1 + m/2r}\right)^2 dt^2 - \left(1 + \frac{m}{2r}\right)^4 (dr^2 + r^2\, d\theta^2 + r^2 \sin^2\theta\, d\phi^2). \tag{4.17}$$

It is interesting to note that from (4.16) we have the equation

$$r^2 + (m - R)r + \frac{m^2}{4} = 0,$$

---

* We showed the possibility of transforming (4.1) to the form (4.2) for the static case. However, such a diagonalization is possible for the general spherically symmetric metric.

and the solution of r is given by

$$r = \frac{-(m - R) \pm \sqrt{(m - R)^2 - m^2}}{2}. \tag{4.18}$$

Thus for $0 \leq R < 2m$, r is complex, and hence the so-called black hole region is not covered in the real domain of r. Again, for $R > 2m$, any value of R gives two real positive values of r, so that this region is mapped out twice in the r space, once in the region $\frac{1}{2}m \leq r < \infty$ and again in the region $0 \leq r \leq m/2$.

The peculiar feature of the new coordinate system is that although there exists a singularity at $R = 0$, it is excluded in the space–time region covered by the new set of coordinates. However, the absence of singularity in the space–time covered by the new coordinates poses a problem for finding a criterion for the regularity of space–time. One way to define a space–time to be nonsingular is to demand that all timelike and null geodesics are complete in the sense that they can be extended to arbitrary values of their affine parameter. The regularity in space–time means that the space–time must contain the complete history of free test particles.

Problems

1. Calculate the scalar $R_{\mu\nu\alpha\beta}R^{\mu\nu\alpha\beta}$ to show that it is finite for all values of R except $R = 0$.

2. Calculate the value of m for a neutron whose mass is 1.66×10^{-24} g.

3. Set up the field equation in isotropic coordinates and thereby directly obtain the metric (4.17).

4. Using the field equation in the nonstatic case with isotropic coordinates, prove Birkhoff's theorem.

4.3. Three Crucial Tests

As in the weak field limit, the general theory of relativity gives the Newtonian equation for the motion of test particles in a gravitational field, we may say that any observation, which is consistent with Newtonian motion, gives indirect support to the general theory of relativity. However, a crucial observational test will be one which will be able to discriminate between the predictions of the general theory of relativity and that of Newtonian theory. In the past, three such tests decided unequivocally in favor of the general theory of relativity.

All three tests are based on studies in the spherically symmetric field, i.e., they are basically tests of the Schwarzschild metric and the geodesic postulate.

Test I

Perihelion Motion of a Planet
Consider the motion of a planet. Although in reality it is a two-body problem, as the mass of the planet is much smaller than that of the Sun, we can regard the planet as a test particle moving in the field of the Sun. Again the Sun is, at least to a high degree of approximation, a spherical body. So the geometry around it is given by the Schwarzschild metric and the planet will move in geodesics in this field.

The geodesics are given by the Euler–Lagrange variational principle

$$\delta \int \left[\left(1 - \frac{2m}{R} \right) \left(\frac{dt}{ds} \right)^2 - \left(1 - \frac{2m}{R} \right)^{-1} \left(\frac{dR}{ds} \right)^2 \right.$$
$$\left. - R^2 \left(\frac{d\theta}{ds} \right)^2 - R^2 \sin^2 \theta \left(\frac{d\phi}{ds} \right)^2 \right] ds = 0. \quad (4.19)$$

The variation (4.19) yields the following equations of motion:

$$\frac{d}{ds} \left(R^2 \frac{d\theta}{ds} \right) = R^2 \sin \theta \cos \theta \left(\frac{d\phi}{ds} \right)^2, \quad (4.20)$$

$$\frac{d}{ds} \left(R^2 \sin^2 \theta \frac{d\phi}{ds} \right) = 0, \quad (4.21)$$

$$\frac{d}{ds} \left[\left(1 - \frac{2m}{R} \right) \frac{dt}{ds} \right] = 0. \quad (4.22)$$

We have not written out the R equation. It is complicated but its first integral is simply given by the metric, namely,

$$\left(1 - \frac{2m}{R} \right) \left(\frac{dt}{ds} \right)^2 - \left(1 - \frac{2m}{R} \right)^{-1} \left(\frac{dR}{ds} \right)^2 - R^2 \left(\frac{d\theta}{ds} \right)^2 - R^2 \sin^2 \theta \left(\frac{d\phi}{ds} \right)^2 = 1. \quad (4.23)$$

If we take $\theta = \pi/2$ where $d\theta/ds = 0$, then from (4.20) θ is constant $= \pi/2$. Equation (4.21) then gives, on integration,

$$R^2 \dot{\phi} = h \quad \text{(constant)}. \quad (4.24)$$

Here and in what follows we denote derivative d/ds by means of an overdot. Equation (4.22) gives

$$\left(1 - \frac{2m}{R} \right) \dot{t} = \gamma \quad \text{(constant)}. \quad (4.25)$$

In (4.24) and (4.25) h and γ are constants and are associated with the angular momentum and energy of the particle, respectively. These are conservation principles corresponding to the cyclic nature of the coordinates ϕ and t.

Utilizing these results, (4.23) may be written as

$$\dot{R}^2 = \gamma^2 - \left(1 - \frac{2m}{R}\right)\left(1 + \frac{h^2}{R^2}\right). \tag{4.26}$$

Now since $(dR/d\phi)^2 = (\dot{R}/\dot{\phi})^2$ it is possible to eliminate the variable S from (4.26) with the help of (4.24). We get, writing $u \equiv 1/R$,

$$\left(\frac{du}{d\phi}\right)^2 = \frac{\gamma^2 - 1}{h^2} + \frac{2m}{h^2}u - u^2 + 2mu^3. \tag{4.27}$$

Differentiating (4.27) with respect to ϕ, we finally get

$$\frac{d^2u}{d\phi^2} + u = \frac{m}{h^2} + 3mu^2. \tag{4.28}$$

Comparing (4.28) with the corresponding Newtonian equation of motion we see that the difference is the appearance of the last term $3mu^2$. However, the term $3mu^2$ is quite small compared to the "Newtonian" term m/h^2, as may easily be seen,

$$\frac{3mu^2}{(m/h^2)} = 3h^2u^2 = 3R^2\left(\frac{d\phi}{ds}\right)^2 \sim \frac{3v^2}{c^2} \ll 1,$$

where v is the velocity of the planet. Since the term $3mu^2$ is very small we can adopt an approximation procedure for studying (4.28). In the zeroth approximation the equation is

$$\frac{d^2u}{d\phi^2} + u = \frac{m}{h^2}, \tag{4.29}$$

which is the Newtonian equation and has the solution $u = m/h^2(1 + e \cos \phi) = 1/l(1 + e \cos \phi)$.

Here l is the semi latus rectum of the orbit and is equal to h^2/m. Substituting this for u in the small term $3mu^2$ we get from (4.28)

$$\frac{d^2u}{d\phi^2} + u \approx \frac{m}{h^2} + \frac{3m}{l^2}(1 + e^2 \cos^2 \phi + 2e \cos \phi). \tag{4.30}$$

Equation (4.30) is of a familiar form met with in the theory of vibrations, ϕ replacing the time variable, so that the periodic terms involving $\cos^2 \phi$ and $\cos \phi$ give rise to "forced vibration" of the same period. Thus, the solution of this equation may be written as $u = u_0 + u_1 + u_2$, i.e.,

$$u_1 = \frac{3m}{l^2}e\phi \sin \phi, \tag{4.31}$$

$$u_2 = \left[\frac{3m}{l^2}\left(1 + \frac{e^2}{2}\right) - \frac{m}{l^2}\frac{e^2}{2}\cos 2\phi\right], \tag{4.32}$$

where u_1 and u_2 are both small. However, while the aperiodic factor in u_2

causes a small change in the latus rectum, the periodic term averaged over a large number of periods vanishes. Contrary to this, u_1 is aperiodic because of the occurrence of ϕ and hence its average effect over a large time does not vanish. To study its effect, we write $u = u_0 + u_1$, i.e.,

$$u = u_0 + u_1 = \frac{1}{l}(1 + e\cos\phi) + \frac{3m}{l^2}e\phi\sin\phi$$

$$= \frac{1}{l}[1 + e\cos(1 - \varepsilon)\phi],\qquad(4.33)$$

where ε is written for the small quantity $3m/l$. Hence

$$\varepsilon = \frac{3m}{l} = \frac{3m}{(h^2/m)} = \frac{3m^2}{h^2} = \frac{3m^2}{R^4(d\phi/ds)^2}$$

$$\approx \frac{3m^2}{R^2}\frac{c^2}{v^2}.$$

For planets on the solar system, $m/R < 10^{-5}$ and $v/c \sim 10^{-3}$, hence $\varepsilon < 10^{-4}$.

The point in the orbit nearest to the Sun is called the perihelion, i.e., R is minimum or u is maximum there. While in the Newtonian case the maxima of u occur after intervals of 2π in ϕ corresponding to closed orbits; with (4.33) the perihelions occur for $\phi(1 - \varepsilon) = 2\pi n$, i.e., at intervals of $\phi = 2\pi/(1 - \varepsilon) \approx 2\pi(1 + \varepsilon)$ and the orbits are open (Fig. 4.1). For most planets ε is so small that the perihelion motion due to general relativity cannot be observed accurately. Only in the case of the planet Mercury it has been accurately determined. The perihelion of Mercury's orbit shows a precession of 5599.74 ± 0.41 arc seconds per century. Of this 5557.18 ± 0.85 arc seconds can be explained as due to Newtonian effects, namely, the influence of other planets, the general precession of equinoxes, etc., leaving an unaccounted perihelion motion of 42.56 ± 0.94 arc seconds per century. To compute the general relativity effect, note that there are 415 revolutions of Mercury per century and the latus rectum l has the value 5.53×10^{12} cm and the mass of the Sun $= 1.476 \times 10^5$ cm (in $G = c = 1$ unit). Thus the precession per century due to the relativity effect is $2\pi\varepsilon \times 415 = 2\pi \cdot 3m/l \cdot 415 = 43.03$ arc seconds with an estimated error limit of ± 0.03 arc seconds. Thus the unaccounted part seems nicely explained.

The corresponding periastronic shift in the case of the Hulse–Taylor bi-

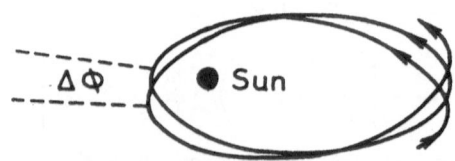

Figure 4.1. Showing perihelion motion.

nary pulsar is much larger and we shall discuss that later on. In fact, the observations of that pulsar period have apparently led to a verification of quite a number of results of the general theory of relativity, like the peri-astronic motion, the gravitational redshift, the transverse Doppler effect, and the rate of loss of energy due to gravitational waves from a revolving binary system.

Guinan and Maloney (1985) have apparently found a very significant difference between observation and theory in the case of a binary system in the constellation Hercules (observed periastronic precession $-0°.65 \pm 0°.18$ per century and the theoretically expected value is $-4°.27 \pm 0°.30$ per century), the reason for the discrepancy is not clear.

Test II

The Bending of Light

The equations of the null geodesics are identical in form to (4.20)–(4.22), except that the independent variable is a parameter λ. Also (4.23) becomes

$$\left(1 - \frac{2m}{R}\right)\dot{t}^2 - \left(1 - \frac{2m}{R}\right)^{-1}\dot{R}^2 - R^2(\dot{\theta}^2 + \sin^2\theta\dot{\phi}^2) = 0, \qquad (4.34)$$

where the overdot now represents $d/d\lambda$. Proceeding further in the same manner we get, in place of (4.28),

$$\frac{d^2u}{d\phi^2} + u = 3mu^2. \qquad (4.35)$$

If the deflecting source S were absent ($m = 0$), (4.35) would become

$$\frac{d^2u}{d\phi^2} + u = 0,$$

the solution of which is $u = u_0 \cos\phi$. This is the equation of a straight line $X'XX''$ in plane polar coordinates, where u_0 corresponds to the distance of closest approach X to the origin S and ϕ is zero for SX (see Fig. 4.2). Substituting the solution $u = u_0 \cos\phi$ in the small term $3mu^2$ (the approxima-

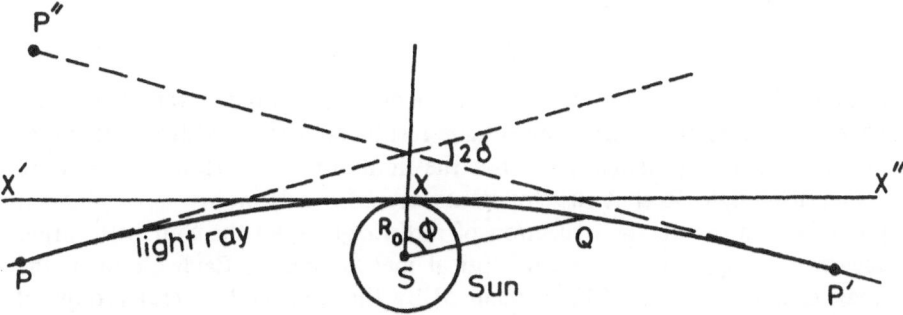

Figure 4.2. The deflection of light.

tion is quite justified because $3mu = 3m/R$ is sufficiently small), (4.35) becomes

$$\frac{d^2u}{d\phi^2} + u = \tfrac{3}{2}mu_0^2(1 + \cos 2\phi). \tag{4.36}$$

Equation (4.36) has the solution $u = u_0 \cos \phi + u_1$ where

$$u_1 = mu_0^2(2 - \cos^2 \phi).$$

The angle XSQ, where SQ is parallel to the asymptote of the light path, is given by putting $u = 0$, i.e.,

$$0 = u_0 \cos \phi + 2mu_0^2 - \tfrac{3}{2}mu_0^2 \cos^2 \phi,$$

which is a quadratic in $\cos \phi$, so that

$$\cos \phi = \tfrac{1}{2}\left[\frac{1}{mu_0} \pm \left\{\left(\frac{1}{mu_0}\right)^2 + 8\right\}^{1/2}\right]. \tag{4.37}$$

Since $mu_0 \equiv m/R_0$ is usually very small, $(1/mu_0)^2 \gg 8$ and we have approximately

$$\cos \phi = \tfrac{1}{2}\left[\frac{1}{mu_0} \pm \frac{1}{mu_0}(1 + 4m^2u_0^2)\right].$$

The positive sign is to be ignored since $\cos \phi$ cannot exceed unity. Thus

$$\cos \phi = -2mu_0,$$

or, again because $2mu_0$ is small,

$$\phi = \left(\frac{\pi}{2} + \delta\right),$$

where $\delta = 2mu_0$. The above results clearly indicate that the net deviation of the light ray $P'' \times P$ is $2\delta = 4mu_0$. Thus

$$\phi = 4mu_0 = \frac{4GM}{c^2R_0}. \tag{4.38}$$

Putting in the value of c, G, M, and R_0, corresponding to a ray grazing the surface of the Sun, we find

$$\phi = 8.48 \times 10^{-6} \text{ radians.}$$

Expressed in seconds, it is $1.75''$. Thus the change in angle is very small, and observations during solar eclipses gave results consistent with prediction of the general theory of relativity, but the limits of observational error were rather too high. Recent developments in radio astronomy have made it possible to measure the deflection of radio signals by the Sun with larger accuracy compared to those in optical measurements. Each October, the quasi-steller source 3C 279 is occulted by the Sun, and several groups of astronomers measured its apparent location during the occulation and at other times. The direction to another radio source 3C 273 is used as a standard for the measurement. Further, we have to correct for the deflection of the radio waves in the solar corona. As this deflection is frequency-dependent, observation at two different frequencies makes it possible to correct for this effect. The latest observation due to Fromalont and Sramek, taking account of all these, gave results in excellent agreement with the general theory of relativity (Fromalont and Sramek, 1975, 1976).

The latest test on the bending of electromagnetic waves in a gravitational field has been provided by the measurement of the deflection of radio signals in the solar gravitational field with VLBI (Robertson and Carter, 1984). The general relativity formula has been nicely confirmed.

Test III

The Gravitational Redshift

A light ray emitted from the surface of a star will appear to have its frequency altered when it reaches an observer on the Earth. This may be explained by the general theory of relativity in the following way. In a general static field the metric is

$$ds^2 = g_{00}\, dt^2 + g_{ik}\, dx^i\, dx^k,$$

with the $g_{\mu\nu}$'s independent of t. The coordinate time interval along a light track is

$$\int dt = \int \frac{|(g_{ik}\, dx^i\, dx^k)|^{1/2}}{|g_{00}|^{1/2}}. \tag{4.39}$$

The right-hand side of (4.39) is independent of t, so that two pulses emitted from the source at interval Δt will reach the observer also at interval Δt. However, the coordinate interval Δt corresponds to the different proper times at the source and the observer, because of different values of g_{00}. Hence the wavelengths λ_e (of emission) and λ_o (of observation) will be related as

$$\frac{\lambda_e}{\lambda_o} = \frac{c\, ds_e}{c\, ds_o} = \frac{|g_{00}|_e^{1/2}\Delta t}{|g_{00}|_o^{1/2}\Delta t} = \frac{|g_{00}|_e^{1/2}}{|g_{00}|_o^{1/2}}, \tag{4.40}$$

which, in the case of the Schwarzschild field, gives

$$\frac{\lambda_e}{\lambda_o} = \frac{(1 - 2m/R_e)^{1/2}}{(1 - 2m/R_o)^{1/2}} \approx \left(1 - \frac{2m}{R_e}\right)^{1/2} \approx 1 - \frac{m}{R_e}, \tag{4.41}$$

where the subscripts e and o refer to emission and observation, respectively. The above approximation is justified if the observer is situated sufficiently far from the source. The fractional redshift factor can therefore be written to a sufficient degree of accuracy as $z = \delta\lambda/\lambda \simeq m/R_e$, which for the Sun is of the order of 10^{-6}.

This, being small, can be easily swamped by the Doppler broadening by the random velocity of the emitting atom due to the large thermal effect on the surface of the Sun. The effect is somewhat larger when measured for the light from a white dwarf star.

Terrestrial Measurement of the Gravitational Redshift

After the discovery of the Mössbauer effect it was possible to measure the small shift in frequency produced by the Earth's gravitational field. Pound and Rebka (1960) measured the gravitational redshift in a terrestrial experiment using this principle. In this experiment, Fe^{57} samples were used both as the emitter and the absorber of the γ-radiation of energy 14.4 keV. The emitter and absorber were placed at the bottom and at the top of a vertical 22.6 m tower. Gamma rays emitted at the bottom suffered a gravitational redshift in traveling to the observer at the top where the gravitational field is comparitively weak. To attain the resonant absorption it was necessary to move the emitter upwards with a small appropriate velocity to result in a Doppler blue shift to compensate for the gravitational redshift. A measurement of the emitter's velocity would then give the magnitude of the redshift caused by the gravitational field. The whole experiment was, however, possible because the emitter was embedded in a crystal. This crystal prevented the recoil while emission took place by absorbing the momentum so generated, and thus could avoid any uncertain correction in the measurement. The calculations are simple.

If the height of the tower is h above the ground then

$$(g_{00})_o = 1 - \frac{2m}{R+h} \quad \text{and} \quad (g_{00})_e = \left(1 - \frac{2m}{R}\right), \quad (4.42)$$

where m and R are the mass and radius of the Earth. Since $2m \ll R$ we have from (4.41)

$$\frac{\lambda_o}{\lambda_e} = \frac{(1 - 2m/(R+h))^{1/2}}{(1 - 2m/R)^{1/2}} \approx 1 - \frac{m}{R+h} + \frac{m}{R}.$$

Hence the redshift factor is

$$z = \frac{\lambda_o - \lambda_e}{\lambda_e} = \frac{m}{R} - \frac{m}{R+h} \approx \frac{mh}{R^2}. \quad (4.43)$$

The approximation made in (4.43) could be justified, considering the smallness of the height h of the tower compared to the radius of the Earth. Recalling the relation $m \equiv GM/C^2$, we get from (4.43)

$$z = \frac{gh}{C^2} \approx 2.44 \times 10^{-15}. \quad (4.44)$$

The observations of Pound and Rebka agreed with the predictions of general relativity within the accuracy of the experimental results. A comparison of the frequencies of signals from hydrogen masers located in a space craft (altitude 10,000 km) and on the surface of the Earth was made by Vesset et al. (1980). The frequency change involved the gravitational redshift (a general relativity effect) and the second-order Doppler shift (a time dilation, i.e., special relativity effect). The results were in excellent agreement with the theory. A very recent verification of the gravitational redshift formula comes from a test involving the effect of Saturn. Voyager 1 was equipped with an ultrastable crystal oscillator (U.S.O.) with a precise frequency. As it moves in and out of the gravitational field of Saturn, there is a "dip" in the frequency of the radio signal transmitted. Assuming that the radiation in the Saturn magneto-sphere has had a negligible effect on the U.S.O., the gravitational radiation formula of general relativity is verified to an accuracy of 1% (Krisher et al., 1990).

More Recent Test

The Radar Echo Delay
The idea of this test is to measure the time required for radar signals to travel to an inner planet or satellite and back to the Earth in two circumstances:

(1) when the ray passes very near the Sun; and
(2) when the ray does not go near the Sun.

The null geodesic equations in the Schwarzschild field is (as before $\theta = \pi/2$ with a proper choice of coordinates)

$$0 = \left(1 - \frac{2m}{R}\right) dt^2 - R^2 \, d\phi^2 - \left(1 - \frac{2m}{R}\right)^{-1} dR^2,$$

or

$$\left(\frac{dR}{dt}\right)^2 = \left(1 - \frac{2m}{R}\right)^2 - R^2 \left(1 - \frac{2m}{R}\right)\left(\frac{d\phi}{dt}\right)^2, \tag{4.45}$$

$$R^2 \left(\frac{d\phi}{d\lambda}\right) = \text{constant} \quad \text{and} \quad \left(1 - \frac{2m}{R}\right)\left(\frac{dt}{d\lambda}\right) = \text{constant}.$$

The last two relations together give

$$R^2 \left(1 - \frac{2m}{R}\right)^{-1}\left(\frac{d\phi}{dt}\right) = \text{constant} = b \quad \text{(say)}.$$

Hence (4.45) becomes

$$\left(\frac{dR}{dt}\right)^2 = \left(1 - \frac{2m}{R}\right)^2 - \frac{b^2}{R^2}\left(1 - \frac{2m}{R}\right)^3. \tag{4.46}$$

Let PQE be the light ray from a planet P to the Earth E, Q being the point of closest approach to the Sun (Fig. 4.3). At P, Q, and E the distances are $R = R_1$, R_0, and R_2, respectively. The time taken by the light ray to travel from P to E is, from (4.48),

$$t = \int_{R_0}^{R_1} \frac{dR}{(1 - 2m/R)\{1 - b^2/R^2(1 - 2m/R)\}^{1/2}}$$

$$+ \int_{R_0}^{R_2} \frac{dR}{(1 - 2m/R)\{1 - b^2/R^2(1 - 2m/R)\}^{1/2}}.$$

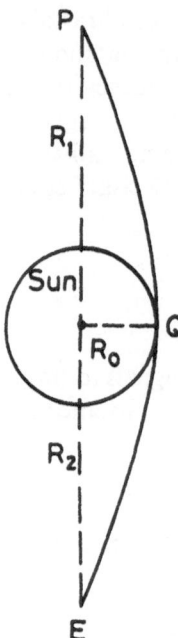

Figure 4.3

Now from (4.46)

$$\frac{dR}{dt} = \left(1 - \frac{2m}{R}\right)\left\{1 - \frac{b^2}{R^2}\left(1 - \frac{2m}{R}\right)\right\}^{1/2}, \tag{4.47}$$

and since $dR/dt = 0$ at $R = R_0$, i.e., at the point of closest approach, we have

$$1 - \frac{b^2}{R_0^2}\left(1 - \frac{2m}{R_0}\right) = 0,$$

or

$$b^2 = \frac{R_0^2}{(1 - 2m/R_0)}. \tag{4.48}$$

Therefore

$$1 - \frac{b^2}{R^2}\left(1 - \frac{2m}{R}\right) = 1 - \frac{R_0^2}{R^2}\frac{(1 - 2m/R)}{(1 - 2m/R_0)}. \tag{4.49}$$

Now, for the solar field, m/R_0 is a small quantity even if the ray grazes the surface. Hence, we approximate (4.49)

$$1 - \frac{b^2}{R^2}\left(1 - \frac{2m}{R}\right) = \left(1 - \frac{R_0^2}{R^2}\right)\left\{1 - \frac{2mR_0}{R(R + R_0)}\right\}.$$

By using the same approximation procedure further we finally get, for the transit time of the light ray from the planet to the Earth,

$$t = \int_{R_0}^{R_1} \left(1 - \frac{R_0^2}{R^2}\right)^{-1/2} \left\{1 + \frac{mR_0}{R(R + R_0)} + \frac{2m}{R}\right\} dR$$

$$+ \int_{R_0}^{R_2} \left(1 - \frac{R_0^2}{R^2}\right)^{-1/2} \left\{1 + \frac{mR_0}{R(R + R_0)} + \frac{2m}{R}\right\} dR.$$

In the absence of the gravitational deflection of light all terms in m will vanish and the above formula reduces to

$$t_0 = \int_{R_0}^{R_1} \left(1 - \frac{R_0^2}{R^2}\right)^{-1/2} dR + \int_{R_0}^{R_2} \left(1 - \frac{R_0^2}{R^2}\right)^{-1/2} dR. \tag{4.50}$$

Performing the elementary integrations, the time delay for a round trip is

$$2(t - t_0) \equiv \Delta t$$

$$= 4m \ln\left[\frac{(R_1 + \sqrt{R_1^2 - R_0^2})(R_2 + \sqrt{R_2^2 - R_0^2})}{R_0^2}\right]$$

$$+ 2m\left(\sqrt{\frac{R_1 - R_2}{R_1 + R_0}} + \sqrt{\frac{R_2 - R_0}{R_2 + R_0}}\right).$$

In particular, if R_0 is much less than both R_1 and R_2, the formula simplifies to

$$\Delta t \approx 4m\left(\ln\frac{4R_1 R_2}{R_0^2} + 1\right).$$

However, the above expression for the coordinate time delay is not very useful directly, for it requires the knowledge of $(R_1^2 - R_0^2)^{1/2}$, etc., to a high degree of accuracy, say to within 1 km; however, the differential systems of radial coordinates themselves differ by larger amounts. Besides, the electrons in the solar corona affect the time delay by an amount which shows considerable variation with time.

In the Viking mission two transponders landed on Mars and two others continued to orbit round it. The latter two transmitted two distinct bands of frequencies and thus the coronal effect could be corrected for. The final results agreed remarkably well with the prediction of general relativity.

4.4. The PPN Formalism

It has become customary to present the results in terms of the "parametrized post-Newtonian formalism" (shortened as PPN formalism). The basic ideas are:

(1) Just as in the general theory of relativity the gravitational field is a metric field in a four-dimensional space–time.

(2) In the space of the spherically symmetric field expressed in isotropic space coordinates, the $g_{\mu\nu}$'s can be expressed in the form of a power series in $1/r$, r being the radial coordinate, i.e.,

$$ds^2 = \left[1 - \frac{2m}{r} + 2\beta\left(\frac{m}{r}\right)^2 + \cdots\right]dt^2$$
$$- \left(1 + \frac{2\gamma m}{r} + \cdots\right)(dr^2 + r^2\,d\theta^2 + r^2\sin^2\theta\,d\phi^2).$$

The Geodetic Precession—Proposed Test

In 1966 Schiff suggested that the precession of the spin of a gyroscope falling freely in the field of the Earth could provide another test of the general theory of relativity. A free fall can be attained by placing the gyroscope in a satellite orbiting round the Earth. However, as yet the experiment has not been performed.

Here we present an outline of the derivation of the spin precession formula on the basis of the general theory of relativity. We first recall some results of Newtonian mechanics. We define a spin vector S^i related with the antisymmetric angular momentum tensor J_{kl}

$$S^i = \tfrac{1}{2}\varepsilon^{ikl}J_{kl}. \tag{4.51}$$

In the absence of external torques, the conservation of angular momentum gives

$$\frac{dS^i}{dt} = 0. \tag{4.52}$$

In general relativity (4.51) and (4.52) are modified to read

$$S^\alpha = \tfrac{1}{2}\varepsilon^{\alpha\beta\mu\sigma}J_{\beta\mu}v_\sigma, \tag{4.53}$$

and

$$S^\mu_{;\nu}v^\nu = 0, \tag{4.54}$$

where $\eta^{\alpha\beta\mu\sigma}$ is the Levi-Civita tensor and v^α is the velocity vector. From (4.53), we get

$$S^\alpha v_\alpha = 0. \tag{4.55}$$

In view of the above relation, we may express S_0 in terms of the 3-space components S_i and the velocity vector v^α. Again, if the spatial velocity of the gyroscope is small compared to the velocity of light, we can take $v^0 = 1$ and $v^i = dx^i/dt$, so that (4.54) written out explicitly gives

$$\frac{dS_i}{dt} - \Gamma^k_{i0}S_k + \Gamma^0_{i0}S_jv^j - \Gamma^k_{il}S_kv^l + \Gamma^0_{ik}v^kv^jS_j = 0. \tag{4.56}$$

As the gyroscope is in the field of the Earth, we may evaluate the Γ's occurring above by using the asymptotic metric for a body with axial symmetry and the angular momentum given in (3.42).

We thus obtain in three-dimensional vector notation

$$\frac{d\mathbf{S}}{dt} = \tfrac{1}{2}\mathbf{S} \times (\nabla \times \mathbf{S}) - 2(\mathbf{V} \cdot \mathbf{S})\nabla\varphi - \mathbf{S}(\mathbf{V} \cdot \nabla\varphi) + \mathbf{V}(\mathbf{S} \cdot \nabla\varphi), \quad (4.57)$$

where we have written

$$\phi = \frac{GM}{C^2 r}, \tag{4.58a}$$

$$\mathbf{S} = \frac{2G}{C^2 r^3}(\mathbf{r} \times \mathbf{J}), \tag{4.58b}$$

M and $\mathbf{J}$ being the mass and angular momentum of the central body.

Again from (4.54), $S^\mu S_\mu$ is conserved. In view of (4.55), $(S^0 S_0)/(S^i S_i) \sim O(v^2)$, hence to a high degree of approximation $S^i S_i$ is conserved. This indicates that the vector S^i only undergoes a change of orientation, hence the term spin precession. To give the spin precession equation an elegant form, we introduce a new spin vector $\mathbf{S}'$

$$\mathbf{S}' = (1 + \phi)\mathbf{S} - \tfrac{1}{2}\mathbf{v}(\mathbf{v} \cdot \mathbf{S}). \tag{4.59}$$

Note that $\mathbf{S}'$ differs from $\mathbf{S}$ by terms of the order of v^2, as ϕ is also of the same order in the case of an orbiting satellite. We then obtain, correct to the first order of approximation

$$\frac{d\mathbf{S}'}{dt} = \boldsymbol{\Omega} \times \mathbf{S}', \tag{4.60}$$

with

$$\boldsymbol{\Omega} = -\tfrac{1}{2}\nabla \times \mathbf{S} - \tfrac{3}{2}\mathbf{v} \times \nabla\phi. \tag{4.61}$$

Using (4.58a) and (4.58b), we get

$$\boldsymbol{\Omega} = \frac{3Gr(\mathbf{r} \cdot \mathbf{J})}{r^5} - \frac{GJ}{r^3} + \frac{3GM(\mathbf{r} \times \mathbf{v})}{2r^3}. \tag{4.62}$$

Obviously, the precession $\boldsymbol{\Omega}$ consists of two parts:

(a) The part represented by the last term, called the geodetic precession. It does not involve $\mathbf{J}$ and would be present in the case of the Schwarzschild field.
(b) The other terms involving $\mathbf{J}$ are referred to as the Lense–Thirring effect. This part does not involve the linear velocity $\mathbf{v}$ of the gyroscope.

For a satellite orbiting the Earth, the geodetic precession is much greater than the Lense–Thirring precession. For a circular orbit at an altitude of about 800 km, the geodetic precession is about 7 s per year while the Lense–Thirring precession is only about 0.1 s.

The actual determination of the spin precession depends on the measurement of the change of angle between the spin of the gyroscope and the light rays from distant stars. The expected precessional velocity is measureable, but elimination of spurious precessions due to gravitational torques (arising from nonhomogeneity of the spherical gyroscope) and the determination of the rest frame of distant stars are difficult problems.

Alternatively, we can present the result in the PPN formalism. In that case, the geodetic precession comes out as $(2\gamma + 1)/3$ times the general relativistic value. Any departure of γ from unity would indicate a defect of the general theory of relativity (or rather the Schwarzschild metric).

Problem

1. Show that for a gyroscope in a circular orbit about the Earth, the value of the geodetic precession is given by

$$\frac{3(GM_\oplus)^{3/2}}{2C^2 r^{5/2}}.$$

Hence check the numerical values given in the test for the geodetic and Lense–Thirring precession (take $G = 6.7 \times 10^{-8}$, $M_\oplus = 6 \times 10^{27}$, $R_\oplus = 6.4 \times 10^8$ all in e.g.s.).

Parametrized Post-Newtonian (PPN)

With the advance in technology, it has become possible to increase the accuracy of observations to such an extent that we can try to decide between various theories of gravitation (including the general theory of relativity) whose predictions differ only slightly. Many of these theories are based on the Riemannian metric, and for these a formalism evolved involving ten different parameters corresponding to the ten components of the metric tensor. This formalism was named the Parametrized Post-Newtonian or, in short, PPN formalism. Most of the experiments concern the solar gravitational field where the field is weak as indicated by the value of the Newtonian gravitational potential, i.e., $(|U|/c^2 \le 10^{-6})$. Now, if the theory implies the conservation principles of energy–momentum and angular momentum and there is no "preferred frame effects," then the number of parameters may be reduced from ten to two and the metric may be written as

$$g_{00} = 1 - 2U + 2\beta U^2 - (2\gamma + 2)\Phi_1 - 2(3\gamma - 2\beta + 1)\Phi_2 - 6\gamma\Phi_4,$$

$$g_{0j} = \tfrac{1}{2}(4\gamma + 3)V_j,$$

$$g_{jk} = -(1 + 2\gamma U)\delta_{jk},$$

with

$$U \equiv \int \frac{\rho'}{|\mathbf{x} - \mathbf{x}'|} d^3 x', \qquad \Phi_2 = \int \frac{\rho' U'}{|\mathbf{x} - \mathbf{x}'|} d^3 x', \qquad \Phi_1 \equiv \int \frac{p' v'^2}{|\mathbf{x} - \mathbf{x}'|} d^3 x',$$

$$V_j \equiv \int \frac{\rho' v_j'}{|\mathbf{x} - \mathbf{x}'|} d^3 x', \qquad \Phi_4 = \int \frac{p' d^3 x'}{|\mathbf{x} - \mathbf{x}'|}.$$

In the above ρ' and v_j' are the energy density and velocity of the source at X', and $d^3 x'$ is the volume element there. Heuristically, γ may be regarded as indicating the amount of space curvature produced by unit mass, and β as indicating the amount of nonlinearity in the superposition law for gravity. As the observational verifications are mostly concerned with the solar field, we are interested in the case of spherical symmetry. There the metric in PPN form is

$$ds^2 = \left[1 - \frac{2m}{r} + 2\beta \left(\frac{m}{r} \right)^2 \right] dt^2 - \left(1 + \frac{2\gamma m}{r} \right)(dr^2 + r^2 \, d\theta^2 + r^2 \sin^2 \theta \, d\phi^2).$$

For the general theory of relativity $\beta = \gamma = 1$. The observational test thus consists in determining β and γ and then examining how far they differ from unity. Thus Hellings et al. (1983), analyzing the Viking radar data, found

$$\beta - 1 = (-2.9 \pm 3.1) \times 10^{-3},$$

$$\gamma - 1 = (-0.7 \pm 1.7) \times 10^{-3},$$

which may be considered to be in excellent agreement with the general theory of relativity. Alternative metric theories like that of Brans–Dicke may be made consistent with these values only by introducing rather unappealing values of the same parameters.

4.5. The Schwarzschild or the Spherically Symmetric Black Hole

The Schwarzschild metric shows some peculiar behavior at $R = 2m$. There g_{tt} vanishes and g_{rr} blows up, and for $R < 2m$ they change sign. However, the requirement that the signature should be $+2$ is satisfied everywhere as is evident from the negative value of the determinant $|g_{\mu\nu}| = -R^4 \sin^2 \theta$. Are then the peculiarities at $R = 2m$ simply characteristic of the particular coordinate system or is $R = 2m$ a physically singular surface? One way of identifying coordinate peculiarities is to calculate the value of scalars, any singular behavior of scalars would obviously indicate singularity of the field itself. In the present case all such scalars are found to be finite and well behaved at $R = 2m$. However, the geometry cannot be said to be static in the inner region $R < 2m$, as in this region the coordinate t is no longer timelike. The timelike coordinate is R and the metric tensor components are not independent of R. Further, we cannot have $dR = 0$ in this region as then $dS^2 < 0$, i.e., the world-line is spacelike, which is not allowed for ordinary bodies. As the R lines are timelike, the question arises of fixing the arrow of time, the direction of time flow. Thus, if the body with boundary radius $R_0 < 2m$ moves towards the center (i.e., if it is a collapsing object), then decreasing R is in the direction of future and all particles inside the horizon have to move in the same direction with no possibility of reversing their direction of motion, for in that case the motion would correspond to a movement toward the past. A "black hole" is thus formed.

Alternatively, if the source is an exploding object, then the direction of future is outwards and everything proceeds in that direction, and we have the so-called "white hole."

In what follows we investigate the behavior of falling test particles in a Schwarzschild field and study the nature of singularity. For the radial motion of a freely falling particle

$$\left(1 - \frac{2m}{R}\right)\dot{t}^2 - \left(1 - \frac{2m}{R}\right)^{-1}\dot{R}^2 = 1, \tag{4.63}$$

where the overdot indicates derivatives with respect to the proper time s. Also for a geodesic

$$\left(1 - \frac{2m}{R}\right)\dot{t} = \gamma, \tag{4.64}$$

γ being a constant of motion. Combining (4.63) and (4.64) we have

$$\left(\frac{dR}{ds}\right)^2 = (\gamma^2 - 1) + \frac{2m}{R}. \tag{4.65}$$

Assuming that the particle starts from rest at $R = R_0$ we get from (4.65) $\gamma^2 = 1 - 2m/R_0$ and then the integration of (4.65) yields

$$\int ds = \int \frac{dR}{(2m/R - 2m/R_0)^{1/2}}.$$

The integral in the parametric form is

$$R = \frac{R_0}{2}(1 + \cos \eta), \qquad s = \frac{R_0}{2}\left(\frac{R_0}{2m}\right)^{1/2}(\eta + \sin \eta). \tag{4.66}$$

We have at $s = 0$, $R = R_0$, i.e., $\eta = 0$ and at $s = s_1$, $R = 0$, i.e., $\eta = \pi$. Finally, we get, for the proper time elapsed between the particle starting from rest at $R = R_0$ and reaching the origin $R = 0$,

$$s_1 = \frac{R_0}{2}\left(\frac{R_0}{2m}\right)^{1/2}, \qquad \pi = \frac{\pi}{2}\left(\frac{R_0^3}{2m}\right)^{1/2}. \tag{4.67}$$

In fact, the proper time is the time measured by an observer falling with the particle. It is interesting to note that no singular behavior is observed at $R = 2m$ and the body goes even up to $R = 0$ in a finite proper time.

A description of the motion in terms of the coordinate time t is strikingly different. Combining (4.64) and (4.65) and assuming that the particle falls freely from rest at infinity so that $\gamma = 1$

$$\frac{dR}{dt} = -\left(\frac{2m}{R}\right)^{1/2}\left(1 - \frac{2m}{R}\right). \tag{4.68}$$

The negative sign is chosen in (4.68) to indicate the motion of a falling particle. Equation (4.68) on integration yields, for the time to fall from radial coordinate R_0 to R $(R_0 > R > 2m)$,

$$t_0 - t \approx \frac{2}{3(2m)^{1/2}} [(R^{3/2} - R_0^{3/2}) + 6m(R^{1/2} - R_0^{1/2})]$$

$$- 2m \ln \frac{[R^{1/2} + (2m)^{1/2}]}{[R_0^{1/2} + (2m)^{1/2}]} \times \frac{[R_0^{1/2} - (2m)^{1/2}]}{[R^{1/2} - (2m)^{1/2}]}. \qquad (4.69)$$

For $R \to 2m$, the expression on the right obviously diverges, hence for R approaching $2m$, we may retain only the diverging term and with $2m \ll R_0$ we get

$$(t_0 - t) \approx - 2m \ln \frac{[(R)^{1/2} + (2m)^{1/2}]}{[R^{1/2} - (2m)^{1/2}]} \approx - 2m \ln \frac{[R^{1/2} + (2m)^{1/2}]^2}{(R - 2m)}. \qquad (4.70)$$

Thus for $R \to 2m$, $R^{1/2} + (2m)^{1/2} \approx 2 \cdot (2m)^{1/2}$, and (7.8) gives

$$R - 2m \simeq 8m \times e^{-(t-t_0)/2m}. \qquad (4.71)$$

Substituting for m the mass of the Sun, for example, we find that the exponential factor in (4.71) is $e^{-\{(t-t_0)10^5\}}$ where $(t - t_0)$ is measured in seconds. Thus, given a time of the order of a millisecond (i.e., $t - t_0 \sim 10^{-3}$ s), the particle will be extremely close to the Schwarzchild surface of radius $2m$. Nevertheless, R will never be less than $2m$ and will reach $2m$ only after the lapse of infinite t. However, t is the measured time according to an observer at rest at $R \to \infty$, and for any observer at rest at any value of $R > 2m$ the measured time will differ from t by a multiplying factor $(1 - 2m/R)^{1/2}$. Thus all observers in the region $R > 2m$ will agree in saying that no test particle on the surface of a collapsing sphere will ever reach $R = 2m$.

How would the body itself, which crosses into the region $R = 2m$ in a finite proper time, behave as it crosses $R = 2m$? Would it, for example, experience large tidal forces? The answer is in the negative, for as we have seen, the tidal forces are determined by the Riemann–Christoffel tensor components (recall the geodesic deviation formula) and they remain quite well behaved at $R = 2m$. Indeed, rather surprisingly, the tidal forces will be less, the larger the mass m. (Verify this by considering the geodesic deviation equation.)

The next question in which we may feel interested is: What exactly will the external observer see if the test particle/collapsing surface goes on emitting light signals as it approaches and crosses $R = 2m$? To answer this we first calculate the frequency change of these signals when received outside.

4.6. Frequency Shift of Spectral Lines of Light Emitted by a Collapsing/Exploding Spherical Body

The problem was studied by Banerjee (1966) considering the spherical body to be built up of pressureless dust.* With the isotropic form of the spherically

---

* The collapse of such a system was considered by Oppenheimer.

symmetric line element

$$ds^2 = e^\nu \, dt^2 - e^\mu (dr^2 + r^2 \, d\theta^2 + r^2 \sin^2 \theta \, d\phi^2), \qquad (4.72)$$

where ν and μ are functions of r and t, the field equations are

$$0 = e^{-\mu} \left(\frac{\mu'^2}{4} + \frac{\mu'\nu'}{2} + \frac{\mu'\nu'}{r} \right) - e^{-\nu} \left(\ddot{\mu} + \tfrac{3}{4}\dot{\mu}^2 - \frac{\dot{\mu}\dot{\nu}}{2} \right), \qquad (4.73)$$

$$0 = e^{-\mu} \left(\frac{\mu'' + \nu''}{2} + \frac{\nu'^2}{4} + \frac{\mu' + \nu'}{2r} \right) - e^{-\nu} \left(\ddot{\mu} + \tfrac{3}{4}\dot{\mu}^2 - \frac{\dot{\mu}\dot{\nu}}{2} \right), \qquad (4.74)$$

$$8\pi\rho = -e^{-\mu} \left(\mu'' + \tfrac{1}{4}\mu'^2 + \frac{2\mu'}{r} \right) + \tfrac{3}{4} e^{-\nu}\dot{\mu}^2, \qquad (4.75)$$

$$0 = 8\pi T_{14} = -(\dot{\mu}' - \tfrac{1}{2}\dot{\mu}\nu'), \qquad (4.76)$$

where primes and dots denote differentiations with respect to r and t, respectively. The last equation follows from the assumption $\nu_r = 0$ (comoving coordinates).

Inside the sphere, the Robertson–Walker metric represents a homogeneous dust distribution (see Chapter 16)

$$e^\mu = \frac{e^q}{(1 + kr^2/4)^2}, \quad q = q(t) \quad \text{for } r \le r_0 \text{ (boundary) and } k = 0, \pm 1,$$
$$e^\nu = 1. \qquad (4.77)$$

the density ρ being given by

$$8\pi\rho = 3ke^{-q} + \tfrac{3}{4}\dot{q}^2. \qquad (4.78)$$

In the outside empty space for $r \ge r_0$, the solution is

$$e^\mu r^2 = \xi^4, \qquad e^\nu = \left(\frac{4\dot{\xi}}{\dot{q}\xi} \right)^2, \qquad (4.79)$$

where the variable ξ is a solution of the differential equation

$$\left(\frac{\partial \xi}{\partial x} \right)^2 = \tfrac{1}{4}\xi^2 + \tfrac{1}{16}\dot{q}^2 \xi^6 - \frac{m}{2}, \qquad (4.80)$$

with $x \equiv \log r$ and

$$m = \frac{4\pi\rho}{3} \frac{e^{3q/2} r_0^3}{(1 + kr_0^2/4)^3}, \qquad (4.81)$$

(verify that the outside and inside metrics are continuous at $r = r_0$). The outside metric may be transformed to the Schwarzschild form

$$ds^2 = \left(1 - \frac{2m}{R} \right) dT^2 - \left(1 - \frac{2m}{R} \right)^{-1} dR^2 - R^2(d\theta + \sin^2 \theta \, d\phi^2),$$

by the transformation

$$R = \xi^2,$$

$$dT = \frac{8r\dot{\xi}\xi'}{|\dot{q}|(\xi^2 - 2m)} dt + \frac{\xi^6|\dot{q}|}{2r(\xi^2 - 2m)} dr. \tag{4.82}$$

(That such a transformation would exist follows from Birkhoff's theorem.) A straightforward calculation then gives

$$\frac{ds}{ds_0} = \frac{1}{(1 - 2m/\xi_0^2)^{1/2}} \frac{(1 - \beta)^{1/2}}{(1 + \beta)^{1/2}}, \tag{4.83}$$

where ds_0 is the proper time interval for emission at the surface $r = r_0$, and ds the corresponding interval for reception by an observer at rest in the Schwarzschild coordinate system at infinity. ξ_0 is the value of ξ at r_0 at the instant of emission and

$$\beta = \frac{\xi_0^3(\dot{q})_0}{4(\partial\xi/\partial x)_0}, \tag{4.84}$$

the subscript zero indicating the instant of emission. We may interpret the two parts in the right-hand side easily, the first factor is the familiar gravitational redshift while the second factor involving β is a typical Doppler shift due to the relative motion of the source and observer.

To have an understanding of the nature of things, it is most convenient to take the case $k = 0$ in (4.78) when we get from (4.83) and (4.84)

$$\frac{ds}{ds_0} = \frac{1}{1 \mp (2m/\xi_0^2)^{1/2}}. \tag{4.85}$$

The upper sign corresponds to a contracting sphere while the lower sign corresponds to an expanding sphere. When a contracting sphere reaches the Schwarzschild radius $\xi_0^2 = 2m$, light has infinite redshift and so is cut off. On the other hand, for an expanding sphere the frequency will be doubled when the sphere reaches the Schwarzschild radius and the light reaches the observer with enhanced intensity.

The above treatment may be generalized to include nonuniformity of density.

4.7. Fall in Apparent Luminosity of a Collapsing Body

The degradation of frequency that we have just studied will lead to a fall in the apparent luminosity as observed by a distant observer. There are two other effects which also affect the observed luminosity, namely:

(a) the contraction of the escaping light cone; and
(b) progressively larger times taken by light rays to reach the observer.

As Fig. 4.4 shows schematically, at $R = 2m$, only the radial rays escape while

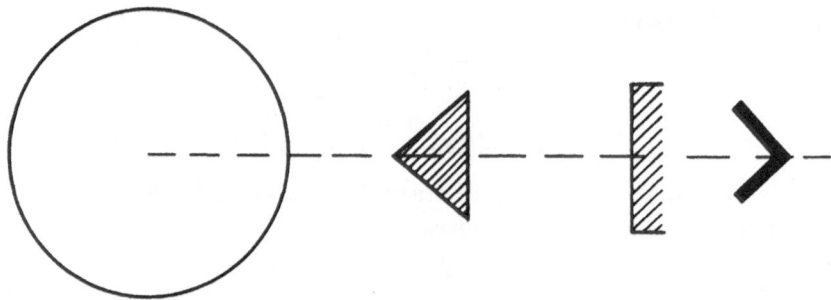

Figure 4.4

at larger values of R, photons emitted within the shaded cone escape. Obviously, this will lead to a decrease of luminosity as the collapsing surface approaches the black hole stage. Figure 4.5 shows three light rays, one proceeding radially and the other two proceeding along slightly longer paths being bent by the gravitational field. Thus the different rays reaching the observer P at the same instant start from the surface at different stages of collapse. Considering all these factors, Podurets (1965) gives, for the evolution of the luminosity with time (when the surface has gone near the Schwarzschild radius), the following relation:

$$L(t) = L_0 e^{-4t/(3\sqrt{3})2m}.$$

Thus the collapsing object quickly passes out of observation.

4.8. Kruskal–Szekeres Coordinates

There are a number of reasons which necessitate a change from the Schwarzschild coordinates. First, although the surface $R = 2m$ is not a singularity, the metric tensor components in these coordinates become singular

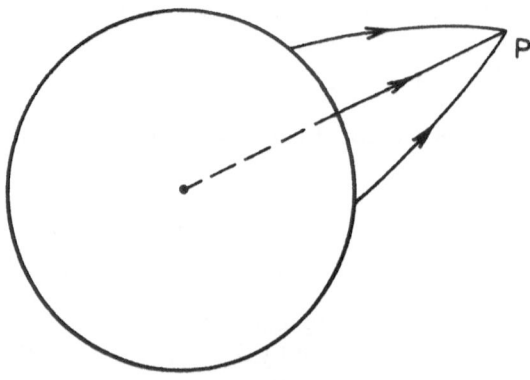

Figure 4.5

there. Of course, to remove this objectionable feature, the transformation formulas must have a singularity at $R = 2m$. Second, the notation R and t, when they no longer represent spatial and temporal coordinates, are rather deceptive. Third, as we have seen in this coordinate system, two distinct physical space–times, namely the black and white holes, have been so fused together that they camouflage as being indistinguishable. A coordinate system which effectively meets all these conditions was discovered independently by Kruskal and Szekeres. While θ and ϕ remain unchanged, the transformation relations for the new coordinates u and v are

$$\left.\begin{aligned} u &= (R/2m - 1)^{1/2}e^{R/4m}\cosh(t/4m) \\ v &= (R/2m - 1)^{1/2}e^{R/4m}\sinh(t/4m) \end{aligned}\right\} \quad \text{for } R \geq 2m, \qquad (4.86)$$

and

$$\left.\begin{aligned} u &= (1 - R/2m)^{1/2}e^{R/4m}\sinh(t/4m) \\ v &= (1 - R/2m)^{1/2}e^{R/4m}\cosh(t/4m) \end{aligned}\right\}, \quad 0 \leq R < 2m. \qquad (4.87)$$

The inverse transformation is given by the formulas

$$(R/2m - 1)e^{R/2m} = u^2 - v^2 \qquad \text{for all } R,$$

and

$$t = \begin{cases} 4m\tanh^{-1}(v/u) & \text{for } R > 2m, \\ 4m\tanh^{-1}(u/v) & \text{for } R < 2m, \end{cases} \qquad (4.88)$$

or, equivalently,

$$v/u = \begin{cases} \cosh(t/4m) & \text{for } R < 2m, \\ \tanh(t/4m) & \text{for } R > 2m, \\ 1 & \text{for } R = 2m. \end{cases} \qquad (4.89)$$

The Schwarzschild metric becomes, in the new coordinate system,

$$ds^2 = \left(\frac{32m^3}{R}\right)e^{-R/2m}(dv^2 - du^2) - R^2(d\theta^2 + \sin^2\theta\, d\phi^2). \qquad (4.90)$$

Figure 4.6 presents the situation in the u, v space (corresponding to the R, t space in Schwarzschild). Obviously, v is now a timelike coordinate while u is spacetime. The $u = \pm v$ lines are null geodesics and thus these world lines of light are inclined at 45° to the axes, as in Minkowski space. Again the event horizon $R = 2m$ is now represented by the very same pair of lines $u = \pm v$, and the singularity $r = 0$ by $(v^2 - u^2) = 1$, which give two hyperbolas with the axis in the direction of v. These double representations of the singularity and the horizon may at first sight seem confusing but we shall now see that this is because the white hole and the black hole have been separated out. In Fig. 4.6, the null lines AOA' and BOB' (for both, $R = 2m$) divide the u–v space into four quadrants. For the unshaded regions in the quadrants II and IV, the Schwarzschild coordinate R does not exist and hence these regions do not belong to physical space. In the quadrant I, $u > v$, hence $R > 2m$, this is the region outside a black hole where there is nothing peculiar. The region III is

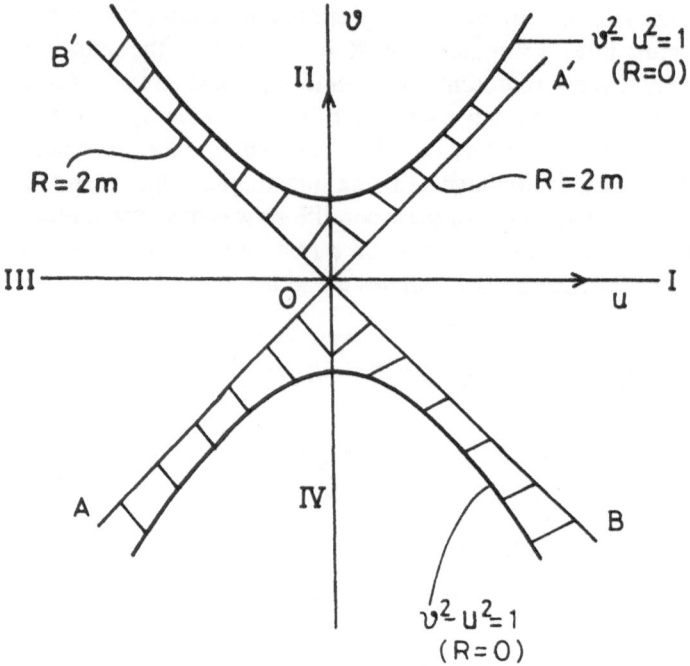

Figure 4.6

similar. The shaded regions in II and IV contain the interesting domains $0 \leq R \leq 2m$. As the direction of the time coordinate points upward, in region IV, a passage from $R = 0$ to $R = 2m$ and then on to region I (or region III) is possible. This is the case of the motion "outwards" and corresponds to a white hole. In region II, the future points from $r = 2m$ toward $r = 0$ (i.e., from the null lines to the hyperbola), and the movement must be from $R = 2m$ towards $R = 0$, this therefore represents a black hole. We thus say that for any ordinary observer (who is in region I or III, i.e., at $R > 2m$) the black hole is always in the future and the white hole is in the past (see the result that the collapsing surface or in-falling test particle ever approaches the surface $R = 2m$, but can reach it only in the infinite future by the reckoning of an outside observer and the corresponding result for a body exploding from a singularity).

Lastly, the regions I and III both represent the space $R > 2m$. There is no way of passing from one to the other. We are in one of them and the other one is simply a mathematical fiction.

4.9. Historical Note on the Schwarzschild Black Hole

The possibility of a one-way membrane for light propagation was first envisaged by Laplace. Laplace based his consideration on the following hypothesis:

(1) Light consists of particles which are affected by gravitation in the same way as material particles; hence, light velocity can increase or decrease under the influence of gravitation.

(2) At emission, light has an invariant velocity c relative to the source.

Obviously, on these assumptions, light will not be able to escape from the surface of a star if c is less than the escape velocity $v = \sqrt{2GM/R}$, where M is the mass and R is the radius of the star. However, there is no difficulty for light particles to enter from outside, as the velocity of incoming light particles will simply be increased by the star's gravitation. Laplace's ideas were of course wrong, and indeed Newtonian physics can never give a one-way passage because of the general principle of dynamical reversibility.

The peculiarity of the Schwarzschild surface was noted quite early, but the general belief was that this signified that, in Nature, we can never compress a mass to its Schwarzschild radius. Einstein (1939) also subscribed to this idea and put forward a model to justify this conjecture. His model was a cluster of particles moving in concentric circles of all radii, orientation, and phase. Thus, the system as a whole had spherical symmetry and was in equilibrium, as for each particle there was a balance between gravitational and centrifugal forces. However, the model was extremely artificial. Curiously, in the very same year the papers of Oppenheimer and Snyder and Oppenheimer and Volkhoff were published, and established the result that a spherically symmetric pressureless dust distribution experiences an uncontrolled collapse right up to a singularity in finite proper time, and in the process a black hole is produced.

Howevery, much later in the 1960s, after the discovery of quasars, ideas of gravitational collapse and black holes gained respectibility and their various aspects were very thoroughly investigated.

Still, the curious point remains that as the collapse up to the horizon takes infinite time in the reckoning of an outside observer and the universe has only a finite age, logically, no true black hole exists for anybody in the universe unless perhaps he himself is within the black hole itself.

Problems

1. Show that for a test particle in the Schwarzschild field, the equation of radial motion can be reduced to a Newtonian equation of motion in a potential field $V(R)$

$$V(R) = \left(1 - \frac{2m}{R}\right)\left(1 + \frac{l^2}{R^2}\right),$$

where l is a constant of motion. Hence discuss the circular orbits, showing that they exist up to $R = 3m$ but are stable up to $R = 6m$.

2. Show that any particle whose impact parameter is less than $4M/v$ (v being the velocity of the particle for $R \to \infty$), will be absorbed by the black hole of mass M.

3. A photon is emitted at an angle α to the direction of increasing R at the coordinate

R ($R > 2m$). Show that it will escape the black hole if

$$\sin \alpha < \frac{3\sqrt{3m}}{R}\left(1 - \frac{2m}{R}\right)^{1/2}.$$

4. Vaidya (1951), in discussing the field due to a radiating star, considered a radial flux of massless particles in a spherically symmetric space–time which is otherwise empty. The energy–stress tensor is then of the form $T_\rho^\mu = \rho k^\mu k_\nu$, where k^μ is a radically directed null vector. Set up Einstein's field equations with the metric

$$ds^2 = e^\nu \, dt^2 - e^\lambda \, dr^2 - r^2(d\theta^2 + \sin^2 \theta \, d\phi^2),$$

where λ and ν are functions of r and t and show that for the case considered by Vaidya

$$e^\nu = \frac{\dot\mu^2}{\mu'(1 - 2\mu/r)}, \qquad e^\lambda = \left(1 - \frac{2\mu}{r}\right)^{-1},$$

with μ ($\equiv \int_0^r 4\pi\rho r^2 \, dr$) obeying the differential equation

$$\frac{\dot\mu'}{\dot\mu} - \frac{\mu''}{\mu'} = \frac{2\mu}{r^2(1 - 2\mu/r)}.$$

5. Electromagnetism in General Relativity

5.1. Introduction

In Chapter 3 the action function for the electromagnetic field was spelt out and its energy–momentum tensor was derived. We obtained there Maxwell's field equations in the form

$$F^{\mu\nu}_{;\nu} = 4\pi J^{\mu}. \tag{5.1}$$

We now define the dual tensor $*F^{\mu\nu}$ by the relation (see (2.22))

$$*F^{\mu\nu} = \tfrac{1}{2}\eta^{\mu\nu\alpha\beta}F_{\alpha\beta}, \qquad *F_{\mu\nu} = \tfrac{1}{2}\eta_{\mu\nu\alpha\beta}F^{\alpha\beta}, \tag{5.2}$$

where $\eta^{\mu\nu\alpha\beta}$ is the permutation tensor.

From the relation $F_{\mu\nu} = A_{\nu,\mu} - A_{\mu,\nu}$ we have

$$F_{\mu\nu,\alpha} + F_{\nu\alpha,\mu} + F_{\alpha\mu,\nu} = 0. \tag{5.3}$$

The relation (5.3) may be written as

$$*F^{\mu\nu}_{;\nu} = 0. \tag{5.4}$$

We recall the energy–momentum tensor given by relation (3.60)

$$4\pi\tau_{\mu\nu} = -F_{\mu\alpha}F^{\alpha}_{\nu} + \tfrac{1}{4}g_{\mu\nu}F_{\alpha\beta}F^{\alpha\beta}. \tag{5.5}$$

Obviously, the trace of this tensor τ^{μ}_{μ} vanishes. Taking the covariant derivative of (5.5) we get, using (5.1),

$$4\pi\tau^{\mu\nu}_{;\nu} = -F^{\mu\alpha}_{;\nu}F^{\nu}_{\alpha} + F^{\mu}_{\alpha}4\pi J^{\alpha} + \tfrac{1}{2}g^{\mu\nu}F_{\alpha\beta;\nu}F^{\alpha\beta}, \tag{5.6}$$

and (5.6) in turn, after some readjustment of the terms, reduces to

$$\begin{aligned}
4\pi(\tau^{\mu\nu}_{;\nu} - F^{\mu\alpha}J_{\alpha}) &= -g_{\alpha\beta}F^{\mu\alpha}_{;\nu}F^{\nu\beta} + \tfrac{1}{2}g^{\mu\nu}F_{\alpha\beta;\nu}F^{\alpha\beta} \\
&= -g^{\mu\nu}(F_{\nu\beta;\alpha})F^{\alpha\beta} + \tfrac{1}{2}g^{\mu\nu}F_{\alpha\beta;\nu}F^{\alpha\beta} \\
&= \tfrac{1}{2}g^{\mu\nu}F^{\alpha\beta}(F_{\alpha\beta;\nu} - 2F_{\nu\beta;\alpha}) \\
&= \tfrac{1}{2}g^{\mu\nu}F^{\alpha\beta}(F_{\alpha\beta;\nu} + F_{\beta\nu;\alpha} + F_{\nu\alpha;\beta}).
\end{aligned} \tag{5.7}$$

The right-hand side of (5.7) vanishes in view of (5.3) and so we finally obtain the result

$$\tau^{\mu\nu}_{;\nu} = F^{\mu\alpha}J_{\alpha}. \tag{5.8}$$

The right-hand expression is the Lorentz force density.

5.2. The Field of a Charged Particle

Here we may also assume spherical symmetry as in the Schwarzschild field for a particle. Thus the metric can be written in the form

$$ds^2 = e^\nu \, dt^2 - e^\lambda \, dr^2 - r^2(d\theta^2 + \sin^2 \theta \, d\varphi^2). \tag{5.9}$$

However, the space is no longer empty, it is the seat of the electric field of the particle. Hence the appropriate field equations would be

$$R_\nu^\mu - \tfrac{1}{2} R \delta_\nu^\mu = -8\pi \tau_\nu^\mu, \tag{5.10}$$

with τ_ν^μ given by (5.5).

Besides, we have Maxwell's equations (5.1) and (5.3) with $J^\mu = 0$, as we are considering the field outside the charge distribution. Consistent with the physical situation, we may assume that the field in this coordinate system is purely electrostatic, i.e., only $F_{10} = -F_{01} \neq 0$, and all other components of the electromagnetic field vanish. (The Einstein–Maxwell equations in this case also admit solutions having $F_{23} = -F_{32} \neq 0$, but that would mean the presence of a magnetic monopole at the origin.) We number the coordinates t, r, θ, ϕ as 0, 1, 2, 3, respectively. Hence we may take $A_0 = \phi \neq 0$, $A_1 = A_2 = A_3 = 0$. Equation (5.1) now gives only one nontrivial relation

$$(g^{00}g^{11}\phi_{,1}\sqrt{-g})_{,1} = 0,$$

or

$$e^{-(\lambda+\nu)/2}\phi_{,1}r^2 = \text{constant} = e \quad \text{(say).} \tag{5.11}$$

Writing out (5.10) explicitly, using (5.5) and (5.11), we get

$$e^{-\lambda}\left(\frac{\nu'}{r} + \frac{1}{r^2}\right) - \frac{1}{r^2} = -8\pi\tau_1^1 = -\frac{e^2}{r^4}, \tag{5.12}$$

$$e^{-\lambda}\left(\frac{\nu''}{2} - \frac{\lambda'\nu'}{4} + \frac{\nu'}{4} + \frac{\nu' - \lambda'}{2r}\right) = -8\pi\tau_2^2 = +\frac{e^2}{r^4}, \tag{5.13}$$

$$e^{-\lambda}\left(\frac{\lambda'}{r} - \frac{1}{r^2}\right) + \frac{1}{r^2} = 8\pi\tau_0^0 = \frac{e^2}{r^4}, \tag{5.14}$$

where primes indicate differentiation with respect to r. Adding (5.12) and (5.14) we readily obtain

$$\lambda' + \nu' = 0,$$
$$\lambda = -\nu. \tag{5.15}$$

The constant of integration being put equal to zero, as for $r \to \infty$, the field becomes asymptotically flat with $\lambda \to 0$, $\nu \to 0$.

Equation (5.14) can be directly integrated to give

$$e^{-\lambda} = 1 - \frac{2m}{r} + \frac{e^2}{r^2},$$

where m is an integration constant. Thus the metric known, after Reissner–Nordström, for a charged particle is

$$ds^2 = \left(1 - \frac{2m}{r} + \frac{e^2}{r^2}\right) dt^2 - \left(1 - \frac{2m}{r} + \frac{e^2}{r^2}\right)^{-1} dr^2 - r^2(d\theta^2 + \sin^2\theta \, d\phi^2).$$

(5.16)

For large values of r, the term e^2/r^2 falls off and the Reissner–Nordström metric (5.16) reduces to the Schwarzschild metric showing that m is the total gravitational mass of the system. However, the electrostatic field energy also contributes to this mass (see Problem 2 below). It is fairly easy to identify the constants m and e. As $r \to \infty$ the geometry tends to the Euclidean form, and $F_{01} \to e/r^2$ showing that e is the electric charge. Alternatively, we may consider the equation for a continuous distribution of charge

$$(F^{0\nu}\sqrt{-g})_{,\nu} = 4\pi J^0 \sqrt{-g} = 4\pi\sigma \left(\frac{dt}{ds}\right)\sqrt{-g},$$

where σ is the charge density and the current is purely convectional so that

$$J^\mu = \sigma v^\mu.$$

Integrating, we obtain

$$\oint F^{0\nu}\sqrt{-g} \, ds_\nu = 4\pi \int \sigma \frac{dt}{ds} \sqrt{g_{00}} \sqrt{|h|} \, d^3x,$$

(5.17)

where $|h|$ is the determinant of h_{ik}, the static metric being expressible in the form

$$ds^2 = g_{00} \, dt^2 + h_{ik} \, dx^i \, dx^k \qquad (i, k \text{ ranging from 1 to 3}).$$

Equation (5.17) is Gauss's flux theorem and with the charges at rest $\sqrt{g_{00}} \, dt/ds = 1$, so that (5.17) yields

$$\oint F^{0\nu}\sqrt{-g} \, ds_\nu = 4\pi Q,$$

(5.18)

where Q is the total charge in the field and the surface integral on the left is over the sphere at infinity. Putting the value of $F^{0\nu}$ in our case, we find $Q = e$. Note that (5.18) does not assume spherical symmetry.

The Reissner–Nordström solution shows beautifully the interaction of two fields—gravitational and electromagnetic. The gravitational field is altered as shown by the appearance of e in the metric and the electric field is also altered, for although the field apparently is still e/r^2, r cannot be identified with the spatial distance in the ordinary sense. Note that $g_{00} = e^\nu$ vanishes at $r = m \pm \sqrt{m^2 - e^2}$. With $m > |e|$, this gives two real values of r, say r_+, and r_- ($r_+ > r_-$). For $r_- < r < r_+$, $g_{00} < 0$ and $g_{11} > 0$ so that t is a spacelike coordinate and r timelike. Thus no reversal of motion within this region is possible as in the case of the Schwarzschild black hole. However, a reversal is possible in the region $r < r_-$. A simple calculation shows that, as in the Schwarzschild case, the time required for any body to reach the surface r_+ is infinite in the reckoning of an outside observer. Thus we may have the following disturbing situation. A body moving toward the singularity $r = 0$, crosses the surfaces r_+ and r_- and then turns back at some $r < r_-$ and appears back in the region $r > r_+$, the entire process occurring in a finite time in the reckoning of the body itself, but an outside observer can only see it approaching r_+ for ever; there is no question of seeing it ever moving outward at $r > r_+$. Such a situation has indeed been constructed by Israel.

5.3. Static Electrovac

We next consider a more general case, the vacuum exterior to a distribution of charged matter at rest. Such a vacuum has been called the static electrovac by Synge. The line element may be written in the form

$$ds^2 = f^2\, dt^2 + h_{ij}\, dx^i\, dx^j, \tag{5.19}$$

mentioned earlier. As before, we write $A_0 = \phi$, $A_i = 0$, so that

$$F_{i0} = \phi_{,i} \quad \text{and} \quad F_{ij} = 0,$$

where the index zero refers to the time coordinate. The components of the energy–momentum tensor are then

$$\tau_{ij} = f^{-2}(\tfrac{1}{2}h_{ij}\Delta_1\phi - \phi_{,i}\phi_{,j}),$$
$$\tau_{i0} = 0, \qquad \tau_{00} = -\tfrac{1}{2}\Delta_1\phi. \tag{5.20}$$

Einstein's field equations then read

$$R_{ij} = -8\pi f^{-2}(\tfrac{1}{2}h_{ij}\Delta_1\phi - \phi_{,i}\phi_{,j}), \tag{5.21}$$
$$R_{00} = f\Delta_2 f = 4\pi\Delta_1\phi. \tag{5.22}$$

The operators Δ_1 and Δ_2 introduced above are defined by

$$\Delta_1\phi = h^{ij}\phi_{,i}\phi_{,j}, \qquad \Delta_2\phi = h^{ij}\phi_{\|ij}, \tag{5.23}$$

where the special symbol $\|$ denotes covariant differentiation with respect to the 3-space metric h_{ij}. The only nontrivial Maxwell's equation is

$$f\Delta_2\phi - h^{ij}f_{,i}\phi_{,j} = 0. \tag{5.24}$$

An interesting result about the electrovac follows if we assume that g_{00} and ϕ are functionally related, so that the level surfaces of g_{00} and ϕ coincide (Majumdar, 1947).

Then (5.22) and (5.24) reduce, respectively, to

$$ff'\Delta_2\phi + (ff'' - 4\pi)\Delta_1\phi = 0, \tag{5.25}$$

and

$$f\Delta_2\phi - f'\Delta_1\phi = 0, \tag{5.26}$$

where the prime denotes the derivative with respect to ϕ. Eliminating $\Delta_2 \phi$ from the above two equations it is easy to obtain

$$ff'' + f'^2 = 4\pi$$

which then integrated yields

$$f^2 = A + B\phi + 4\pi\phi^2, \tag{5.27}$$

where A and B are arbitrary constants and the above form of g_{00} holds irrespective of any symmetry condition. If, following Majumdar, the constants are chosen in such a way that (5.27) reduces to

$$g_{00} = f^2 = (1 \pm \phi')^2, \tag{5.28}$$

where $\phi' = 2\pi^{1/2}\phi$. Then using the field equations we can show that the line element (5.19) assumes the form

$$ds^2 = f^2 \, dt^2 - f^{-2}(dx^2 + dy^2 + dz^2) \tag{5.29}$$

and the whole set of field equations now reduces to a simple equation

$$\nabla^2 V = 0,$$

where $f^{-1} = (1 + V)$ and ∇^2 stands for the ordinary Laplacian in the Euclidean 3-space. So any solution for V of Laplace's equation can, in principle, generate a solution for the static electrovac.

5.4. The Already Unified Field Theory

We may regard Maxwell's equations and Einstein's equations as coupled systems, the coupling being achieved through the stress–energy tensor. A different approach was developed by Misner and Wheeler (1957) based on some earlier work by Rainich. They found a set of necessary and sufficient conditions that the geometry must satisfy in order to be the seat of an electromagnetic field. However, there were two limitations:

(1) that the space is otherwise empty; and
(2) the field is nonnull.

An electromagnetic field is called null when both the invariants $F^{\mu\nu}F_{\mu\nu}$ and $*F^{\mu\nu}F_{\mu\nu}$ vanish. In what follows we give an outline of the already unified theory, a name coined by the authors.

Using the definitions of dual electromagnetic field tensors (5.2) we may verify the following identities:

$$F_{\mu\alpha}F^{\nu\alpha} - *F^*_{\mu\alpha}F^{\nu\alpha} = \tfrac{1}{2}\delta^\nu_\mu F_{\alpha\beta}F^{\alpha\beta}, \tag{5.30}$$

$$F^*_{\mu\alpha}F^{\nu\alpha} = \tfrac{1}{4}\delta^\nu_\mu F_{\alpha\beta}F^{\alpha\beta}, \tag{5.31}$$

$$\omega_{\mu\nu}\omega^{\mu\sigma} = \tfrac{1}{4}\delta^\sigma_\nu \omega_{\alpha\beta}\omega^{\alpha\beta}, \tag{5.32}$$

where

$$\omega_{\mu\nu} = F_{\mu\nu} + i*F_{\mu\nu}.$$

So the electromagnetic energy–momentum tensor (5.5) can be written in an equivalent way

$$4\pi\tau_{\mu\nu} = -(F_{\mu\alpha}F_\nu^\alpha + *F_{\mu\alpha}*F_\nu^\alpha) = 2\omega_{\mu\alpha}\bar{\omega}_\nu^{\ \alpha}, \tag{5.33}$$

where $\bar{\omega}_{\mu\nu}$ is the complex conjugate of $\omega_{\mu\nu}$. It is now not difficult to write out Rainich's algebraic relations between the τ_ν^μ's

$$\tau_\mu^\mu = 0, \tag{5.34}$$

$$\tau_\mu^\alpha \tau_\alpha^\nu = \tfrac{1}{4}\tau_{\alpha\beta}\tau^{\alpha\beta}\delta_\mu^\nu = \rho^2\delta_\mu^\nu, \tag{5.35}$$

where

$$\rho^2 = \tfrac{1}{4}\tau_{\alpha\beta}\tau^{\alpha\beta} = I_1^2 + I_2^2. \tag{5.36}$$

The two electromagnetic invariants I_1 and I_2 are defined as

$$I_1 = \tfrac{1}{2}F_{\mu\nu}F^{\mu\nu} \quad \text{and} \quad I_2 = \tfrac{1}{2}F_{\mu\nu}*F^{\mu\nu}.$$

Again, since in the Minkowski frame $\tau_{00} = E^2 + H^2$ is nonnegative, it is demanded that for any timelike vector u^α we must have

$$\tau_{\alpha\beta}u^\alpha u^\beta \geq 0. \tag{5.37}$$

Relations (5.34)–(5.37) are not only necessary consequences of (5.33), but are also sufficient for the existence of an antisymmetric tensor $F_{\mu\nu}$ satisfying (5.33). The remaining question is: Will the $F^{\mu\nu}$'s satisfy Maxwell's equations? To answer this question, we note that the energy–stress tensor $\tau^{\mu\nu}$ does not uniquely determine the field tensor $F^{\mu\nu}$. In fact, if θ be any real variable then $F'_{\mu\nu}$ and $F_{\mu\nu}$ give the same value of $\tau^{\mu\nu}$, when they are related by

$$F_{\mu\nu} = F'_{\mu\nu} \cos\theta + *F'_{\mu\nu} \sin\theta, \tag{5.38}$$

or

$$*F_{\mu\nu} = -F'_{\mu\nu} \sin\theta + *F'_{\mu\nu} \cos\theta,$$

which in complex form is

$$\omega_{\mu\nu} = \omega'_{\mu\nu}e^{-i\theta}. \tag{5.39}$$

The transformation from $F'_{\mu\nu}$ to $F_{\mu\nu}$ is called the duality rotation (Misner and Wheeler, 1957).

We may now reduce any nonnull field to a simple electric or magnetic field (according as $F^{\mu\nu}F_{\mu\nu} < 0$ or > 0) or a combination of the two in the same direction by a suitable Lorentz transformation. Again, from (5.38),

$$F^{\mu\nu}F_{\mu\nu} = F'^{\mu\nu}F'_{\mu\nu}(\cos 2\theta + \sin 2\theta),$$

and

$$*F^{\mu\nu}F_{\mu\nu} = *F'^{\mu\nu}F'_{\mu\nu}.$$

Thus, by a duality rotation, a simple electric field may be converted into a magnetic field and vice versa. A superposition of duality rotation and Lorentz transformation, therefore, may reduce any nonnull field to a simple electric field. This electric field is called an extremal field by Misner and Wheeler. The actual field $F_{\mu\nu}$ at the point can be obtained from the extremal field by a duality rotation through an angle θ.

It has already been shown that the Rainich algebraic conditions on the energy–momentum tensor enable us to find an antisymmetric tensor at each point in space–time, subject to a duality rotation. Our next task is to find the condition that the complexion θ, which is so far arbitrary, can be chosen so that the resultant field tensor $F_{\mu\nu}$ satisfies Maxwell's equations everywhere in space–time.

From (5.39) we have

$$\omega^{\mu\nu}_{;\nu} = \omega'^{\mu\nu}_{;\nu}e^{-i\theta} - i\omega'^{\mu\nu}e^{-i\theta}\theta_{,\nu}.$$

Hence $\omega^{\mu\nu}$ will satisfy Maxwell's equations if there exists a θ satisfying

$$\theta_{,\nu} = -i\frac{\omega'^{\mu\nu}_{,\nu}}{\omega'^{\mu\nu}}.$$

The complexion θ above is found to satisfy the relation

$$\theta_{,\nu} = \frac{(-g)^{1/2}\varepsilon_{\nu\alpha\beta\mu}\tau^{\beta\sigma;\mu}}{\rho^2}\tau^{\alpha}_{\sigma}. \tag{5.40}$$

Now according to Einstein's theory of gravitation the electromagnetic energy momentum tensor acts as a source of the gravitational field and so the field equations can be written as

$$R_{\mu\nu} = -8\pi\tau_{\mu\nu},$$

the trace τ^{μ}_{μ} being zero.

So, finally, we observe that in order that an otherwise empty Reimannian space may be understood as the seat of an electromagnetic field the necessary and sufficient conditions are, in view of (5.34), (5.35), (5.37), and (5.40), as follows:

$$R = 0, \tag{5.41a}$$

$$R^{\alpha}_{\mu}R^{\nu}_{\alpha} = \rho^2\delta^{\nu}_{\mu} = \tfrac{1}{4}R_{\alpha\beta}R^{\alpha\beta}\delta^{\nu}_{\mu}, \tag{5.41b}$$

$$R_{00} \leq 0, \tag{5.41c}$$

and

$$\alpha_{\mu,\nu} - \alpha_{\nu,\mu} = 0, \tag{5.41d}$$

where

$$\alpha_{\nu} \equiv \frac{(-g)^{1/2}\varepsilon_{\nu\alpha\beta\mu}R^{\beta\sigma;\mu}}{\rho^2}R^{\alpha}_{\sigma}. \tag{5.42}$$

Equations (5.41a–d), along with definition (5.42), are the equations of the already unified theory. Note that only the geometric objects, the Ricci tensor, and $g_{\mu\nu}$ occur in the equations, the electromagnetic field tensor being completely eliminated. For null fields, however, the equation for α is undefined and the already unified theory cannot be formulated.

Problems

1. Considering a charged dust distribution, show that $\dot{v}^{\mu} = (\sigma/\rho)F^{\mu\alpha}v_{\alpha}$ where σ and ρ are the charge and matter densities, respectively. Set up the equation of motion of a charged particle moving in the Reissner–Nordström metric. Integrate for the special cases of purely radial and circular motion.

2. For large distances, the gravitational field is given as $-m/r^2 + 2e^2/r^3$. Explain the paradox that the electrostatic field energy appears to reduce the attraction. (*Hint*: Study the identity $R_0^0 \sqrt{|g|} = [g^{00}g^{ik}g_{00,i}\sqrt{|g|}]_{,k}$).

3. Defining the electric and magnetic field vectors by the expressions

$$E_{\mu} = F_{\mu\nu}v^{\nu} \qquad \text{and} \qquad B_{\mu} = \tfrac{1}{2}\eta_{\mu\nu\sigma\rho}F^{\nu\sigma}v^{\rho},$$

where v^{μ} is a unit timelike vector, express the components of τ_{ν}^{μ} for an electromagnetic field in terms of E_{μ} and B_{μ}.

4. A charged dust distribution is in equilibrium and the metric is written the form

$$ds^2 = g_{00}\,dt^2 + g_{ik}\,dx^i\,dx^k.$$

Show that g_{00}, the electrostatic potential ϕ, and the ratio of the charge density to mass density are functionally related (De and Raychaudhuri 1968).

5. For the electromagnetic energy–stress tensor, prove the Rainich relations

$$\tau_\alpha^\alpha = 0, \qquad \tau_\beta^\alpha\tau_\sigma^\beta = \tfrac{1}{4}\delta_\sigma^\alpha(\tau_{\mu\nu}\tau^{\mu\nu}).$$

6. *$F_{\mu\nu}$ is the tensor dual to the electromagnetic field tensor $F^{\mu\nu}$. Express the electromagnetic energy–stress tensor in terms of *$F_{\mu\nu}$.

7. An electromagnetic field is called a null field if $F^{\mu\nu}F_{\mu\nu} = {}^*F^{\mu\nu}F_{\mu\nu} = 0$. Show that the necessary and sufficient condition for an electromagnetic field to be null is that $F_{\mu\nu}$ must be in the form

$$F_{\mu\nu} = k_\mu a_\nu - k_\nu a_\mu,$$

where k_μ is a null vector and $a^\mu k_\mu = 0$.

8. Show that for a null field to satisfy Maxwell's equations, k_μ must be geodesic.

6. Axially Symmetric Fields

6.1. The Lie Derivative and the Killing Equation

Consider a transformation from the unprimed coordinates x^α to primed coordinates x'^β. In general, x'^β will be functions of x^α and a number of parameters a_r

$$x'^\beta = f^\beta(x^\alpha, a_1, a_2, \ldots, a_r). \tag{6.1}$$

(To fix ideas, take the case of a rotation about the z-axis through an angle θ; then

$$x' = x \cos \theta + y \sin \theta,$$

$$y' = y \cos \theta - z \sin \theta, \tag{6.2}$$

$$z' = z,$$

here θ is the parameter.) Let us, for simplicity, take only one parameter, say t, and choose it in such a way that $t = 0$ gives the identity transformation, then

$$x'^\beta = f^\beta(x^\alpha, t), \tag{6.3}$$

with

$$x^\beta = f^\beta(x^\alpha, 0).$$

Then, assuming differentiability of the function, we have

$$x'^\beta(dt) = x^\beta + \left(\frac{\partial x'^\beta}{\partial t}\right)_{t=0} dt + \cdots.$$

The vector $\xi^\beta \equiv (\partial x'^\beta/\partial t)_{t=0}$ is called the generator of the transformation. Relation (6.3) can also be viewed as a shift in the position of points P (recall that (6.2) can be viewed either as a rotation of the space or a rotation of the coordinates). As the parameter t changes continuously P will trace out a curve.

A vector field U^α in a transformation will change to U'^β,

$$U'^\beta = U^\alpha \frac{\partial x'^\beta}{\partial x^\alpha} = U^\alpha(\delta_\alpha^\beta + \xi^\beta_{,\alpha} dt + \cdots). \tag{6.4}$$

This is the value of the transformed vector at x'^β. However, the untransformed

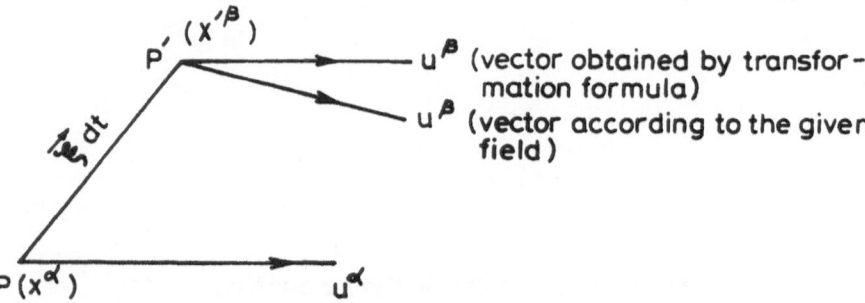

Figure 6.1

vector has the value at x'^β

$$U^\beta(x'^\beta) = U^\beta + U^\beta_{,\alpha}\delta x^\alpha$$
$$= U^\beta + U^\beta_{,\alpha}\xi^\alpha\, dt + \cdots . \tag{6.5}$$

Both (6.4) and (6.5) are vectors at the same point, hence their difference is also a vector. We now divide the difference by the scalar dt, and proceed to the limit $dt \to 0$, to get

$$U^\alpha_{,\beta}\xi^\beta - U^\beta\xi^\alpha_{,\beta}.$$

The above expression is called the Lie derivative of U^α with respect to ξ^β and written

$$\mathscr{L}_\xi U^\alpha = U^\alpha_{,\beta}\xi^\beta - U^\beta\xi^\alpha_{,\beta} = U^\alpha_{;\beta}\xi^\beta - U^\beta\xi^\alpha_{;\beta}. \tag{6.6}$$

Diagrammatically, we represent the situation in Fig. 6.1.

It is easy to extend the idea to tensors of higher rank. For a tensor $C^{\alpha\beta}$ of rank 2, we have, by the transformation law,

$$C'^{\alpha\beta} = C^{\mu\nu}\frac{\partial x^\alpha}{\partial x^\mu}\frac{\partial x^\beta}{\partial x^\nu}$$
$$= C^{\mu\nu}(\delta^\alpha_\mu + \xi^\alpha_{,\mu}\, dt\ldots)(\delta^\beta_\nu + \xi^\beta_{,\nu}\, dt\ldots)$$
$$= C^{\alpha\beta} + (C^{\mu\beta}\xi^\alpha_{,\mu} + C^{\alpha\nu}\xi^\beta_{,\nu})\, dt + \cdots,$$

and from the functional dependence of the field $C^{\alpha\beta}$ on the coordinates,

$$C'^{\alpha\beta} = C^{\alpha\beta} + C^{\alpha\beta}_{,\mu}\xi^\mu\, dt.$$

The Lie derivative is thus

$$\mathscr{L}_\xi C^{\alpha\beta} = C^{\alpha\beta}_{,\mu}\xi^\mu - C^{\mu\beta}\xi^\alpha_{,\mu} - C^{\alpha\nu}\xi^\beta_{,\nu}$$
$$= C^{\alpha\beta}_{;\mu}\xi^\mu - C^{\mu\beta}\xi^\alpha_{;\mu} - C^{\alpha\nu}\xi^\beta_{;\nu}. \tag{6.7}$$

(The last step may be obtained by direct calculation or arrived at from the consideration that the object is a tensor.)

The Lie derivative $\mathscr{L}_\xi U^\alpha$ is also called the commutator of the vectors U^α, ξ^β and sometimes written as (U, ξ). Obviously, $(U, \xi) = -(\xi, U)$. When the Lie derivative vanishes we say that the vectors commute. The vanishing means that the transformation does not alter the intrinsic vector (or tensor) field. As the geometry of space is determined by the metric tensor $g_{\mu\nu}$, a vanishing of the Lie derivative of $g_{\mu\nu}$ with respect to ξ^α would indicate that the transformation generated by ξ^α does not alter the intrinsic geometry. We then say that the transformation is a motion or isometry of the space. Explicitly, the condition for this is

$$\mathscr{L}_\xi g_{\mu\nu} = 0 \quad \rightarrow \quad g_{\mu\nu,\alpha}\xi^\alpha + g_{\mu\alpha}\xi^\alpha_{,\nu} + g_{\alpha\nu}\xi^\alpha_{,\mu} = 0, \tag{6.8}$$

which again can be reduced to

$$\xi_{\alpha;\beta} + \xi_{\beta;\alpha} = 0. \tag{6.9}$$

Equation (6.9) is called a Killing equation and is the condition for ξ^α to generate a motion (ξ^α is then called a Killing vector). (In (6.7) we have deduced the expression for the Lie derivative of a contravariant tensor, the reader may check the expression for the Lie derivative of a covariant tensor used in (6.8) by direct calculation.)

6.2. Static and Stationary Metrics

Any continuous symmetry property indicates the existence of a Killing vector and vice versa. We now show how the metric can be reduced to a form in which the $g_{\mu\nu}$'s are independent of a particular coordinate if a Killing vector exists.

Consider the differential equation

$$\xi^\mu \frac{\partial\phi}{\partial x^\mu} = 0.$$

This is a first-order differential equation and has $(n-1)$ independent solutions say $\phi^2, \phi^3, \phi^4, \ldots, \phi^n$. Let ϕ^1 be any solution of the differential equation

$$\xi^\mu \frac{\partial\phi}{\partial x^\mu} = 1.$$

If we take $\phi^1, \phi^2, \ldots, \phi^n$ to be a new coordinate system then the vector ξ^μ transforms to

$$\xi'^\alpha = \xi^\mu \frac{\partial\phi^\alpha}{\partial x^\mu} = \delta^\alpha_1. \tag{6.10}$$

Thus any vector ξ^α may be transformed to have only one nonvanishing component and that one is unity. Now if ξ^μ is a Killing vector and if it is transformed to the form (6.10), then in the new coordinate system we get from (6.8)

$$g_{\mu\nu,1} = 0. \qquad (6.11)$$

In particular, if ζ^μ is a timelike vector, then writing $\phi^1 \equiv t$, $\phi^i = x^i$ ($i = 2$, $3, \ldots, n$) the metric is

$$ds^2 = g_{tt}\, dt^2 + g_{ik}\, dx^i\, dx^k + 2g_{it}\, dx^i\, dt, \qquad (6.12)$$

where the Latin indices run only over the space coordinates. In the case $g_{it} = 0$, i.e., the timelike Killing vector is orthogonal to a family of three spaces, we say the space–time is static, while if $g_{it} \neq 0$, the space–time is named as stationary. We shall presently see that these nondiagonal elements in the metric tensor correspond to the existence of rotation in the system (see (3.42)).

Problems

1. Show that the metric

$$dS^2 = dx^{3^2} + f(x^3)(dx^{1^2} + dx^{2^2})$$

admits the Killing vectors ζ^μ where

$$\zeta^\mu = (1, 0, 0);\ (0, 1, 0);\ (x^2, x^1, 0).$$

2. Find the Killing vectors admitted by the spherically symmetric metric

$$dS^2 = e^\nu\, dt^2 - e^\lambda\, dr^2 - r^2(d\theta^2 + \sin^2\theta\, d\phi^2),$$

where $\nu = \nu(r, t)$; $\lambda = \lambda(r, t)$.

3. In the metric

$$dS^2 = dt^2 - dr^2 - dz^2 + 2m\, d\phi\, dt - l\, d\phi^2.$$

where m and l are functions of r alone. Find the condition that it may admit four linearly independent Killing vectors. (*Hint:* Three Killing vectors correspond to translation along t, z, and ϕ. Take the fourth Killing vector as having $\zeta^z = \zeta^a_{,z} = 0$.)

6.3. The Axially Symmetric Static Metric

We have already discussed, in the weak field approximation, the field due to a rotating stationary system and seen how the nondiagonal element $g_{\phi t}$ is connected with the angular momentum. We shall now investigate cases of axial or rotational symmetry, i.e., the situation is such that a rotation of the system through any angle about an axis does not bring about any essential change in the same. We shall further assume that the situation is stationary; we can describe the situation by saying that there are two Killing vectors, one timelike and the other spacelike, and the orbits of the spacelike Killing vector are closed. Two situations can arise, the timelike and the spacelike Killing vectors may or may not be orthogonal to one another according to whether the system has or has not an angular momentum. We first consider the case where they are orthogonal. The metric, in this case, can be written in the form

$$dS^2 = g_{00}\, dt^2 + g_{\phi\phi}\, d\phi^2 + \sum g_{ab}\, dx^a\, dx^b,\qquad(6.13)$$

with the $g_{\mu\nu}$'s independent of ϕ and t and the Latin indices run over 1 and 2. Note that the metric is invariant with respect to the transformation $t \to -t$, and ϕ to $-\phi$.

We shall first show that any 2-space metric can be transformed to the conformally flat form. Consider a transformation from x^1, x^2 to $\bar{x}^1$, $\bar{x}^2$. If the transformed metric is to be conformally flat, we must have

$$\bar{g}^{11} = \bar{g}^{22}$$

or

$$g^{11}\left(\frac{\partial\bar{x}^1}{\partial x^1}\right)^2 + 2g^{12}\left(\frac{\partial\bar{x}^1}{\partial x^1}\right)\left(\frac{\partial\bar{x}^1}{\partial x^2}\right) + g^{22}\left(\frac{\partial\bar{x}^1}{\partial x^1}\right)^2$$

$$= g^{11}\left(\frac{\partial\bar{x}^2}{\partial x^1}\right)^2 + 2g^{12}\left(\frac{\partial\bar{x}^2}{\partial x^1}\right)\left(\frac{\partial\bar{x}^2}{\partial x^2}\right) + g^{22}\left(\frac{\partial\bar{x}^2}{\partial x^2}\right)^2,$$

$$\bar{g}^{12} = 0,$$

or

$$g^{11}\left(\frac{\partial\bar{x}^1}{\partial x^1}\right)\left(\frac{\partial\bar{x}^2}{\partial x^1}\right) + g^{12}\left[\left(\frac{\partial\bar{x}^1}{\partial x^1}\right)\left(\frac{\partial\bar{x}^2}{\partial x^2}\right) + \left(\frac{\partial\bar{x}^2}{\partial x^1}\right)\left(\frac{\partial\bar{x}^1}{\partial x^2}\right)\right]$$

$$+ g^{22}\left(\frac{\partial\bar{x}^1}{\partial x^2}\right)\left(\frac{\partial\bar{x}^2}{\partial x^2}\right) = 0.$$

Thus, we have only two partial differential equations to determine the two functions $\bar{x}^1\,(x^1,\,x^2)$ and $\bar{x}^2\,(x^1,\,x^2)$ and a solution can be found. (In fact, there are infinite solutions as we have only two equations amongst the four first-order derivatives $\partial\bar{x}^i/\partial x^k$.) Hence, we can reduce the metric (6.13) to the form

$$dS^2 = \gamma^2\, dx^{0^2} - \alpha^2(dx^{1^2} + dx^{2^2}) - \beta^2\, dx^{3^2},\qquad(6.14)$$

where α, β, and γ are functions of x^1 and x^2 only.

6.4. Weyl's Canonical Form

With the above metric, the nonvanishing components of the Ricci tensor are

$$R_{11} = \left(\frac{\alpha_1}{\alpha}\right)_1 + \left(\frac{\alpha_2}{\alpha}\right)_2 + \frac{\beta_{11}}{\beta} + \frac{\gamma_{11}}{\gamma} + \frac{\alpha_2}{\alpha}\left(\frac{\beta_2}{\beta} + \frac{\gamma_2}{\gamma}\right) - \frac{\alpha_1}{\alpha}\left(\frac{\beta_1}{\beta} + \frac{\gamma_1}{\gamma}\right),\qquad(6.15a)$$

$$R_{22} = \left(\frac{\alpha_1}{\alpha}\right)_1 + \left(\frac{\alpha_2}{\alpha}\right)_2 + \frac{\beta_{22}}{\beta} + \frac{\gamma_{22}}{\gamma} + \frac{\alpha_1}{\alpha}\left(\frac{\beta_1}{\beta} + \frac{\gamma_1}{\gamma}\right) - \frac{\alpha_2}{\alpha}\left(\frac{\beta_2}{\beta} + \frac{\gamma_2}{\gamma}\right),\qquad(6.15b)$$

$$R_{12} = \frac{\beta_{12}}{\beta} + \frac{\gamma_{12}}{\gamma} - \frac{\alpha_2}{\alpha}\left(\frac{\beta_1}{\beta} + \frac{\gamma_1}{\gamma}\right) - \frac{\alpha_1}{\alpha}\left(\frac{\beta_2}{\beta} + \frac{\gamma_2}{\gamma}\right),\qquad(6.15c)$$

$$R_{33} = \frac{\beta}{\alpha^2}\left\{\beta_{11} + \beta_{22} + \frac{1}{\gamma}(\beta_1\gamma_1 + \beta_2\gamma_2)\right\}, \qquad (6.15d)$$

$$R_{00} = -\frac{\gamma}{\alpha^2}\left\{\gamma_{11} + \gamma_{22} + \frac{1}{\beta}(\beta_1\gamma_1 + \beta_2\gamma_2)\right\}, \qquad (6.15e)$$

where the subscripts 1 and 2 indicate partial derivatives with respect to the corresponding coordinate. We have, from the above,

$$R_3^3 + R_0^0 = -\frac{1}{\alpha^2\beta\gamma}\{(\beta\gamma)_{11} + (\beta\gamma)_{22}\}. \qquad (6.16)$$

In the case of empty space, $R_{\mu\nu} = 0$ and, consequently, from (6.16)

$$(\beta\gamma)_{11} + (\beta\gamma)_{22} = 0. \qquad (6.17)$$

Again consider a transformation from x^1, x^2 to ξ, η where ξ, η are, respectively, the real and imaginary parts of an analytic function of the complex argument $(x^1 + ix^2)$, i.e.,

$$\xi + i\eta = f(x^1 + ix^2). \qquad (6.18)$$

The Cauchy–Riemann conditions are

$$\frac{\partial\xi}{\partial x^1} = \frac{\partial\eta}{\partial x^2}, \qquad (6.19a)$$

$$\frac{\partial\xi}{\partial x^2} = -\frac{\partial\eta}{\partial x^1}. \qquad (6.19b)$$

Writing out the transformation formulas, we see that the conformally flat form of the metric is preserved in the (ξ, η) coordinates. Again from (6.19)

$$\frac{\partial^2\xi}{\partial x^{1^2}} + \frac{\partial^2\xi}{\partial x^{2^2}} = 0. \qquad (6.20)$$

We now identify ξ with $(\beta\gamma)$, this is permissible in view of (6.17) and (6.20). Writing r for ξ and Z for η, the metric then assumes the form

$$dS^2 = e^{2\lambda}\, dt^{2(\nu-\lambda)}\,(dr^2 + dZ^2) - r^2 e^{-2\lambda}\, d\phi^2. \qquad (6.21)$$

This form of the metric is known as Weyl's canonical form of the axially symmetric metric. It is to be noted that the only equation we have used is $R_3^3 + R_0^0 = 0$. The metric (6.21) therefore holds not only for empty space but also whenever $T_1^1 + T_2^2 = 0$. In particular, this is the case for incoherent dust at rest as also for electromagnetic fields if the field tensor has only F^{14}, F^{13} (or F^{24}, F^{23}) nonvanishing. However, it does not hold for fluids with non-vanishing pressure.

We rewrite (6.15) in terms of λ and ν of the metric (6.21)

$$\tfrac{1}{2}(R_{11} + R_{22}) = \nu_{11} + \nu_{22} - \left(\lambda_{11} + \lambda_{22} + \frac{\lambda_1}{r}\right) + \lambda_1^2 + \lambda_2^2, \qquad (6.22a)$$

$$\tfrac{1}{2}(R_{11} - R_{22}) = \lambda_1^2 - \lambda_2^2 - \frac{v_1}{r}, \tag{6.22b}$$

$$R_{12} = 2\lambda_1 \lambda_2 - \frac{v_2}{r}, \tag{6.22c}$$

$$R_3^3 - R_0^0 = 2e^{2(\lambda - v)}\left(\lambda_{11} + \lambda_{22} + \frac{\lambda_1}{r}\right), \tag{6.22d}$$

$$R_3^3 + R_0^0 = 0, \tag{6.22e}$$

where the coordinates t, r, Z, ϕ have been numbered 0, 1, 2, 3, respectively. For the vacuum, the left-hand sides vanish and the resulting equations are

$$\lambda_{11} + \lambda_{12} + \frac{\lambda_1}{r} = 0, \tag{6.23}$$

$$v_1^2 = r(\lambda_1^2 - \lambda_2^2), \qquad v_2 = 2r\lambda_1\lambda_2, \tag{6.24}$$

$$v_{11} + v_{22} + \lambda_1^2 + \lambda_2^2 = 0. \tag{6.25}$$

Of the four equations (6.23)–(6.25), equation (6.25) follows from the other three, while (6.23) is the integrability condition for (6.24), i.e., the former follows from $v_{12} = v_{21}$ using (6.24). A method of solving the field equation will be to assume a solution of (6.23). This being Laplace's equation in two dimensions, there are well-known solutions and we will then integrate (6.24) directly.

6.5. The Case of Two Mass Particles

We shall now prove that there does not exist any solution subject to the following conditions:

(1) The metric is static and $R_{\mu v} = 0$ everywhere and the space–time is asymptotically flat.
(2) The geometry is regular everywhere except on two points on the axis of symmetry.

This result can be interpreted as indicating that in the general theory of relativity, two free particles cannot be in equilibrium, this is as we would demand from a complete theory. To make the idea clear, consider Maxwell's equation in electromagnetism. We have a static solution, the electrostatic potential

$$\phi(\mathbf{r}, t) = \frac{e_1}{|\mathbf{r} - \mathbf{r}_1|} + \frac{e_2}{|\mathbf{r} - \mathbf{r}_2|},$$

representing two point changes e_1, e_2 at positions $\mathbf{r}_1$ and $\mathbf{r}_2$. Unless we make the additional inputs of Lorentz force and Newton's equations of motion, there is nothing to tell us that this solution is in reality invalid. Thus, "force" and "motion" are not included in Maxwell's equations. On the contrary, the nonexistence of the solution of this type, representing two particles in equilibrium in general relativity, indicates that the interaction between particles and their consequent motion are in a way contained within the field equations of general relativity. In fact, this idea has led to a successful deduction of the equation of motion in a two-body system.

Let us give a proof of the theorem. Introduce the variables ρ_1 and ρ_2 by the relations

$$\rho_1^2 = r^2 + (Z - Z_1)^2, \qquad \rho_2^2 = r^2 + (Z - Z_2)^2,$$

where Z_1 and Z_2 are two constants. The reader will recognize that ρ_1 and ρ_2 are simply the distances of the "field point" (r, Z) from the "source point" $(0, Z_1)$ and $(0, Z_2)$ in Euclidean space.

Take the solution of Laplace's equation (6.23) in the form

$$\lambda = -\frac{m_1}{\rho_1} - \frac{m_2}{\rho_2}, \tag{6.26}$$

where m_1 and m_2 are constants and λ has poles at the source points. Equation (6.24) is now readily integrated to give

$$v = -\frac{m_1^2 r^2}{2\rho_1 4} - \frac{m_2^2 r^2}{2\rho_2 4} + \frac{2m_1 m_2}{(Z_1 - Z_2)}\left[\frac{r^2 + (Z - Z_1)(Z - Z_2)}{\rho_1 \rho_2} - 1\right]. \tag{6.27}$$

In (6.27), the integration constant has been chosen so that $v \to 0$ at spatial infinity. We may be tempted to interpret the solution given by (6.26) and (6.27) as presenting two particles of masses m_1 and m_2 at $(0, Z_1)$ and $(0, Z_2)$, respectively, and come to the conclusion that the two particles are at rest in spite of their gravitational interaction. However, such an interpretation is viable only if the field is regular everywhere except at $\rho_1 = 0$ and $\rho_2 = 0$, as we shall see right now this is not so.

Note that the metric (6.21) has an apparent singularity at $r = 0$ (because $g_{\phi\phi}$ vanishes). This can be dismissed as a mere coordinate singularity if we can justify ϕ as an angular coordinate. There are a number of ways to test this, we could, for example, calculate the scalars formed from the Riemann–Christoffel tensor (recall that we could, in this way, discriminate between the horizon $R = 2m$ from the true singularity at $R = 0$ in the Schwarzschild metric). However, the computation of scalars is a laborious procedure. Instead, we can use the following criterion (called the "elementary flatness" requirement).

Compute the following ratio:

$$\chi \equiv Lt_{r_0 \to 0} \frac{\text{proper length of } \phi \text{ line at } r = r_0 \text{ from } \phi = 0 \text{ to } 2\pi}{\text{proper length of } r \text{ line for } r = 0 \text{ to } r_0}$$

$$= 2\pi (e^{-v})_{r \to 0}.$$

If ϕ is really an angular coordinate and the geometry is regular at $r = 0$, then χ is the ratio of the length of the circumference to its radius in Euclidean space, i.e., we should have $\chi = 2\pi$. Conversely, if $\chi = 2\pi$, we can by a suitable transformation reduce (at $r = 0$) $g_{rr}\,dr^2 + g_{\phi\phi}\,d\phi^2$ to the Euclidean form, $dx^2 + dy^2$. But from the above $\chi = 2\pi$ requires ν to vanish at $r = 0$ for all values of Z. However, (6.27) shows that $e^\nu \neq 1$ if $(Z - Z_1)$ and $(Z - Z_2)$ are of opposite signs, i.e., if the point lies on the axis anywhere between the two poles (note that ρ_1, ρ_2 are the positive roots of ρ_1^2 and ρ_2^2). Thus the entire Z-axis between the two source points is singular. We interpret this as indicating that in order to maintain the two gravitating particles in equilibrium, we need stresses all along the joining line.

A third method of examining the regularity of the field is to note that ν_2 vanishes for $r = 0$. Hence the line integral $\int \nabla \nu\, dl$ over any line with ends on the Z-axis should vanish. It is easy to see that this is not so when the ends lie between Z_1 and Z_2.

6.6. The Schwarzschild Metric in the Form (6.21)

Obviously, spherical symmetry includes axial symmetry and so we can expect to express the Schwarzschild metric in the form (6.21). In fact,

$$e^{2\lambda} = \frac{\rho_1 + \rho_2 - 2m}{\rho_1 + \rho_2 + 2m}, \tag{6.28a}$$

$$e^{2\nu} = \frac{(\rho_1 + \rho_2)^2 - 4m^2}{4\rho_1\rho_2}, \tag{6.28b}$$

where

$$\rho_1^2 = r^2 + (Z - m)^2,$$
$$\rho_2^2 = r^2 + (Z + m)^2, \tag{6.29}$$

give a solution of the vacuum field equations. The solution may be transformed to the Schwarzschild metric

$$dS^2 = \left(1 - \frac{2m}{R}\right)dt^2 - \left(1 - \frac{2m}{R}\right)^{-1} dR^2 - R^2(d\theta^2 + \sin^2\theta\, d\phi^2),$$

if

$$R = m + \tfrac{1}{2}(\rho_1 - \rho_2) \quad \text{and} \quad \cos\theta = \frac{1}{2m}(\rho_1 - \rho_2). \tag{6.30}$$

With (6.29) and (6.30), $(\rho_1 + \rho_2) > 2m$, so that the black or white hole region is not mapped in the axially symmetric form.

Problems

1. Verify by direct substitution that (6.28) is a solution of the vacuum field equations.

2. Consider the variables $x = R/m - 1$ and $y = \cos\theta$. Show that x constant lines are ellipses and y constant lines are hyperbolas in the Euclidean $r - Z$ space (x, y are known as prolate spheroidal coordinates).

3. Write out (6.23) and (6.24) in terms of derivatives with respect to x and y introduced in Problem 2, and hence obtain the solution (6.28).

4. Calculate the Newtonian gravitational potential due to a thin rod along the Z-axis and point out its relation with the metric tensor components (6.28).

6.7. Stationary Axisymmetric Vacuum Solutions (Ernst, 1968)

We shall quote some results without giving formal proofs. First, the stationary axisymmetric vacuum metric can in general be reduced to the form

$$dS^2 = f(dt - \omega \, d\phi)^2 - f^{-1}\{e^{2\gamma}(d\rho^2 + dz^2) + \rho^2 \, d\phi^2\}, \qquad (6.31)$$

where f, ω, and γ are functions of ρ and Z only. Note that if $\omega = 0$, the metric passes over to the Weyl form. An elegant method of solution in due to Ernst. Introduce a function $\chi(\rho, Z)$ defined by the equation

$$-\rho^{-1}\mathbf{e}_\phi \times \nabla\omega = f^{-2}\nabla\chi, \qquad (6.32)$$

where $\mathbf{e}_\phi$ is the unit vector in the ϕ direction. Ernst then shows that the field equations reduce to

$$\gamma_1 = \frac{1}{4\rho}\left\{\frac{\rho^2}{f^2}(f_1^2 - f_2^2) - f^2(\omega_1^2 - \omega_2^2)\right\} \qquad (6.33)$$

$$= \frac{1}{2\rho}\left\{\frac{\rho^2}{f^2}f_1 f_2 - f^2\omega_1\omega_2\right\}, \qquad (6.34)$$

and

$$(\text{Re } \zeta)\nabla^2\zeta = \nabla\zeta \, \nabla\zeta, \qquad (6.35)$$

where the subscripts 1 and 2 indicate partial differentiation with respect to ρ and Z, respectively, and $\zeta = f + i\chi$. In (6.35), Re ζ indicates the real part of ζ, i.e., f and the Laplacian and the gradient operations are in the three-dimensional flat space.

Using spheroidal coordinates (see Problem 2 above for the definition) the equation system can be solved fairly easily. In particular, taking ζ real, corresponding to $\omega = 0$ when the metric reduces to the static form, and we can obtain a transformed Schwarzschild metric. With ζ complex we can obtain a host of stationary metrics of which the two are known after Kerr and Tomimaksu–Sato. We shall not go into the details of derivation, this is straightforward but involves rather lengthy calculations. In the next chapter we study the Kerr metric in some detail. In the following problems we give the steps leading to the Kerr metric.

Problems

1. Show that if we put $\zeta = (\eta - 1)/(\eta + 1)$ then (6.35) reduces to

$$(\eta\eta^* - 1)\nabla^2\eta = 2\eta^*\nabla\eta\nabla\eta.$$

2. If η, in the above problem $= -e^{i\alpha} \coth \psi$ where α is a constant, show that $\nabla^2\psi = 0$.

3. In spheroidal coordinates, x and y show that the equation for η becomes

$$(\eta\eta^* - 1)\left[(x^2 - 1)\frac{\partial^2\eta}{\partial x^2} + 2x\frac{\partial\eta}{\partial x} + (1 - y^2)\frac{\partial^2\eta}{\partial y^2} - 2y\frac{\partial\eta}{\partial y}\right]$$
$$= 2\eta^*\left[(x^2 - 1)\left(\frac{\partial\eta}{\partial x}\right)^2 + (1 - y^2)\left(\frac{\partial\eta}{\partial y}\right)^2\right].$$

4. Verify that $\eta = x \cos \lambda + iy \sin \lambda$ is a solution of the equation in Problem 3 and use this solution to obtain the metric. Show that this metric can be transformed to the Kerr metric given in the next chapter.

7. The Kerr Metric or the Rotating Black Hole

7.1. The Kerr Metric in Boyer–Lindquist Coordinates

In many ways the Kerr metric, which satisfies the vacuum field equation $R_{\mu\nu} = 0$, is next in importance to the Schwarzschild metric; indeed, in a sense, it is more realistic as it incorporates an angular momentum of the source and, as we all know, rotation is almost a universal property of stars. We present the metric in Boyer–Lindquist (1967) coordinates (the original form in which Kerr presented his metric is somewhat more complicated)

$$ds^2 = \frac{\Delta}{\rho^2}(dt - a \sin^2 \theta \, d\phi)^2 - \frac{\sin^2 \theta}{\rho^2}[(r^2 + a^2) \, d\phi - a \, dt]^2$$

$$- \rho^2 \left(\frac{dr^2}{\Delta} + d\theta^2 \right), \tag{7.1}$$

where

$$\Delta \equiv r^2 - 2mr + a^2, \tag{7.2}$$

$$\rho^2 \equiv r^2 + a^2 \cos^2 \theta. \tag{7.3}$$

The metric involves two arbitrary constants m and a, their interpretation becomes easy if we write (7.1) explicitly

$$dS^2 = \frac{r^2 + a^2 \cos^2 \theta - 2mr}{r^2 + a^2 \cos^2 \theta} dt^2 + \frac{4mar \sin^2 \theta}{r^2 + a^2 \cos^2 \theta} d\phi \, dt$$

$$- \frac{[(r^2 + a^2)(r^2 + a^2 \cos^2 \theta) + 2mra^2 \sin^2 \theta]}{r^2 + a^2 \cos^2 \theta} \sin^2 \theta \, d\phi^2$$

$$- (r^2 + a^2 \cos^2 \theta)\left(d\theta^2 + \frac{dr^2}{r^2 + a^2 - 2mr} \right). \tag{7.4}$$

Comparing (7.4) with the expression for the axially symmetric field at large distances (3.41) that we have found earlier, we see that m represents the gravitational mass of the system and $(m \cdot a)$, the angular momentum, is in the direction $\theta = 0$. A straightforward calculation shows that

$$\det |g_{\mu\nu}| = -\sin^2 \theta (r^2 + a^2 \cos^2 \theta) \le 0,$$

so that the signature requirement is satisfied everywhere.

7.2. The Black Hole Property

Obviously, at large distances r, t is a timelike coordinate and the metric is stationary, i.e., it admits a Killing vector which is timelike but not hypersurface orthogonal (mark the term $d\phi\, dt$). There is also another Killing vector corresponding to axial symmetry, the metric admits a translation of the angular coordinate ϕ, i.e., $\phi \to \phi + \psi$ leaves the metric unchanged, if ψ be an arbitrary constant.

It is essential for axial symmetry that ϕ should be an angular coordinate, or the path traced out by any point under the translation along ϕ is closed. That ϕ is an angular coordinate, at least for $r \to \infty$, follows from the asymptotic form of the metric

$$ds^2 = dt^2 - dr^2 - r^2\, d\theta^2 - r^2 \sin^2 \theta\, d\phi^2.$$

For another type of criterion for an angular coordinate, see the discussion later in connection with the Gödel metric. However, that test cannot be applied here as the metric has a singularity at $r = 0$.

Now g_{tt} vanishes for two values of r

$$r_\pm = m \pm \sqrt{m^2 - a^2 \cos^2 \theta},$$

and in between these two values g_{tt} is negative, i.e., t is spacelike. Of course, r_+ will have real values only if $a < m$. Light emitted from the $r = r_+$ surface would suffer infinite redshift in going outside, and this surface is also the stationary limit as within this surface the Killing vector along the t-axis is no longer timelike.

However, unlike the Schwarzschild metric, r does not change its spacelike character in crossing the infinite redshift surface and hence dr for a particle motion may vanish and be of either sign even for $r < r_+$. The change of r to timelike character takes place for $r < r_h$ where

$$r_h = m + \sqrt{m^2 - a^2}.$$

This represents a spherical surface and below this value of r_h (and above r_-) dr cannot vanish for any body as that would make $ds^2 < 0$. The $r = r_h$ surface is thus the horizon or one-way membrane. Figure 7.1 shows the sections of the surface $r = r_+$ and $r = r_h$. The shaded region between the two surfaces is known as the ergosphere.

What will be the nature of motion of a particle in this region? Clearly, no particle can have $d\phi = 0$, for in that case in this region $ds^2 < 0$. Indeed, the condition for a world-line to be timelike or null in this region is

$$g_{tt} + 2g_{\phi t}\left(\frac{d\phi}{dt}\right) + g_{\phi\phi}\left(\frac{d\phi}{dt}\right)^2 \geq 0,$$

or $d\phi/dt$ must lie between the two roots of $g_{\phi\phi}(d\phi/dt)^2 + 2g_{\phi t}(d\phi/dt) + g_{tt} = 0$

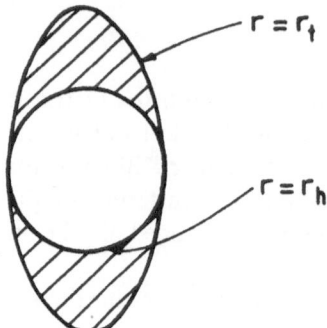

Figure 7.1

(because $g_{\phi\phi}$ is negative). The roots are

$$[-g_{\phi t} \pm \sqrt{(g_{\phi t}^2 - g_{tt}g_{\phi\phi})}]g_{\phi\phi}^{-1}.$$

Both the roots are positive and thus we have the result that in this region all particles are constrained to rotate in the same sense as the angular momentum of the source, a phenomenon referred to as "the dragging of the inertial frames." As we go from r_+ toward r_h, the two roots approach one another and finally at the horizon as $g_{\phi t}^2 = g_{tt}g_{\phi\phi}$, the roots are identical and the permitted range of $d\phi/dt$ collapses to a single value. This is the limit of the range up to which motions in either direction of r, as also with $dr = 0$, are possible.

7.3. Locally Nonrotating Observers

We may write the Kerr metric in the form

$$ds^2 = \left(g_{tt} - \frac{g_{\phi t}^2}{g_{\phi\phi}}\right)dt^2 + g_{\phi\phi}\left(d\phi + \frac{g_{\phi t}}{g_{\phi\phi}}dt\right)^2 + g_{\theta\theta}\,d\theta^2 + g_{rr}\,dr^2. \quad (7.5)$$

If now for an observer $d\phi/dt = -(g_{\phi t}/g_{\phi\phi})$, then along his world-line

$$ds^2 = \left(g_{tt} - \frac{g_{\phi t}^2}{g_{\phi\phi}}\right)dt^2 + g_{\theta\theta}\,d\theta^2 + g_{rr}\,dr^2,$$

i.e., he will not experience any change of the angular coordinate ϕ. Such an observer is called a locally nonrotating observer and for him $d\phi/ds$ vanishes, his angular momentum, defined as the component of the four momentum in the direction of ϕ, also vanishes.

7.4. The Horizon as a Null Surface

We have seen that in the Kruskal–Szekeres diagram, the horizon $R = 2m$ is represented by two null lines orthogonal to one another. They might have appeared to be accidental but we show below that there is an inherent relation between horizons (i.e., hypersurfaces which allow only one-way passage) and null surfaces (i.e., hyersurfaces the normal to which at every point is a null vector).

We start our consideration with a flat Minkowski space. If we take a 3-space spanned by three orthogonal spacelike congruences, then the normal will be everywhere timelike, and we may choose our time coordinate suitably so that the equation of the surface becomes $t = $ constant $= t_0$ (say); obviously, the 3-space separates the space–time into two halves with $t > t_0$ and $t < t_0$, respectively. With the idea that all particle world-lines must be future directed, we see that the world-lines can cross the surface only in the direction from $t < t_0$ to $t > t_0$. Thus, this surface is a one-way membrane. In the limiting case when the normals become null vectors, the interesting thing that happens is that the normal also lies on the surface (recall the curious fact that a null vector is self-orthogonal), and the future directed null cone still lies entirely on one side of the hypersurface, touching the hypersurface on one side. Beyond this, when the hypersurface contains timelike lines (the normal being spacelike in that case), the separation into the two half-spaces is no longer temporal and passage through the hypersurface can occur in either direction. These results may be taken over directly to the space–time of general relativity, for at any point the curved space–time may be replaced by the tangent flat space–time.

Taking the Kerr metric, we see that a null hypersurface would have an equation independent of ϕ and t. This is because of the symmetry. Perform the symmetry operation, i.e., translation along ϕ and t on an element of null hypersurface and we develop the entire subspace. Hence, if its equation is $f(r, \theta) = 0$, the normal would be null if

$$g^{rr}\left(\frac{\partial f}{\partial r}\right)^2 + g^{\theta\theta}\left(\frac{\partial f}{\partial \theta}\right)^2 = 0,$$

or

$$(r^2 - 2mr + a^2)\left(\frac{\partial f}{\partial r}\right)^2 + \left(\frac{\partial f}{\partial \theta}\right)^2 = 0. \tag{7.6}$$

Also,

$$\left(\frac{\partial f}{\partial r}\right)dr + \left(\frac{\partial f}{\partial \theta}\right)d\theta = 0,$$

or

$$\left(\frac{\partial f}{\partial r}\right) + \left(\frac{\partial f}{\partial \theta}\right)\left(\frac{d\theta}{dr}\right)_f = 0, \tag{7.7}$$

where dr, $d\theta$ is a vector lying on the surface and $(d\theta/dr)_f$ signifies that the derivative is to be taken subject to f being held constant. From (7.6) and (7.7), obviously, one solution is $(\partial f/\partial\theta) = 0$ giving $f(r) \equiv r^2 - 2mr + a^2 = 0$. This gives the horizons as we have already noted. To investigate other possible solutions, we eliminate $(\partial f/\partial r)$ and $(\partial f/\partial\theta)$ from (7.6) and (7.7) to obtain

$$(r^2 - 2mr + a^2)\left(\frac{d\theta}{dr}\right)^2 + 1 = 0,$$

or

$$d\theta = \frac{dr}{\sqrt{2mr - r^2 - a^2}}.$$

Integrating we obtain the equation of the null surface

$$\sin(\theta - \theta_0) = \frac{r - m}{\sqrt{m^2 - a^2}}.$$

Obviously, this surface lies entirely inside the horizon, i.e., $r \leq r_h$, hence this null surface does not play any important part. Any surface having an equation of the form $f(r, \theta) = 0$ will have spacelike normals for $r > r_h$, as both the r and θ coordinates are spacelike in that region.

The null lines, which are at once the normals and tangents to the horizon, will be referred to as generators of the horizon. Indeed, we can proceed further and show that these generators are geodesic. A general theorem about the null geodesic property of the horizon of any black hole, dynamic or static, has been shown to be true.

7.5. The Kerr–Newmann Metric

Just as the Reissner–Nordstrom metric can be looked upon as a generalization of the Schwarzschild metric with an electromagnetic field replacing the vacuum, the so-called Kerr–Newmann metric considers an electromagnetic field in vacuo where, like the Kerr metric, the source has rotation. The metric is of the same form as the Kerr metric, only the definition of Δ has to be altered by the addition of a charge term

$$\Delta = r^2 - 2mr + a^2 + e^2,$$

where e is the total charge of the source. Again if a is put to zero, the Kerr–Newmann metric passes over to that of Reissner and Nordstrom.

It is conjectured that the most general black hole solution is the Kerr–Newmann metric or, in other words, a black hole field in empty space is characterized by only three parameters: mass, angular momentum, and charge, all others being last or made unobservable in the process of collapse. This is the so-called no-hair theorem which we shall consider in greater detail later.

7.6. The Penrose Process

While according to the classical ideas (as distinct from the quantum ideas introduced by Hawking), nothing can escape from a Schwarzschild black hole and there is no way to extract any energy from such a situation, Penrose has conceived a process by which energy can be extracted from a Kerr black hole. The idea is to utilize the peculiar situation in the ergosphere where the following conditions hold:

(i) the coordinate t is spacelike; and

(ii) a motion in either direction of r is possible as r is also spacelike.

Condition (i) allows a particle to move in the negative direction of t in the ergosphere, and the energy of such a particle will be negative as reckoned by the observer at rest at infinity whose velocity vector is $v^\mu = (dt/ds, 0, 0, 0)$. Consider now a body to go down to the ergosphere and there break up into two fragments, one with negative energy (i.e., proceeding along the negative direction of t) and the other with correspondingly enhanced energy. If now the negative energy part goes down into the black hole and the other part escapes into the outside vacuum, we shall have derived energy from the black hole.

However, there is a severe restriction that the two split parts must have a high relative velocity $> c/2$. Again, another way of obtaining energy may be by the scattering of electromagnetic waves after their entry into the ergosphere. In principle, the entire rotational kinetic energy of the Kerr black hole can be extracted, making it pass into a Schwarzschild black hole condition when it becomes "dead" in the sense that it cannot yield any further energy.

An interesting series of investigations has been made by Dadhich and collaborators (see the review by Wagh and Dadhich, 1989) on the extraction of energy from a rotating black hole situated in an external electromagnetic field. The requirement of the high relative velocity then no longer exists. Further, the effective ergosphere now extends much beyond $r = 2M$ and the efficiency of the energy extraction is also increased. It is claimed that, with the external magnetic field $B \sim 10^{-6}$ G, $a/M = 0.998$, $M = 10^8 \, M_\odot$, we can have 100% efficiency of the energy extraction. (For a detailed review of the works, see Wagh and Dadhich, 1989.)

Problems

1. Using the Hamilton–Jacobi method show that $p^2_{\,\theta} + \cos^2 \theta [a^2(m_0^2 - E^2) + p_\phi^2 \sin^{-2} \theta]$ is a constant of motion for a test particle in the Kerr metric.

2. Study the motion of test particles in the Kerr metric and obtain the condition for circular orbits.

3. Show that there exists within the ergosphere a timelike Killing vector K^μ of the form $K^\mu = \delta_t^\mu + \lambda \delta_\phi^\mu$ where λ is a constant and find λ. In view of this Killing vector, is it unjustified to consider the outer boundary of the ergosphere as the static limit?

4. Corresponding to the ignorable coordinates t and ϕ in the Kerr–Newmann metric, show that the conserved energy and angular momentum of a particle of rest mass m_0 and charges q are

$$E = m_0 v_0 + q A_\phi,$$

$$p_\phi = m_0 v_\phi + q A_\phi,$$

where A^μ is the electromagnetic potential for the Kerr–Newmann field. Hence, using the condition $v^\mu v_\mu = 1$, show that

$$E = \frac{\beta + \sqrt{\beta^2 - \alpha\gamma}}{\alpha},$$

where α, β, and γ are given by

$$\alpha = (r^2 + a^2)^2 - \Delta a^2 \sin^2 \theta,$$

$$\beta = (p_\phi a + eqr)(r^2 + a^2) - p_\phi a\Delta$$

$$= (p_\phi a + eqr)^2 - \Delta(p_\phi/\sin^2 \theta)^2 - m_0^2 \Delta \rho^2 - \rho^2[(p^r)^2 + \Delta(p^\theta)^2].$$

Hence show that an extraction of energy leads to a reduction of at least one of the two constants a and e.

8. The Energy–Momentum Pseudotensor of the Gravitational Field and Loss of Energy by Gravitational Radiation

8.1. The Pseudo-Energy–Momentum Tensor

As is well known, when we consider an interaction present, the kinetic enegy is no longer conserved. In the case of electromagnetic fields or Newtonian gravitation, we recover an energy conservation by introducing the concept of field energy. In general, for any conserved entity, an equation of the form

$$\frac{\partial}{\partial t} \int X \, dv = -\oint \mathbf{Y} \, d\mathbf{s} \qquad (8.1)$$

holds, where the surface integral on the right-hand side is over the boundary of the volume of integration on the left. We then say that the change of the entity X within the volume V is exactly accounted for by the flux through the boundary. As an example, we recall the equation derived from Maxwell's equations

$$\frac{\partial}{\partial t} \int \frac{E^2 + B^2}{8\pi} \, dv = -\frac{c}{4\pi} \oint (\mathbf{E} \times \mathbf{H}) \, d\mathbf{s}. \qquad (8.2)$$

We recognize the integrand on the left-hand side as the energy density and identify $(c/4\pi)(\mathbf{E} \times \mathbf{H})$ as the energy flux vector. The form (8.1) is readily arrived at if we have an equation

$$S^{\alpha\beta}_{,\beta} = 0. \qquad (8.3)$$

However, the gravitational field in general relativity poses some peculiar problems. First, there is no role of field energy in the Einstein equations

$$R^{\mu}_{\gamma} - \tfrac{1}{2}\delta^{\mu}_{\gamma}R = -8\pi T^{\mu}_{\gamma}, \qquad (8.4)$$

T^{μ}_{γ} arising entirely from nongravitational fields.

Second, in analogy with the case of electromagnetic fields, we would expect that the field energy–momentum complex would involve only the field variables $g_{\mu\gamma}$ and their first derivatives $g_{\mu\gamma,\alpha}$. However, because of the equivalence principle, gravitational fields locally can be transformed away, i.e., in the local Lorentz coordinates $g_{\mu\gamma} = \eta_{\mu\gamma}$ and $g_{\mu\gamma,\alpha} = 0$, so that the field energy–momentum would vanish. Such vanishing in a particular coordinate system is possible only for nontensorial entities. Hence we have to accept the situation that the energy–momentum complex that we are seeking will not be tensors. However, globally, gravitational fields do have significance and therefore the volume integrals of our field energy, etc., should have a coordinate independent meaning at least in the case of asymptotically flat space–time. (The last condition comes from the fact that the energy–momentum, etc., cannot be defined operationally unless the field is asymptotically flat.)

Thus we demand the following for the $S^{\alpha\beta}$ in gravitational fields:

(a) $S^{\alpha\beta}$ should contain the energy–momentum tensor $T^{\alpha\beta}$ and another part involving $g_{\mu\gamma}$ and their first derivatives.
(b) $S^{\alpha\beta}$ should be symmetric, i.e., $S^{\alpha\beta} = S^{\beta\alpha}$, for only then can we define an angular momentum which is conserved.
(c) In asymptotically flat space–time, $\int S^{\alpha\beta}\, dv$, taken over the entire space section at a constant time coordinate t_0, should have a coordinate independent meaning consistent with our ideas of energy and momentum of an isolated system.

The most commonly used $S^{\alpha\beta}$ is due to Landau and Lifshitz who define

$$S^{\alpha\beta} \equiv h^{\alpha\beta\gamma}_{,\gamma}$$

with

$$h^{\alpha\beta\gamma} \equiv \frac{1}{16\pi}[|g|(g^{\alpha\beta}g^{\gamma\delta} - g^{\alpha\gamma}g^{\beta\delta})]_{,\delta}.$$

As $h^{\alpha\beta\gamma}$ is antisymmetric with respect to β and γ, it follows that (8.3) is satisfied. Further, it can readily be verified that

$$S^{\alpha\beta} = S^{\beta\alpha}.$$

Using the field equations, Landau and Lifshitz showed that condition (a) is also satisfied

$$S^{\alpha\beta} = |g|(T^{\alpha\beta} + t^{\alpha\beta}),$$

where $t^{\alpha\beta}$ involved only $g_{\mu\nu}$ and their first derivatives. (The second derivatives cancel out where $T^{\alpha\beta}$ is replaced by the Einstein equation.) The explicit expression for $t^{\alpha\beta}$ can be worked out and comes out to be

$$
\begin{aligned}
t^{\alpha\beta} = \frac{1}{16\pi}\{ & (2\Gamma^{\mu}_{\nu\sigma}\Gamma^{\rho}_{\mu\rho} - \Gamma^{\mu}_{\nu\rho}\Gamma^{\rho}_{\sigma\mu} - \Gamma^{\mu}_{\nu\mu}\Gamma^{\rho}_{\sigma\rho})(g^{\alpha\nu}g^{\beta\sigma} - g^{\alpha\beta}g^{\nu\sigma}) \\
& + g^{\alpha\nu}g^{\sigma\mu}(\Gamma^{\beta}_{\nu\rho}\Gamma^{\rho}_{\sigma\mu} + \Gamma^{\beta}_{\sigma\mu}\Gamma^{\rho}_{\nu\rho} - \Gamma^{\beta}_{\mu\rho}\Gamma^{\rho}_{\sigma\nu} - \Gamma^{\beta}_{\nu\sigma}\Gamma^{\rho}_{\mu\rho}) \\
& + g^{\beta\nu}g^{\sigma\mu}(\Gamma^{\alpha}_{\nu\rho}\Gamma^{\rho}_{\sigma\mu} + \Gamma^{\alpha}_{\sigma\mu}\Gamma^{\rho}_{\nu\rho} - \Gamma^{\alpha}_{\mu\rho}\Gamma^{\rho}_{\sigma\nu} - \Gamma^{\alpha}_{\nu\sigma}\Gamma^{\rho}_{\mu\rho}) \\
& + g^{\nu\sigma}g^{\mu\rho}(\Gamma^{\alpha}_{\nu\mu}\Gamma^{\beta}_{\sigma\rho} - \Gamma^{\alpha}_{\nu\sigma}\Gamma^{\beta}_{\mu\rho})\}.
\end{aligned}
\tag{8.5}
$$

The expression, while undoubtedly complicated, simplifies enormously in the case of plane gravitational waves where it is commonly used.

Regarding condition (c) we note that if we identify the enegy–momentum vector p^{α} with $\int S^{\alpha 0}\, dV$ then, as for an asymptotically flat space, for $r \to \infty$

(zero referring to the time coordinate)

$$g_{00} = 1 - \frac{2m}{r} + \frac{2m^2}{r^2} + O\left(\frac{1}{r^3}\right),$$

$$g_{0i} = O\left(\frac{1}{r^2}\right),$$

$$g_{ik} = \left[1 + \frac{2m}{r}\right]\delta_{ik} + O\left(\frac{1}{r^3}\right),$$

(where i and k run over 1 to 3, the space coordinates, and we assume stationary conditions) we have

$$p^\alpha = \int S^{\alpha 0} \, dV = \int h^{\alpha 0\gamma}_{,\gamma} \, dV = \int h^{\alpha 0i}_{,i} \, dV$$

$$= \oint h^{\alpha 0i} \, dS_i,$$

and

$$p^0 = \frac{1}{16\pi} \oint \frac{4m}{r^2} r^2 \, d\omega = m,$$

$$p^i = 0.$$

These results are consistent with our ideas of the energy and momentum of the system and are also coordinate independent, insofar as the weak field values of $g_{\mu\nu}$ are not altered by any coordinate transformation which alters the $g_{\mu\nu}$ at large distances only infinitesimally.

It is usual to call $t^{\alpha\beta}$ the energy–momentum pseudotensor of the field, but it is to be emphasized that it has no local significance and even the volume integral can be used meaningfully only for asymptotically flat fields. Last, we define the angular momentum $J^{\mu\nu}$ as an antisymmetric object

$$J^{\mu\nu} = \int (x^\mu S^{\nu 0} - x^\nu S^{\mu 0}) \, dv. \tag{8.6}$$

8.2. Historical Note

The failure to obtain a tensor expression for the energy–momentum system of the gravitational field, at one time seemed disturbing to many relativists. However, eventually, it became clear that the equivalence principle excluded such a possibility. Einstein was the first to consider a "pseudotensor" for the field energy–momentum–stress.

With the Lagrangian $\mathscr{L}\sqrt{|g|}$ introduced in the problem, standard field-theoretic procedure led to the entity t_α^ρ defined by

$$t_\alpha^\rho = \frac{1}{16\pi\sqrt{|g|}}\left(\frac{\partial \mathscr{L}}{\partial g_{,\rho}^{\mu\nu}}g_{,\alpha}^{\mu\nu} - \delta_\alpha^\rho \mathscr{L}\right),$$

such that

$$[(T_\alpha^\beta + t_\alpha^\beta)\sqrt{|g|}]_{,\beta} = 0.$$

Einstein identified t_α^β with the energy–stress–momentum of the field. However, while satisfactory from other points of view, the Einstein $t^{\beta\alpha}$'s were not symmetric and hence did not allow any consideration of angular momentum. This defect is remedied in the Landau–Lifshitz formulation.

8.3. Loss of Energy by Gravitational Radiation

The observationally important situation has the following characteristics:

(a) The observation is made on Earth where we are justified in considering the radiation field to be weak.
(b) The source is far away (compared to, say, the wavelength of the radiation and the dimensions of the source), but the gravitational field is presumably quite intense at the location of the source and velocities there may be comparable to that of light. To fix our ideas, we may consider the source to be a binary, one or both members being a compact object.

Clearly, under this situation, we cannot take the field to be weak everywhere and we base our calculation on the linearized field equations. Unfortunately, a fully satisfactory way of calculating the radiation flux in this situation is not available and we do go through some approximations, the validity of which is highly suspect. However, it is in the spirit of the present time to believe in the final result, although the steps towards that are often questionable.

We start by recalling the linearized field equations ((3.19) and (3.20))

$$\gamma_{\nu,\mu}^\mu = 0, \tag{8.7}$$

$$\Box\gamma_{\mu\nu} = \frac{16\pi G}{c^4} T_{\mu\nu}, \tag{8.8}$$

where

$$\left.\begin{aligned}\gamma_{\mu\nu} &= h_{\mu\nu} - \tfrac{1}{2}\eta_{\mu\nu}h, \\ g_{\mu\nu} &= \eta_{\mu\nu} + h_{\mu\nu},\end{aligned}\right\} \tag{8.9}$$

which assumes implicitly that $T_{\mu\nu}$, the energy–stress components, are also small. Note that we have restored G and c whereas in the earlier discussions we have taken $G = c = 1$.

In the intense field region (8.8) will obviously break down; nevertheless, we assume an equation of the form

$$\Box \gamma_{\mu\nu} = \frac{16\pi G}{C^4} S_{\mu\nu},$$ (8.10)

where $S_{\mu\nu}$ contains, besides $T_{\mu\nu}$, the terms that come from the departure of the Ricci tensor from the relation

$$R_{\mu\nu} = \tfrac{1}{2}\Box h_{\mu\nu}.$$ (8.11)

Equation (8.10) has the well-known retarded solution

$$\gamma_{\mu\nu}(\mathbf{r}, t) = -\frac{4G}{C^4} \int \frac{S_{\mu\nu}(\mathbf{r}', t')}{|\mathbf{r} - \mathbf{r}'|} d^3x',$$ (8.12)

where

$$t' = t - \frac{|\mathbf{r} - \mathbf{r}'|}{C}.$$

With our assumption that the linear dimensions of the source are small compared to the distance of the field point, we may take $|\mathbf{r} - \mathbf{r}'| = r$, so that (8.12) becomes

$$\gamma_{\mu\nu}(\mathbf{r}, t) = -\frac{4G}{C^4 r} \int S_{\mu\nu}(\mathbf{r}, t') d^3x'.$$ (8.13)

In view of (8.7) and (8.13)

$$S_{\mu\nu,\nu} = 0,$$ (8.14)

which, written explicitly, gives (zero indicating the time coordinate, and the Latin indices i, j indicating the space coordinates)

$$S_{00,0} - S_{0j,j} = 0,$$ (8.15)

$$S_{i0,0} - S_{ij,j} = 0.$$ (8.16)

Using (8.16), we have

$$\frac{\partial}{\partial x^0} \int S_{i0} x^1 \, d^3x = \int \frac{\partial}{\partial x^j} (S_{ij} x^1) \, d^3x - \int S_{il} \, d^3x.$$ (8.17)

The first integral on the right can be converted into a surface integral over the sphere at infinity, and vanishes as $S_{ij} \to 0$ faster than r^{-3} as $r \to \infty$. Hence

$$\int S_{il} \, d^3x = -\frac{\partial}{\partial x^0} \int S_{i0} x^l \, d^3x$$

$$= -\frac{\partial}{\partial x^0} \int S_{l0} x^i \, d^3x$$ (8.18)

(because the left-hand side is symmetric in i and l). Again, using (8.15),

$$\frac{\partial}{\partial x^0} \int S_{00} x^i x^l \, d^3x = \int \frac{\partial}{\partial x^j} (S_{0j}) x^i x^l \, d^3x$$

$$= \int \frac{\partial}{\partial x^j} (S_{0j} x^i x^l) \, d^3x - \int (S_{0i} x^l + S_{0l} x^i) \, d^3x.$$ (8.19)

From (8.18) and (8.19), we finally have (the divergence integral in (8.19) vanishing because of the behavior of S_{0j} at infinity)

$$\int S_{il}\, d^3x = \frac{1}{2}\frac{\partial^2}{\partial x_0^2}\int S_{00} x^i x^l\, d^3x. \qquad (8.20)$$

Identifying S_{00} with T_{00} (as is justified) and taking it equal to ρC^2, where ρ is the energy density we find from (8.13) and (8.20)

$$\gamma_{il}(\mathbf{r}, t) = -\frac{2G}{C^4 r}\frac{\partial^2}{\partial t^2}\int \rho x^i x^l\, d^3x, \qquad (8.21)$$

where we have replaced x^0 by ct. Equation (8.21) shows the wave field γ_{il} as arising from the second time derivative of the quadrupole moment of the mass distribution. The fact that the monopole and dipole terms do not give rise to any wave field is essentially due to the conservation of energy and momentum, which has been taken care of in our analysis by (8.14).

At a large distance from a bounded source, we can consider the wave to be plane over small areas, so that we may use the result, that for the wave propagating in the x' direction the only surviving components of $h_{\mu\nu}$ are $h_{22} = h_{33}$ and h_{23}, as discussed in Chapter 3 and from (8.21),

$$h_{22} = -h_{33} = -\frac{2G}{3C^4\pi}(\ddot{D}_{22} - \ddot{D}_{33}),$$

$$h_{23} = -\frac{2G}{3C^4\pi}\ddot{D}_{23},$$

where we have introduced the traceless quadrupole moment tensor

$$D_{ik} = \int \rho(3x_i x_k - \delta_{ik}r^2)\, d^3x.$$

In the far-from-the-source region the flux of radiation is given by the t^{01} component of the energy–momentum pseudotensor of the gravitational field, which in the present case is

$$ct^{01} = \frac{c^3}{16\pi G}\left[(\dot{h}_{23})^2 + \tfrac{1}{4}(\dot{h}_{22} - \dot{h}_{33})^2\right]$$

$$= \frac{G}{36\pi c^5 r^2}\left[\left(\frac{\dddot{D}_{22} - \dddot{D}_{33}}{2}\right)^2 + \dddot{D}_{23}^2\right]. \qquad (8.22)$$

However, to evaluate the total energy lost by the radiating system, we must consider waves proceeding in all possible directions. We can write, for the energy flux in any direction with unit vector n^i,

$$t^{0i}n^i = \frac{G}{36C^5 r^2\pi}\left[\tfrac{1}{2}\dddot{D}_{k1}\dddot{D}_{kl} - \dddot{D}_{kl}\dddot{D}_{km}n^l n^m + \tfrac{1}{4}(\dddot{D}_{kl}n^k n^l)^2\right],$$

so that the power radiated through a solid angle $d\Omega$ is

$$\frac{G}{36\pi C^5} [\tfrac{1}{2}\dddot{D}_{kl}\dddot{D}_{kl} - \dddot{D}_{kl}\dddot{D}_{km}n^l n^m + \tfrac{1}{4}(\dddot{D}_{kl}n^k n^l)^2] \, d\Omega,$$

summed up over all vectors n^i. Using the following formulas:

$$\frac{1}{4\pi} \int n^i n^k \, d\Omega = \tfrac{1}{3}\delta_{ik},$$

$$\frac{1}{4\pi} \int n^i n^k n^l n^m \, d\Omega = \tfrac{1}{15}(\delta_{ik}\delta_{lm} + \delta_{kl}\delta_{ni} + \delta_{il}\delta_{km}),$$

find for the rate of energy loss by the source

$$-\frac{dE}{dt} = \frac{G}{45C^5} \dddot{D}_{ki}\dddot{D}_{ki}. \tag{8.23}$$

This formula, known as the quadrupole radiation formula, was obtained by Einstein and is widely used, but as the reader must have noted the deduction is seriously in doubt for realistic sources because of the assumption involved. However, we shall presently see that it has recently been claimed that this formula has been observationally verified in the case of a binary pulsar system.

8.4. The Case of a Binary Star

We now go on to compute the power loss from a binary star, the masses of the two members being comparable. (Thus, unlike the planetary case, none of the stars can be considered to be a test particle describing a geodesic in the field of the other.) However, the radiation loss is a small quantity of second order ($\sim \dot{h}_{\alpha\beta}^2$), hence assuming that the general relativity correction to Newtonian mechanics will be small, they can be neglected.

We take, for simplicity, the orbits to be circular about the common center of mass. We are considering the two stars to be mass particles. Then if a_1 and a_2 are the distances from the center of mass and $a = a_1 + a_2$, the distance between the two stars,

$$m_1 \cdot a_1 = m_2 \cdot a_2 = \mu \cdot a,$$

where μ, the reduced mass, is given by

$$\mu = \frac{m_1 m_2}{m_1 + m_2},$$

Then the angular velocity ω of either particle is given by

$$\omega^2 = \frac{G(m_1 + m_2)}{a^3}. \tag{8.24}$$

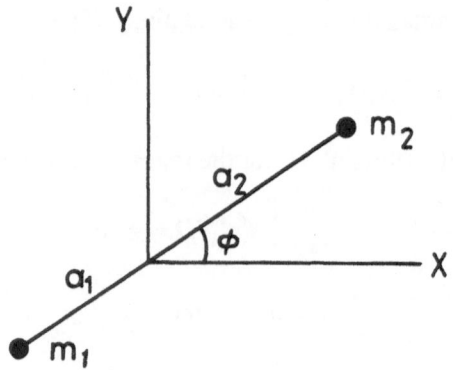

Figure 8.1

If the orbit is in the $X-Y$ plane

$$D_{xx} = 3(m_1 a_1^2 + m_2 a_2^2) \cos^2 \phi - m_1 a_1^2 - m_2 a_2^2,$$

$$D_{yy} = 3(m_1 a_1^2 + m_2 a_2^2) \sin^2 \phi - m_1 a_1^2 - m_2 a_2^2,$$

$$D_{xy} = 3(m_1 a_1^2 + m_2 a_2^2) \cos \phi \sin \phi,$$

and all the other D_{ik}'s are constant. Performing the differentiation and substituting for $\omega = \dot{\phi}$ from (8.24), we get from (8.23)

$$-\frac{dE}{dt} = \frac{32}{5} \frac{G^4}{C^5} \frac{(m_1 + m_2) m_1^2 m_2^2}{a^5}.$$

However, the energy

$$E = -\frac{1}{2} \frac{G\mu(m_1 + m_2)}{a}.$$

Hence a radiation of energy away would lead to a decrease in a and a consequent increase in ω. If T is the time period

$$\frac{\dot{T}}{T} = -\frac{\dot{\omega}}{\omega} = +\frac{3}{2}, \quad \frac{\dot{a}}{a} = -\frac{3}{2}, \quad \frac{\dot{E}}{E} = -\frac{96}{5} \frac{G^3}{C^5} \frac{(m_1 + m_2) m_1 m_2}{a^4}.$$

The calculation proceeds along the same lines as in the case of elliptic motion where the basic formulas are

$$a = -\frac{Gm_1 m_2}{2E},$$

$$e^2 = 1 + \frac{2EL^2(m_1 + m_2)}{G^2 m_1^3 m_2^3},$$

$$r = \frac{a(1 - e^2)}{1 + e \cos \phi},$$

$$m_1 \cdot r_1 = m_2 \cdot r_2 = \mu \cdot r,$$

where E is the energy, L is the angular momentum, r is the separation between masses, and a and e are the semimajor axis and the eccentricity of the orbit, respectively.

A straightforward calculation then gives

$$-\frac{dE}{dt} = \frac{8m_1^2 m_2^2 G^3}{15a^2(1 - e^2)^2 C^5}[12(1 + e\cos\phi)^2 + e^2\sin^2\phi] \cdot \dot{\phi}^2.$$

Eliminating $\dot{\phi}$ we get, for the rate of energy loss averaged over a period,

$$-\left\langle\frac{dE}{dt}\right\rangle = \frac{32}{5}\frac{G^4 m_1^2 m_2^2(m_1 + m_2)}{C^5 a^5(1 - e^2)^{7/2}}(1 + \tfrac{73}{24}e^2 + \tfrac{37}{96}e^4),$$

and finally

$$\frac{\dot{T}}{T} = -\frac{96}{5}\frac{m_1 m_2}{(T/2\pi)^{8/3}(m_1 + m_2)^{1/3}(1 - e^2)^{7/2}}(1 + \tfrac{73}{24}e^2 - \tfrac{37}{96}e^4).$$

Problems

1. Work out the details of the calculation $\dot{T}/T$ in the case of elliptic motion.

2. The decrease of angular momentum due to radiation for a binary is given by

$$\frac{dL}{dt} = -\frac{32}{5}\frac{G^{7/2}}{C^5}\frac{\mu^2 M^{5/2}}{a^{7/2}}\frac{(1 + \tfrac{7}{8}e^2)}{(1 - e^2)^2}$$

(Lightman et al., 1975). Prove this. Assuming the above formula, calculate the rate of change of eccentricity e. Assuming that the Hulse–Taylor binary had started with $e \sim 1$, estimate the time it has taken for e to decrease to its present value ~ 0.6. The mass of either star $= 1.4\ M_\odot$:

$$\text{Period } P = 0.060 \text{ s},$$

$$\dot{P} = 8.6 \times 10^{-18} \text{ s s}^{-1},$$

$$\text{Orbital period} = 27907 \text{ s},$$

$$\text{Semimajor axis } a = 6.5 \text{ light seconds}.$$

3. An ellipsoidal body is spinning about one of its principal axes. If I_1 and I_2 are the principal moments of inertia in the equatorial plane, show that the rate of its radiation is given by

$$\frac{dE}{dt} = -\frac{32}{5}\frac{G}{C^5}(I_1 - I_2)^2\Omega^6.$$

In particular, if the ellipticity in the equatorial plane is

$$\varepsilon \equiv \frac{a - b}{(a + b)^2},$$

$$dE = -\frac{32}{5}\frac{G}{C^5}I^2\varepsilon^2\Omega^6.$$

4. Calculate the rate of emission of energy when two point masses, m_1 and m_2, fall towards each other, assuming Newtonian equations of motion.

9. Analysis of the Observational Data of the Hulse–Taylor Pulsar. Confirmation of the Einstein Quadrupole Radiation Formula

We shall now see how the observations on the Hulse–Taylor pulsar PSR 1913 + 16 (the letters PSR indicate that it is a pulsar and 1913 + 16 indicates its angular position in the sky—the right ascension is 19^h 13^m and the declination is 16^0) have led to a verification of quite a number of predictions of general relativity, including the formula for the energy loss from a binary system due to gravitational radiation. The verification has been of a blanket type in the sense that many theoretical effects tie up to account for the observations; however, other effects have independent verifications so that the observations have been taken to be basically a confirmation of the energy loss formula and the existence of gravitational radiation.

Practically the only quantitative observation that we can make about a pulsar is the period of the pulsar. The pulsed radiation is invariably of a broad-band type, so that no discrete frequency can be identified and the usual type of frequency shift of spectral lines cannot be measured. However, the pulse period in the case of PSR 1913 + 16 undergoes a periodic change. This change is interpreted as due to the presence of a companion star. How can the existence of a companion star affect the observed period when the period is believed to arise from the spin of the pulsar? This can happen in a number of ways, in which both the special and the general theory of relativity play their parts. We proceed to analyze them.

The Doppler Effect

If there is a relative velocity v between the source (in this case, the pulsar) and the observer, any period emission τ in the source appears to be τ' to the observer where

$$\frac{\tau'}{\tau} = \frac{1 + v_1/c}{\sqrt{1 - v^2/c^2}}, \tag{9.1}$$

where v_1 is the longitudinal component of the relative velocity (i.e., the velocity component along the line of sight). The numerator represents a classical (i.e., nonrelativistic) effect while the denominator arises owing to time dilation (a special relativity effect). This factor is sometimes referred to as the transverse Doppler effect, as it occurs even when the motion is normal to the line of sight. Note that we have no direct knowledge to τ, we merely observe τ' so that at first sight (9.1) may appear to be of not much help.

However, the velocity of the pulsar which describes an ellipse (at the Newtonian level of approximation) due to it being a member of a binary, periodically varies and τ' also periodically changes correspondingly. Equation (9.1) can be expanded in a power series of v/c

$$\frac{\tau'}{\tau} \approx 1 + \frac{v_1}{C} + \frac{1}{2}\frac{v^2}{c^2}. \tag{9.2}$$

Under the conditions obtaining at present, terms involving cubes and higher powers of v/c cannot be observed.

The Effect due to the Gravitational Field of the Companion Star

The companion star produces a fairly intense gravitational field at the position of the pulsar, as the masses of the two are $\sim M_\odot$ and the distance r between them is $\sim R_\odot$. The change of period, as we have already studied, is given by

$$\frac{\tau'}{\tau} = \left(1 - \frac{2GM_c}{rc^2}\right)^{-1/2} \approx 1 + \frac{GM_c}{rc^2}, \tag{9.3}$$

where M_c is the mass of the companion star.* We can express the variables v_1, v^2/c^2, and GM_c/rc^2 in terms of the masses of the two stars and the characteristics of the orbit.

From the Newtonian equations of motion of the two stars (considered simply as mass particles)

$$m_1 \ddot{\mathbf{r}}_1 = \frac{Gm_1 m_2}{|\mathbf{r}_1 - \mathbf{r}_2|^3}(\mathbf{r}_2 - \mathbf{r}_1),$$

$$m_2 \ddot{\mathbf{r}}_2 = -\frac{Gm_1 m_2}{|\mathbf{r}_1 - \mathbf{r}_2|^3}(\mathbf{r}_2 - \mathbf{r}_1), \tag{9.4}$$

we get $(m_1 r_1 + m_2 r_2)\ddot{} = 0$. In view of (9.4) we can write out the equation of motion in the rest frame of the center of mass

$$\ddot{\mathbf{r}}_1 = -\frac{Gm_2}{(r_1 + r_2)^2}\frac{\mathbf{r}_1}{r_1}$$

$$= -\frac{Gm_2^3}{(m_1 + m_2)^2}\frac{\mathbf{r}_1}{r_1^3}, \tag{9.5}$$

$$\ddot{\mathbf{r}}_2 = +\frac{Gm_1^3}{(m_1 + m_2)^2}\frac{\mathbf{r}_1}{r_2^2 r_1}.$$

Thus, in general, both the stars will describe ellipses each having a focus at the center of mass. We have the standard equation for the elliptic orbit

$$r_1 = \frac{a_1(1 - e^2)}{(1 + e \cos \phi)},$$

---

* Note that the pulsar's own gravitational field causes a constant change in the period and as such is not observable, although the effect is quite considerable because GM/Rc^2 for a pulsar is estimated to be ~ 0.1.

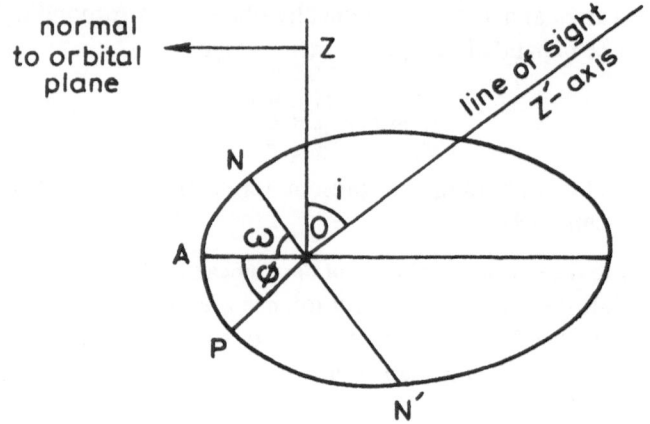

Figure 9.1

OZ'—line of sight.
OZ—perpendicular to the plane of the orbit.
NON'—nodal line, i.e., normal to both *OZ* and *OZ'*.
   A—periastron.
   O—center of mass, a focus of the ellipse.
   P—a point of the ellipse with coordinates r and ϕ.
   ω—$\angle NOA$ (not a constant because of the general relativity effect, see motion of the
      perihelion of Mercury).
   i—$\angle ZOZ'$.

where a_1 is the semimajor axis and the zero of ϕ is taken at the periastron (i.e., minimum of r). The constancy of the areal velocity (Kepler's second law) gives, for the orbital period,

$$T = \frac{\pi a_1^2 \sqrt{1 - e^2}}{r_1^2 \dot{\phi}/2},$$

or

$$\dot{\phi} = \frac{2\pi}{T} \frac{(1 + e \cos \phi)^2}{(1 - e^2)^{3/2}}. \tag{9.6}$$

a straightforward calculation now gives, for the velocity component in the direction of the line of sight (see Fig. 9.1),

$$v_1 = \frac{2\pi a_1 \sin i}{T(1 - e^2)^{1/2}} [e \cos(\omega + \phi) + \cos \omega]. \tag{9.7}$$

We shall presently see that the gravitational shift (9.3) leads to a change of the order of v^2/c^2, hence the first-order effect is simply v_1/c ((9.2)). Thus v_1 (or rather its variation) is determined, and with the help of (9.7) and (9.6) this leads to a determination of T, e, ω, and ϕ (as a function of time) and also $a_1 \sin i$ (see notes at the end of the chapter). Again from dynamics, in view of (9.5) (recall Kepler's third law),

$$T^2 = \frac{4\pi^2 a_1^3}{G} \frac{(m_1 + m_2)^2}{m_2^3}.$$

We write this as

$$f(m_1, m_2, i) \equiv \frac{(m_2 \sin i)^3}{(m_1 + m_2)^2} = \frac{(a_1 \sin i)^3}{G} \left(\frac{2\pi}{T}\right)^2. \tag{9.8}$$

Thus $f(m_1, m_2, i)$, which is known as the mass function, is found. Equation (9.5) on integration gives the energy equation

$$\tfrac{1}{2} v_1^2 - \frac{Gm_2^3}{(m_1 + m_2)^2 r_1} = \text{constant},$$

or

$$\tfrac{1}{2} v_1^2 + \frac{Gm_2}{r} = Gm_2 \left[\frac{1}{r} + \left(\frac{m_2}{m_1 + m_2}\right)^2 \frac{1}{r_1}\right] + \text{constant}$$

$$= Gm_2 \left[\frac{m_2}{m_1 + m_2} + \left(\frac{m_2}{m_1 + m_2}\right)^2\right] \frac{1}{r_1} + \text{constant}$$

$$= Gm_2^2 \frac{(m_1 + 2m_2)}{(m_1 + m_2)^2} \frac{(1 + e \cos \phi)}{a_1(1 - e^2)} + \text{constant}$$

$$= \frac{Ge}{a_1(1 - e^2)} \frac{m_2^2(m_1 + 2m_2)}{(m_1 + m_2)^2} \cos \phi + \text{constant}. \tag{9.9}$$

Thus the transverse Doppler effect and the gravitational effect get tied up and determines a function $g(m_1, m_2, i)$, namely,

$$g(m_1, m_2, i) = \frac{m_2^2(m_1 + 2m_2)}{(m_1 + m_2)^2} \sin i. \tag{9.10}$$

Equations (9.8) and (9.10) provide two relations between the three unknowns m_1, m_2, and i. A further relation between m_2 and i is given by the periastron advance

$$\dot{\omega} = \frac{6\pi Gm_2 \sin i}{T_c^2 (1 - e^2)(a_1 \sin i)}. \tag{9.11}$$

(Equation (9.11) is the same as that derived in the case of Mercury; indeed, as a post-Newtonian calculation by Robertson has shown, the formula for periastron advance is unaffected by the relative masses.) The determination of $\dot{\omega}$ preceeds as follows, so long as the observations extend over, say, a few weeks, ω is effectively a constant and its value determined. However, if we again determine ω after months or years, then a different value of ω comes out and from this we evaluate $\dot{\omega}$ (its observed value is about 4.2° per year). Hence from (9.8), (9.10), and (9.11), we get the values of m_1, m_2, and i. In reality, the matter is much more complicated, for in the observed τ', (9.7) and (9.9) appear in combination.

We now plug in these values in the last equation of the last chapter to evaluate $\dot{T}$ which comes out as

$$(\dot{T})_{\text{calculated}} = (-2.403 \pm 0.005) \times 10^{-12},$$

$$(\dot{T})_{\text{obs}} = (-2.30 \pm 0.22) \times 10^{-12}.$$

The agreement is thus excellent. However, there are some questions which remain, to some extent, unanswered, such as:

(1) Is the Einstein formula for radiation loss correct?
(2) Is it justified to consider the pulsar and the companion stars as mass particles, undeformed by their mutual attraction?
(3) Is it certain that there is no extraneous factor which we have not taken into consideration?

While it is possible to give arguments in favor of our assumption, some degree of uncertainty remains.

It is usually held that it is difficult to believe that Nature has so conspired so that other effects have given rise to a value of $\dot{T}/T$ which is just that required by the radiation loss. However, in the history of science, such coincidences are not altogether unknown. We conclude our discussion by giving a table of data on this pulsar.

Hulse–Taylor Binary Pulsar—PSR 1913 + 16	
Pulsation period, P	0.0590299952709 s
$\dot{P}$	8.628×10^{-18} s s^{-1}
$\ddot{P}$	?
$a \sin i$	2.34186 light seconds
e	0.61714
Orbital period, T	27906.9816 s
$\dot{\omega}$	4.226° yr^{-1}
Pulsar mass	$1.42 \pm 0.06\ M_{\odot}$
Companion mass	$1.41 \pm 0.06\ M_{\odot}$
Sin i	0.72 ± 0.03
Pulsar semimajor axis, a_1	3.24 ± 0.13 light seconds
Companion semimajor axis, a_2	3.26 ± 0.13 light seconds

Note I. In (9.7) the constant term involving $e \cos \omega$ will not be observable. The varying Doppler shift will allow a determination of $(\omega + \phi)$ as a function of time. Hence ϕ is determined. Then (9.6) allows a determination of ϕ and e. Going back to (9.7), $a_1 \sin i$ and ω are determined. The value of T is simply the period of variation of τ'.

Note II. It is particularly satisfactory that the analysis of the Hulse–Taylor

Pulsar data, as done by Taylor and Weisberg (1982), were confirmed by the independent work of Boriakoff et al., (1982). A still later analysis by Taylor and Weisberg (1989) gives

$$\text{pulsar mass} = 1.442 \pm 0.003 \, M_\odot,$$

$$\text{companion} = 1.386 \pm 0.003 \, M_\odot,$$

$$(\dot{T})_{\text{calculated}} = (-2.40216 \pm 0.00021) \times 10^{-12},$$

$$(\dot{T})_{\text{obs}} = (-2.427 \pm 0.026) \times 10^{-12}.$$

Further bounds are set on the temporal variation of the gravitational constant G

$$(\dot{G}/G) = (1.2 \pm 1.3) \times 10^{-11} \, \text{yr}^{-1}.$$

which is consistent with a nonvarying G.

Part II
Relativistic Astrophysics

10. White Dwarf Stars

10.1. Introduction

Relativistic astrophysics are concerned with the applications of the special and general theory of relativity to astrophysical situations. Chandrasekhar's work on white dwarf stars was the first significant application of special relativity to astrophysical theory, and general relativity has made some impact only recently. One way to see the reason for this is to consider the Schwarzschild exterior solution, which is valid in the empty space surrounding a spherical star. The metric differs from the flat space metric by terms of the order of GM/c^2r, and this has a negligibly small value for all astronomical objects except the neutron star and the black hole.

For the Sun, the value of GM/c^2r at the surface is $\sim 10^{-6}$, while for a compact star like the white dwarf it is larger, but still quite small, being $\sim 10^{-4}$; for the neutron stars, the theoretical models lead to a value $\sim 10^{-1}$. In the case of the black hole, it is not quite correct to assign any specific radius; however, the black hole phenomena appear only when the surface has shrunk inside r_0 given by $GM/r_0c^2 = 0.5$ (note that GM/rc^2 is a nondimensional quantity related to the Newtonian gravitational potential).

The reader may wonder that if general relativity effects are significant in astronomy, as is evident from the crucial tests, why they are of no importance in astrophysics. The reason is two-fold: first, the observational data in astronomy have been obtained to a very high degree of accuracy, and second, the crucial tests involved only the dynamics of test particles and photons, while astrophysical theory involves the properties of matter within the stars and their interactions. Both these factors are little affected by considerations of relativity at low densities which prevail in ordinary stars.

Again, while theoretical speculations about neutron stars and black holes can be traced to the 1930s, they have been seriously considered only after the discovery of pulsars and quasars, i.e., during the last three decades.

10.2. The Contraction of a Radiating Star in the Absence of Energy Generation

The standard cosmological theory provides a locale for thermonuclear reaction in the early hot and dense universe. There is a formation of helium from the primeval neutrons and protons, the final abundance by mass being nearly

76% hydrogen and 23% helium. A star of the first generation is believed to have been formed from this mixture at a much later epoch, and then the material of the star becomes heated owing to gravitational compression. At a certain stage, the temperature becomes sufficiently high to trigger off a fresh thermonuclear reaction, leading to a further production of helium by the fusion of protons (hydrogen nuclei). This releases energy and leads to a consequent flux of radiation from the surface of the star. The compression now stops and there is a balance between the energy loss (by radiation) and energy generation (due to the "burning" of hydrogen-forming helium). However, as time goes on, hydrogen burning stops and there is then an uncompensated loss of energy by radiation. This leads to a contraction of the star as the following calculation shows.

The gravitational potential energy of a spherical distribution is (in obvious notation)

$$W_g = -\int_0^R \frac{Gm(r)}{r} 4\pi r^2 \rho \, dr$$

$$= \int_0^R \frac{dP}{dr} 4\pi r^3 \, dr = -3 \int_0^R 4\pi r^2 \rho \, dr, \tag{10.1}$$

where we have used the equation of hydrostatic equilibrium

$$\frac{dP}{dr} = -\frac{Gm(r)}{r^2} \rho, \tag{10.2}$$

and the condition that the boundary pressure $P_R = 0$. Note that

$$m(r) = \int_0^r 4\pi r^2 \rho \, dr. \tag{10.3}$$

We may estimate the internal energy, defined as the work done to compress the system isentropically, from the state of infinite dilution by assuming an equation of state

$$P = K\rho^\Gamma, \tag{10.4}$$

where K and Γ are constants. Thus defined, the internal energy U is given by

$$\int dU = \int_0^P -P d\left(\frac{M}{\rho}\right) = \int_0^P KM\rho^{\Gamma-2} \, d\rho$$

$$= KM \frac{\rho^{\Gamma-1}}{\Gamma-1} = \frac{M}{\rho} \frac{P}{\Gamma-1}.$$

Thus the energy density at any point equals $P/(\Gamma - 1)$ and the total internal energy is

$$\int_0^R \frac{P}{\Gamma-1} 4\pi r^2 \, dr = -\frac{1}{3(\Gamma-1)} W_g.$$

Finally, the sum of the gravitational and internal energies is

$$W_g\left[1 - \frac{1}{3(\Gamma-1)}\right] \sim -\frac{3\Gamma-4}{3(\Gamma-1)} a \frac{GM^2}{R}, \tag{10.5}$$

where we have used the well-known result, that for a spherical distribution,

$$W_g \simeq -a\frac{GM^2}{R}.$$

The constant a depends on the form of density distribution. The above relation shows that any loss of total energy of a spherical star would require a decrease in its radius R if $\Gamma > \frac{3}{4}$. So long as classical nonrelativistic statistics hold, $\Gamma = \frac{5}{3}$ for a gas of noninteracting particles and hence there would be a compression. This decrease in volume means a corresponding decrease in the phase space available for the electron within a definite momentum range. For fermions, a minimum phase space volume h^3 is required for each pair of particles. Thus, with a progressive decrease in volume, a stage is reached when the particles cannot be accommodated within the phase space if the momenta of the particles p correspond to the temperature, i.e.,

$$\frac{p^2}{2m} \sim KT.$$

Thus the fermions then acquire higher momenta and energy and, finally, the maximum momentum, up to which the phase space is filled up, becomes independent of temperature. The fermions are then said to be degenerate; in other words, degeneracy is said to set in when a departure from classical statistics occur).

The situation regarding Γ changes when degeneracy sets in; first, a degenerate system cannot lose energy and, second, Γ in the case of relativistic degeneracy tends towards $\frac{4}{3}$ as we show below. However, the actual history of a star is much more complicated and it has phases of further thermonuclear reactions. We shall go into them later in Chapter 11.

10.3. Degeneracy and the Equation of State

The distribution formula for noninteracting particles is

$$dn = g\frac{4\pi p^2\, dp}{h^3[e^{(\varepsilon-\varepsilon_0)/KT} \pm 1]}, \tag{10.6}$$

where dn is the number density of particles having momenta between p and $p + dp$, with the corresponding energy ε given by

$$\varepsilon^2 = p^2c^2 + m^2c^4,$$

and g is the spin multiplicity factor. The constant m is the rest mass of the particles. The positive and negative signs correspond to Fermi and Bose statistics, respectively. As we are interested in electrons and neutrons, hereafter we shall take the positive sign and $g = 2$.

For nonrelativistic particles (i.e., $p \ll mc$), $\varepsilon = mc^2 + p^2/2m$, it is usual to absorb the constant term mc^2 within ε_0 and we find that if $e^{-\varepsilon_0/KT}$ is large the

Fermi distribution goes over to the classical form

$$dn = \frac{8\pi p^2 \, dp}{h^3} e^{-p^2/(2m\,KT)} e^{+\varepsilon_0/KT}$$

and integration over the entire momentum range yields

$$e^{-\varepsilon_0/KT} = \frac{2(2\pi \cdot mKT)^{3/2}}{nh^3}.$$

Hence the condition for the applicability of classical statistics is $2(2\pi \cdot mKT)^{3/2}/nh^3 \gg 1$. On the other hand, when $(mKT)^{3/2}/nh^3$ becomes much less than unity, we say the system is degenerate. Thus, at the same temperature T and concentration n, the electron system may be degenerate but nucleons with their greater mass may be nondegenerate. This is precisely the case in white dwarfs where the matter density is in the range 10^5–10^7 g cm^{-3} and the temperature $\sim 10^7$ K.

In the extreme limiting case of $T \rightarrow 0$, we get

$$dn = \frac{8\pi p^2 \, dp}{h^3} \qquad \text{for} \quad \varepsilon < \varepsilon_0,$$

$$= 0 \qquad \text{for} \quad \varepsilon > \varepsilon_0. \tag{10.7}$$

The meaning of the above expressions is simple; as $T \rightarrow 0$, there is no temperature excitation, all the particles therefore would try to go to the lowest energy state but this is barred by the Pauli principle. Thus the lowest energy cells become filled, up to the level necessary to accommodate the n particles on the basis of two particles per unit cell. Above this energy, there is no particle whatsoever. The limiting energy is known as the Fermi energy.

Substituting values of m, K, and h we find that electrons become degenerate at $T \sim 10^6$ if $n \gtrsim 10^{23}$ cm^{-3}. Hence for the electron number density $\sim 10^{29}$ (corresponding to the matter density $\sim 10^5$ g cm^{-3}) degeneracy will be quite pronounced. Again from (10.7), the Fermi momentum is

$$P_\mathrm{F} \sim m_0 c n^{1/3} \cdot 2 \cdot 10^{-10}.$$

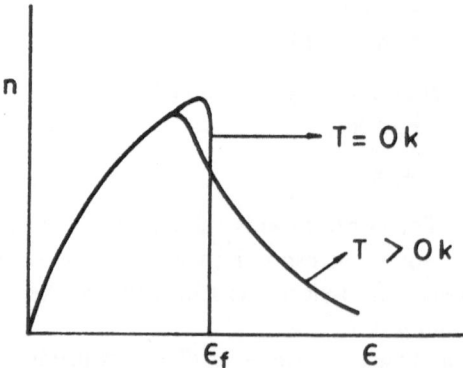

Figure 10.1. Fermi distribution for a nonrelativistic gas.

Hence relativistic correction would be quite significant for $n \gtrsim 10^{31}$ cm^{-3} corresponding to the matter density 10^7 g cm^{-3} and rather negligible at 10^5 g cm^{-3}.

The pressure P of a gas of noninteracting particles is

$$P = \tfrac{1}{3} \int pv \, dn,$$

where v is the velocity corresponding to the momentum p. Thus, for the degenerate system,

$$P = \frac{2}{3h^3} \int_0^{p_f} \frac{p^2 c^2 4\pi p^2 \, dp}{(p^2 c^2 + m^2 c^4)^{1/2}}, \tag{10.8}$$

where p_f is the Fermi momentum. In an actual calculation of a stellar structure we have to compute the above expression to determine the relation between the pressure and density, as was first done by Chandrasekhar in his study of white dwarfs. However, we consider here the two limiting cases and draw some general conclusions. In the extreme nonrelativistic limit, $p_f \ll mc$,

$$P = \frac{8\pi}{15mh^3} p_f^5,$$

and in the same condition the mass density of these fermions is

$$\rho' = \frac{2m}{h^3} \int_0^{p_f} 4\pi p^2 \, dp = \frac{8\pi m}{3h^3} p_f^3.$$

However, ρ' is quite different from the actual mass density prevailing in the case of white dwarfs; for the system which gives rise to pressure consists of electrons while the dominant contributors to matter density consist of nucleons. The condition of charge neutrality means that the number of electrons is the same as the number of protons, hence the number of nucleons is A/Z times the number of electrons (A is the mass number and Z is the atomic number of the nuclei). Thus, the actual mass density is $\rho' \cdot A/Z \cdot m_u/m_e \sim 3680 \, \rho'$ (as $A/Z \sim 2$ and $m_u/m_e \sim 1838$), where m_u is the atomic mass unit, m_e is the electron mass). Hence, finally, the relation between the pressure and density is

$$P = (3\pi^2)^{2/3} \hbar^2 \left(\frac{Z}{Am_u} \right)^{5/3} \frac{\rho^{5/3}}{m_e} \alpha \rho^{5/3}.$$

Thus the adiabatic index is 5/3. In the ultrarelativistic limit $p_f \gg mc$, we get

$$P = \frac{2\pi c}{3h^3} p_f^4.$$

The expression for matter density is the same as before (in spite of the relativistic increase in the mass of the electrons, their masses still do not contribute appreciably to the stellar mass), so that

$$P = \tfrac{1}{4}(3\pi^2)^{1/3}\hbar c\rho^{4/3}\left(\frac{Z}{Am_u}\right)^{4/3},$$

the adiabatic index is, in this case, 4/3.

Problem

1. Examine how far, in the ultrarelativistic condition, electron masses are negligible compared to neutron rest mass.

10.4. Limiting Mass for White Dwarfs

In a realistic star model, we have to take account of the fall in density as we go from the center to the surface. Thus, while a relativistic condition may obtain in the central region, the low density region near the surface will have nonrelativistic electrons. Chandrasekhar integrated the equations of hydrostatic equilibrium

$$\frac{dP}{dr} = -\frac{Gm(r)\rho}{r^2},$$

$$m(r) = \int_0^r 4\pi r^2 \rho\, dr,$$

$$P = f(\rho),$$

the last relation being obtained from integration of (10.8). The integration was performed for different values of the central density ρ_c, and the boundary of the star was identified with the surface where the pressure vanished. In this way, Chandrasekhar found a relation between the mass M of the star and the central density as shown in Fig. 10.2.

It is seen that as ρ_c increases, M increases tending to a limiting value for large ρ_c. Effectively, the limiting value was attained for $\rho_c \sim 10^7$ g cm^{-3} and

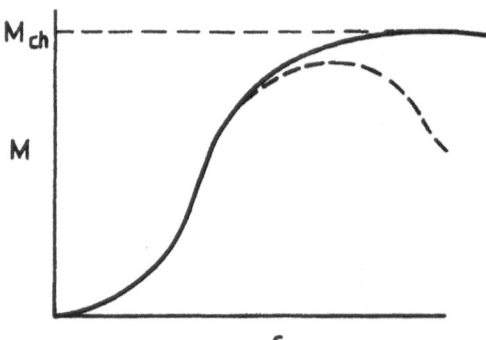

Figure 10.2. Continuous curve showing Chandrasekhar limit. If general relativity is taken into account, there is a maximum mass (as shown by dotted lines)

$$M_{ch} \approx 1.4\, M_\odot \qquad \text{for} \quad \mu_e = 2.$$

was somewhat dependent on the nature of the nuclei that were present in the star. The limiting value M_{ch} is

$$M_{ch} = \frac{5.76 \, M_\odot}{\mu_e^2},$$

where μ_e is the mean molecular weight per electron and is therefore equal to A/Z (or to the average value of A/Z if different nuclear species are present). If the thermonuclear reactions proceed towards completion, then the nuclei present will be that of iron, $^{56}_{26}Fe$, having the highest value of binding energy per nucleon and $M_{ch} = 1.26 \, M_\odot$. However, in white dwarfs, this state is not reached; if the dominant nuclei is carbon, $M_{ch} = 1.44 \, M_\odot$.

10.5. A Simple Argument for the Mass Limit

The condition for the equilibrium of a star can be expressed as a variational principle, $\delta \int \varepsilon \, dV = 0$, where ε is the total energy density. Thus, for equilibrium, the total energy has to be an extremum. We shall use this to arrive at a mass limit for stars in the case of ultrarelativistic degeneracy.

If N is the total number of particles of mass m, which has become degenerate, the Fermi energy is (taking spin-$\frac{1}{2}$ particles of $g = 2$)

$$\varepsilon_f = \frac{p_f^2}{2m} = \left(\frac{9N}{32\pi^2}\right)^{2/3} \frac{h^2}{2mR^2} \qquad \text{(nonrelativistic case)},$$

$$= p_f c = \left(\frac{9N}{32\pi^2}\right)^{1/3} \frac{hc}{R} \qquad \text{(ultrarelativistic case)}.$$

Thus the total kinetic energy is

$$E_{kin} \sim \left(\frac{9}{32\pi^2}\right)^{2/3} N^{5/3} \frac{h^2}{2mR^2} \qquad \text{(nonrelativistic case)},$$

$$\sim \left(\frac{9}{32\pi^2}\right)^{1/3} N^{4/3} \frac{hc}{R} \qquad \text{(ultrarelativistic case)}.$$

The other contribution to the total energy comes from gravitational attraction, which is

$$E_G \sim -\frac{3}{5}, \qquad \frac{GM^2}{R} \sim -\frac{3}{5}\frac{G}{R}(Nm_n)^2,$$

where N is the number of nucleons and is within a factor ~ 2, equal to the number of particles of the degenerate species (electrons or neutrons). Hence the total energy becomes

$$E = E_G + E_{kin} \sim -\frac{3}{5}\frac{GM^2 m_n^2}{R} + \left(\frac{9}{32\pi^2}\right) N^{5/3} \frac{h^2}{2mR^2} \qquad \text{(nonrelativistic case)},$$

$$(10.9a)$$

$$\sim -\frac{3}{5}\frac{GM^2m_n^2}{R} + \left(\frac{9}{32\pi^2}\right)^{1/3}N^{4/3}\frac{hc}{R} \qquad \text{(ultrarelativistic case)}.$$

$$(10.9b)$$

For the nonrelativistic case there is a minimum of E for all N, it occurs at $(\delta E/\delta R = 0, \delta^2 E/\delta R^2 > 0)$

$$R = \frac{5}{3}\left(\frac{9}{32\pi^2}\right)^{2/3}\frac{h^2}{mGN^{1/3}m_n^2} \qquad \text{(nonrelativistic degeneracy)},$$

and thus there exists no limit to N or the mass M. This was the result obtained by Fowler, who first introduced the idea of degeneracy in the discussion on white dwarfs. Chandrasekhar considered the relativistic case. Equation (10.9b) shows that dE/dR can vanish only for

$$N \lesssim \left(\frac{9}{32\pi^2}\right)^{1/2}\left(\frac{5}{3G}\right)^{3/2}\frac{(hc)^{3/2}}{m_n^3}. \qquad (10.10)$$

That (10.10) gives an upper limit to N follows from the observation that the ultrarelativistic formulas are approached asymptotically for $N \to \infty$, smaller values of N lead to a situation between (10.9a) and (10.9b) and hence equilibrium is possible. From (10.10) we get the mass limit

$$M_{\text{lim}} \sim Nm_n \sim \frac{3}{4\pi}\left(\frac{5}{3G}\right)^{3/2}\frac{(hc)^{3/2}}{m_n^2} \sim M_\odot.$$

However, tthis is only an order of magnitude estimate, as we have assumed each of the degenerate particles to have exactly the Fermi energy and used the expression for gravitational potential energy which is true for a uniform sphere, and also neglected the interactions other than gravitational. We may also estimate a bound on the radius R. This must be small enough to push the Fermi momentum above mc so that we may have relativistic degeneracy. Thus

$$R \le \frac{3}{4\pi}\left(\frac{5hc}{3m_n^2 G}\right)^{1/2}\frac{h}{mc},$$

which gives

$R \le 10^9$ cm   for electron degeneracy (white dwarfs),

$R \le 10^6$ cm   for neutron degeneracy (neutron stars).

10.6. Critique of Chandrasekhar's Result and Later Works

Chandrasekhar assumed the electrons to be completely free and thus neglected the Coulomb interaction between the electrons themselves and the nuclei. Assuming a uniform distribution of electrons, Coulomb interaction reduces the pressure from the free electron pressure P_0 by an amount

δP:

$$\frac{\delta P}{P_0} = -\frac{2^{5/3}}{5}\left(\frac{3}{\pi}\right)^{1/3}\alpha Z^{2/3}, \tag{10.11}$$

in this case the nuclei of the atomic number Z are arranged in a lattice and α is the fine structure constant $= 1/137$. The effect is obviously small, even for iron ($Z = 26$). However, this has an appreciable effect on the radius of the star.

Another effect which was not taken into account by Chandrasekhar is the capture of electrons by nuclei. This occurs when the Fermi energy of the electrons is high; in the case of $^{56}_{26}$Fe, the threshold energy for electron capture is 4.2 MeV. Such high energies are attained at densities $\sim 10^9$ g cm^{-3} and the adiabatic index then goes below the value $\frac{4}{3}$. The effect is to give a maximum in the $M - \rho_c$ plot instead of the steady approach to the limit obtained by Chandrasekhar.

10.7. Historical Note

We have already noted that Chandrasekhar was the first to consider relativistic degeneracy and arrive at the conclusion regarding the existence of a limiting mass for equilibrium, and concluded (Chandrasekhar, 1939), "A star of large mass cannot pass into the white dwarf stage and one is left speculating on other possibilities." The "speculating on other possibilites" was done by Eddington (1935) who remarked, "Chandrasekhar shows that a star of mass greater than a certain limit ... has to go on radiating and radiating and contracting until, I suppose, it gets down to a few kilometers radius when gravity becomes strong enough to hold the radiation and the star can at last find peace." While he summarily dismissed such a situation, the present day relativist does indeed believe that under some circumstances the star, indeed, "gets down to a few kilometers radius when gravity becomes strong enough to hold the radiation" but he denies any "peace" to the star even then—it has to go right up to a singularity in space–time.

Landau (1932) was also sceptic about the validity of quantum mechanics and statistics. Thus, he wrote, "For $M > 1.5\ M_\odot$ there exists ... no cause preventing the system collapsing to a point ... we must conclude that all stars heavier than 1.5 $M_\odot$ certainly possess regions in which the laws of quantum mechanics (and therefore of quantum statistics) are violated." Again the present-day physicist believes in "the collapse to a point" (although his mass limit is somewhat higher), but he sees a possibility of escape not in violation of quantum mechanics or quantum statistics but in an as-yet unknown quantum theory of gravitation.

It is interesting to note how the climate has changed, the ideas of a black hole and gravitational collapse to a singularity, which appear in the writings of Eddington and Landau, were considered simply absurd and unacceptable in the 1930s; while today, to the physicist (or at least to a large group amongst them), a black hole is the accepted fate of some stars and many think that the existence of a few black holes have already been confirmed by observations. (See the chapter on black holes for details.)

10.8. Observational Data on White Dwarfs

The main observational data on white dwarfs are on the following lines:

(i) The amount of radiation received in our locale and the distribution of energy amongst different wavelengths. This allows us to have an idea about the temperature of the white dwarfs.

(ii) The distances of the white dwarfs by the parallax method. Combining this with observation (i) we may estimate the total emission from the star and thereby determine an effective temperature and also the radius of the star using the formula for the luminosity

$$L = \sigma \cdot 4\pi R^2 T^4.$$

(iii) If the white dwarf is a member of a binary system, we can estimate its mass.

(iv) A shift in the spectral lines is observed toward longer wavelengths, and interpreting this as a gravitational redshift $\delta\lambda/\lambda = GM/Rc^2$ we can determine M/R and compare it with the values obtained from observations (i)–(iii). We quote the results for the two cases where the relevant data have been obtained with fair accuracy:

$M/M_\odot$	$R/R_\odot$	$v = GM/Rc$ cal. in km s^{-1}	$v = c(\delta\lambda/\lambda)$ obs. km s^{-1}
Sirius B 1.05 ± 0.03	0.0007	89 ± 16	91 ± 8
40 Eri B 0.48 ± 0.02	0.012	24 ± 1	22 ± 1.4

10.9. The Cooling and Age of White Dwarfs

We shall now show that the luminosities of the white dwarfs can be correlated with their ages. A white dwarf has a central core of high density where the electrons are degenerate. The degeneracy leads to two consequences:

(a) a very high mean free path of the electrons leading to a high thermal conductivity so that the core is effectively at a uniform temperature; and

(b) the electron system cannot lose energy, the energy of radiation comes solely from the positive ions.

The core is covered by a surface region where the density and temperature progressively fall, this region has little energy to radiate but it acts as an absorbing region for the radiation that diffuses from the core. The absorption may arise from a variety of causes like photoionization, free–free transition, etc. Without going into the details of these processes, we may write

$$\mathbf{n}\nabla I = Q - k\rho I, \qquad (10.12)$$

where k is the mass absorption coefficient, I is the intensity of radiation defined as the radiation proceeding through a unit solid angle in unit time, $\mathbf{n}$ is the unit vector along the path of the radiation, and Q represents the emission of energy per unit volume per unit time at the point under consideration.

The flux vector $\mathbf{F}$ is obviously $\equiv \int \mathbf{n}I\, d\Omega$, the integration over the solid angle being over the entire domain 4π. Hence we obtain from (10.12)

$$F_i = -\frac{1}{k\rho} \int n_i n_k \frac{\delta I}{\delta x_k}\, d\Omega,$$

where the term in Q has fallen off as the emission is assumed isotropic. We thus get

$$F_i = -\frac{4\pi}{3k\rho} \frac{\delta I}{\delta x_i}.$$

The intensity I is related with the energy density ρ_γ of radiation

$$I = \frac{c}{4\pi}\rho_\gamma = \frac{c}{4\pi}(aT^4),$$

where we assume thermodynamic equilibrium and hence the validity of Planck's formula. Finally, the luminosity L for a spherical star is

$$\text{constant} = L = 4\pi r^2 F_r = -\frac{16\pi r^2 ac T^3}{3k\rho}\frac{dT}{dr}. \qquad (10.13)$$

In reality, k is frequency dependent and hence we should have modified the treatment taking into account the variation of k, so k used here is to be regarded as an averaged-out value known as the Rosseland mean.

We give the relation (without proof) between k, the density of matter ρ, and the temperature T

$$k = k_0 \rho T^{-3/5}, \qquad (10.14)$$

with

$$k_0 = 4.34 \times 10^{24}\, Z(1 + X)\, \text{cm}^2\, \text{g}^{-1},$$

$$Z = \text{mass fraction of all elements of mass number} > 4,$$

$$X = \ldots\ldots\ldots\ldots\ H,$$

$$Y = \ldots\ldots\ldots\ldots\ He.$$

So that $X + Y + Z = 1$. Another relation between T and ρ can be obtained by combining the equation of hydrostatic equilibrium,

$$\frac{dP}{dr} = -\frac{GM}{r^2}\rho, \tag{10.15}$$

with the equation of state. Throughout the surface region, M may be regarded as a constant, as the density in the envelope is so small that the mass of the envelope can be neglected in comparison the mass of the core. The equation of state may be taken to be that due to a perfect gas, thus

$$P = \frac{\rho KT}{\mu m_n} \tag{10.16}$$

where μm_n is the mass per particle, and the atomic mass unit can be taken to be the mass of a nucleon for our present purposes. Equations (10.13) and (10.15) may be used to eliminate r

$$\frac{dP}{dr} = \frac{16\pi ac}{3k_0 L}GM\frac{T^{6.5}}{\rho}, \tag{10.17}$$

where we have used (10.14).

Eliminating ρ between (10.16) and (10.17) yields an equation connecting P and T which is readily integrated. Finally, substituting for P from (10.16) we obtain

$$\rho = \left(\frac{32\pi acGM\mu}{25.5k_0 LK}\right)^{1/2} T^{3.25}. \tag{10.18}$$

However, our motivation is to obtain L as a function of T. To eliminate ρ we note the following. Equation (10.16) holds for the nondegenerate case, i.e., it also holds for electrons in the outer envelope, while in the core the electrons are degenerate. We consider a simplified situation, an abrupt change from a degenerate to a nondegenerate condition at a certain surface. However, at this surface we assume the continuity of electron pressure P_e and matter density ρ. Thus the electron number density is $\rho/\mu_e m_n$, μ_e = number of nucleons/ number of electrons, and m_n is the mass of a nucleon, and the nondegenerate formula for the electron pressure is given by

$$P_e = \frac{\rho KT}{\mu_e m_n}.$$

The degenerate formula (nonrelativistic) gives

$$P_e = 1.0 \times 10^{13}\left(\frac{\rho}{\mu_e}\right)^{5/3}.$$

Hence, we have at the surface

$$\rho = (2.4 \times 10^{-8})\mu_e T^{3/2}. \tag{10.19}$$

From (10.18) and (10.19) we get the desired relation

$$L = (5.7 \times 10^5) \frac{\mu}{\mu_e^2} \frac{1}{Z(1 + X)} \frac{M}{M_\odot} T^{3.5}, \qquad (10.20)$$

where we have introduced the value of k_0 and it should be noted that T is the temperature at the core–envelope interface, and because of the uniformity of temperature in the body of the core it is effectively the tempeature. In most actual cases, (10.20) can be written approximately as

$$L \approx 10^6 \, T^{3.5}.$$

The emission of radiation causes a cooling of the core; of course, this energy comes from the ions. If the temperature is high enough so that the ions may be considered to form a perfect gas, the ionic thermal energy will be

$$(\tfrac{3}{2}KT \times \text{number of ions}) \simeq \tfrac{3}{2}KT \frac{M}{m_{ion}}.$$

The energy conservation relation gives

$$L = -\frac{d}{dt}\left(\tfrac{3}{2}KT \frac{M}{m_{ion}}\right)$$

or

$$-\frac{d}{dt}\left(\tfrac{3}{2}KT \frac{M}{m_{ion}}\right) = BT^{7/2} \qquad (B \sim 10^6).$$

On integration we get

$$\frac{3}{5}\frac{K}{m_{ion}}(T^{-5/2} - T_0^{-5/2}) = B(t - t_0) = B\tau,$$

where T_0 is the temperature at birth (time t_0) and τ is the age of the white dwarf. Finally, neglecting $1/T_0$ in comparison with $1/T$, we get the formula connecting age, temperature, and luminosity

$$\tau = \frac{3}{5}\frac{KT}{m_{ion}}L,$$

which gives a value of $\tau \sim 10^9$ yr for $L \sim 10^{-3} L_\odot$ and $T \sim 10^7$ K. Thus we can hardly expect to observe white dwarfs with $L < 10^{-3} L_\odot$ as the age τ is rather too long for such cases. However, quite a number of dwarfs have been observed with such low luminosity, and an explanation has been sought in the energy expression for ions. As the temperature falls, the perfect gas assumption for the ions no longer holds good, and at low enough temperatures the ions form a solid and we have to calculate the lattice energy; thus, the specific heat at very low temperatures would be given by the Debye T^3 law, and at temperatures higher than the Debye characteristic temperature by the Dulong–Petit law. Figure 10.3 shows qualitatively the variation of C_V with temperature.

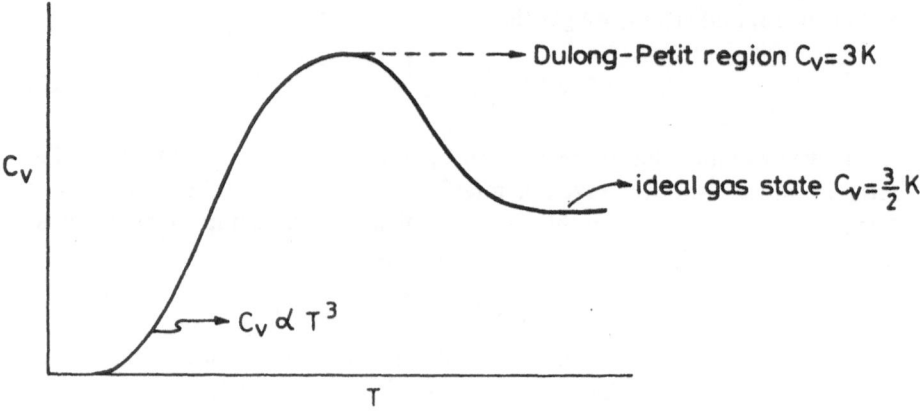

Figure 10.3. Qualitative change of C_V with temperature.

Lamb and Van Horn (1975) and Sweeney (1976) used the observed luminosities to calculate the age of white dwarfs (carbon white dwarfs of mass ~ 1 $M_\odot$). They compared this with the age of clusters in which the white dwarfs occur. Obviously, the white dwarf age should be less than the cluster age if all the stars in the cluster were formed at the same time. Their results are in agreement with this conclusion.

Cluster	Luminosity range of white dwarfs $\log(L/L_\odot)$	Theoretical age for white dwarfs (yr)	Cluster age (yr)
Hyades	-1.0 to -2.4	$4 \times 10^7 - 5 \times 10^8$	$(4–5) \times 10^8$
Praesepe	-2.3 to -2.8	$4 \times 10^8 - 9 \times 10^8$	9×10^8

Observations, so far, show a sharp decrease in the number of white dwarfs of luminosity below $10^{-4}\, L_\odot$. The interpretation would be that the white dwarfs have not yet become old enough to have such low luminosities.

Problems

1. Use (10.2) and (10.3) to prove the following results:
   (a) Both the pressure and the density decrease monotonically from the center to the surface. (Use the condition $dp/d\rho > 0$. Why is it called a stability condition?)
   (b) Given the value of the central density ρ_c, there is a unique equilibrium configuration.

2. Estimate the temperature below which an electron gas with number density 10^{23} cm^{-3} will become degenerate. At this temperature, what is the condition for a neutron gas to be degenerate?

3. If the matter density ρ is 10^7 g cm^{-3}, show that the electrons attain relativistic energy. What should be the minimum matter density for the Fermi energy of the electron to reach the threshold value for the inverse β-decay reaction? (Use the approximate result, neutron mass is greater than parton mass by 2.5 times the electron mass.)

4. If a white dwarf is a carbon sphere of uniform density 10^6 g cm^{-3} and mass equal to the Chandrasekhar limit, what will be the gravitational redshift for light emitted from the surface? Use Newtonian equations and compare with the general relativity result by using Schwarzschild's interior metric given in Chapter 13.

5. Use the condition of hydrostatic equilibrium to show that

$$\frac{1}{r^2}\frac{d}{dr}\left(\frac{r^2}{\rho}\frac{d\rho}{dr}\right) = -4\pi G\rho.$$

Hence derive the Lane–Emden equation

$$\frac{1}{\xi^2}\frac{d}{d\xi}\left(\xi^2\frac{d\theta}{d\xi}\right) = -\theta^n,$$

where

$$\rho = \rho_c\theta^n,$$

$$r = a\xi,$$

$$a^2 = \left[\frac{(n+1)k\rho_c^{(1/n-1)}}{4\pi G}\right],$$

where n is a constant called the polytropic index which appears in the equation of state $p = k\rho^{(1+1/n)}$, and the boundary conditions on θ are $\theta(0) = 1$, $\theta'(0) = 0$.

11. Stellar Evolution, Supernovae, and Compact Objects

11.1. Introduction

In this chapter our object is to give the reader an idea of the processes that take place inside ordinary stars, leading to the birth of compact objects like white dwarfs, neutron stars, and black holes. At the conclusion of this study, we shall have occasion to look at the explosive phenomena, the novae and supernovae. However, a complete history of the evolution of stars is complicated and there are some parts which are not quite clear. We will omit all mathematical details and try to give only a broad outline of what is currently the standard picture believed by most astrophysicists.

11.2. The Evolution of Stars

A star is supposed to be born by condensation of gaseous material inside galaxies. As we shall see later, the big bang model of the universe envisages that at a very early stage, when the temperature was $\sim 10^9$ K, the primordial protons and neutrons united to form helium, leading to a relative abundance (by number) of about 90% hydrogen and 10% helium and just a trace of deuterons ($\sim 10^{-4}$–10^{-5} times proton). A star formed out of such a mixture may be called a first-generation star but as yet such stars have not been observed. The youngest stars, called Population I stars, have a high metal content (and in this group is the Sun) and they occur in the spiral arms of our galaxy. The oldest stars observed are referred to as Population II stars. They are less rich in metals, and occur throughout the galaxy.

During the formation of the star, as the gas contracts due to self-gravitation, the potential energy is converted into heat, and at a certain stage the temperature becomes so high ($\sim 10^7$ K) that the following thermonuclear reactions occur leading to a formation of helium nuclei from protons:

$$p + p \rightarrow D + e^+ + v,$$

$$D + p \rightarrow H_e^3 + \gamma,$$

$$He^3 + He^3 \rightarrow He^4 + 2p.$$

These reactions occur only when the charged nuclei have sufficient energy to come into contact, overcoming the Coulomb barrier; this is why a high temperature is required to trigger off these reactions. The net result of the above reactions is a formation of He^4 nuclei at the cost of protons, and an emission of energy in the form of radiation along with some neutrinos; this process is referred to as the burning of hydrogen into helium, as long as this energy generation lasts we have a quasi-static state for the stars. In the case of second generation stars, there are some other reactions leading to the formation of helium from hydrogen. The carbon and nitrogen present in these stars act as catalysts for the conversion of hydrogen into helium.

In any case, the helium formed, being heavier than hydrogen, sinks toward the center of the star, and after some time a helium-rich central core is formed in which there is little hydrogen left to burn. There should still be some hydrogen burning in the outside shell but, as the energy generation stops in the core, it goes through a stage of contraction due to its own gravitation, while the envelope of the star expands considerably and the star becomes a red giant.

When the contracting core reaches a temperature of $T \sim 10^8$ K, a second stage of thermonuclear reactions sets in with helium "burning" to form carbon. Again, carbon being heavier than helium, a carbon-rich core is formed in which the reaction stops and, outside, there is a helium shell, while still beyond there is the hydrogen-rich envelope. Now the carbon core contracts, temperature rises still further, and then carbon burns to form neon nuclei. The story goes on repeating; a stage of nuclear burning, formation of a non-burning core, its contraction, rise in temperature, and fresh nuclear burning. Finally, we have a central core of iron group nuclei with shells of nuclear species like silicon, oxygen, neon, carbon, helium, and hydrogen—residues of unburnt fuels. It is said that the star now has an onionlike structure. However, every star does not go all the way to the formation of the iron–nickel core. Calculations seem to indicate that this happens only for fairly massive stars ($M \gtrsim 10\ M_\odot$). Again the time scale of the evolutionary processes depend on the stellar mass. The more massive the star is, the processes that we have mentioned also go faster. A rough relation between the hydrogen burning time T and the mass of the star is

$$T = 1.3 \times 10^{10} \left(\frac{M}{M_\odot}\right)^{-2.5} \text{ yr.}$$

Thus stars, whose mass do not exceed the solar mass appreciably, are yet to complete the hydrogen burning (note that the age of the universe $\lesssim 2 \times 10^{10}$ yr).

However, the masses of most white dwarfs seem to be below 1 $M_\odot$, and hence two conclusions seem plausible. First, that in the process of the formation of white dwarfs, a considerable original mass of the progenitor stars has been lost and, second, in the white dwarfs, the nuclear reactions have not gone up to the stage of building up an iron–nickel core. Indeed, it is generally believed that the white dwarfs are composed primarily of carbon and oxygen and the loss of mass is by an ejection of the envelope. We will discuss this again later.

Consider now the case of massive stars in which an iron–nickel core has been formed. As these are the most stable nuclei, in the sense that the binding energy per nucleon has the highest value for them, further fusion reactions would require a supply of energy from outside rather than be associated with an evolution of energy. Hence, such reactions do not occur. However, at this stage, the neutrino luminosity (i.e., the loss of energy by neutrino flux) far exceeds the electromagnetic luminosity. The important processes giving rise to neutrino production are:

(a) Pair annihilation

$$e^+ + e^- \rightarrow \nu + \bar{\nu}.$$

(b) Photo annihilation

$$e^- + \gamma \rightarrow e^- + \nu + \bar{\nu}.$$

(c) Plasma decay

$$\text{plasmon} \rightarrow \nu + \bar{\nu}.$$

A plasmon is a quantized electromagnetic wave propagating in a dense plasma with an effective nonzero rest mass.

Computation shows that for a star of mass 15 $M_\odot$ the neutrino luminosity is $\sim 10^{15}\, L_\odot$, whereas the luminosity of the average galaxies (due to photons) is $\sim 10^{11}\, L_\odot$. An important event occurs at this stage. The iron-rich core attains a quasi-equilibrium state supported by electron degeneracy pressure with a mass close to the Chandrasekhar limit. This is referred to as core convergence.

11.3. The Dynamical Collapse

Two things happen next which bring about a remarkable change:

(a) Photodisintegration of the iron group nuclei by high energy γ-rays. This results in α particles and neutrons

$$_{26}\text{Fe}^{56} + \gamma \rightarrow 13\,_2\text{He}^4 + 4n'.$$

(Later on, at a still higher temperature, the α particles also undergo photo-disintegration into nucleons.)

(b) Neutronization, with the volume decreasing—the degenerate electrons are pushed up to higher energies so that, at a certain stage, the electrons are absorbed, converting both free and bound protons into neutrons.

Both of these cause a fall in pressure and thus the adiabatic index Γ falls below the critical value $\frac{4}{3}$ for stability. General relativity also has a contribution (although small), in bringing about instability as it reduces the critical value of Γ by a factor of the order GM/Rc^2 (the Newtonian potential).

There now ensues a rapid dynamical collapse of the core in contrast to the previous slow quasi-static collapse. The motion of the collapsing core is homologous, which means that the velocity of an element is proportional to its radial distance. However, the homologous condition may obtain only up to the sonic point (where the velocity is equal to the sound velocity), as the material beyond cannot be influenced by the changes in the subsonic core. Indeed, beyond the sonic point, matter falls freely under gravity.

As the homologous core becomes compactified to nuclear density ($\sim 2.8 \times 10^{14}$ g cm^{-3}), the nuclei are broken up, the equation of state stiffens, and Γ exceeds $\frac{4}{3}$ and tends towards the monatomic gas value $\frac{5}{3}$. It is now possible to have an equilibrium configuration but, due to the acquired high velocity during the dynamical collapse, the compactification goes much beyond the equilibrium stage and then the core bounces. This bouncing core produces a shock wave at the sonic point and causes an explosive ejection of the outlying envelope, leading to a supernova explosion and the core settles down to form a neutron star. If the progenitor star mass exceeds ~ 40 $M_\odot$, then explosions apparently do not occur and the ultimate result is a black hole.

11.4. Some Numerical Results

Numerical simulation leads to the general result that if the core mass exceeds 1.5 $M_\odot$ (which is likely to be the case if the progenitor star mass exceeds, say, 12 $M_\odot$), then there is no explosion. For an original mass of 10 $M_\odot$, the core mass is 1.38 $M_\odot$ surrounded by a Ne–O shell of 0.13 $M_\odot$ and a He–H envelope expanding with a velocity 100–300 km s^{-1} up to 10^{16} cm. The explosion ejected the shell through several Ne flashes. At the onset of dynamical collapse, the central density and temperature were 5×10^9 g cm^{-3} and 7.6×10^9 K. The whole explosion took about 50 ms and 0.06 $M_\odot$ was ejected with velocity $\sim 2.5 \times 10^4$ km s^{-1} and the residue was a neutron star of 1.44 $M_\odot$. The total energy released was $\sim 5 \times 10^{50}$ erg which is somewhat less than observed in type II supernovae. The above computational results are taken from a study of Hillebrandt (1982).

11.5. Explosive Processes

There are many explosive processes showing wide a variation in their characteristics. We make a brief mention of some of them.

(a) *Explosions leading to white dwarfs.* In this process, low mass stars quietly eject their envelopes with velocities of a few tens of kilometers per second. The mass ejected may be anything from a fraction of the solar mass to as much as 5–6 $M_\odot$, and the ejection continues for about 10^5 yr. The central star contracts until the degenerate electron pressure brings about equilibrium and a white dwarf is formed.

(b) *Helium flashes.* These are believed to occur in low mass stars after the stage of hydrogen burning. The helium burning core contracts but, unlike normal red giants, the envelope does not expand initially to reduce the

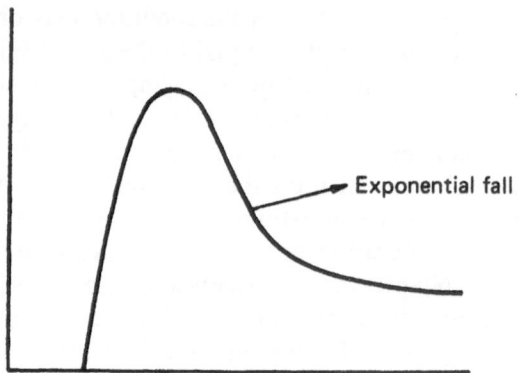

Figure 11.1. Luminosity curve for type I supernova.

temperature. So the heat generated by the core contraction raises the temperature causing a runaway reaction which is called a flash.

(c) *Nova outbursts.* Novae are explosive objects associated with an average energy emission of 10^{42}–10^{44} ergs. But the energy release sometimes is as high as 10^{46} or 10^{47} ergs. Novae occur in binary systems involving a white dwarf and a red giant which entirely fills up its Roche lobe. The white dwarf accretes mass from the red giant to an extent of 10^{-4} or 10^{-3} $M_\odot$ with velocities of $\sim 10^3$ km s^{-1}. Considerable heat is thus generated and the consequent nuclear burning causes the outburst. Generally, novae occur periodically as the explosive reactions occur only at the surface of the white dwarf without disrupting the star itself.

(d) *Supernovae.* Observations of supernova explosions in other galaxies seem to indicate that their rate of occurrence is about one every (70 ± 30) yr in giant spirals like our own. Curiously, in our galaxy, the rate seems to be much lower, about 1 in 300 years.

Supernovae emit about $\sim 10^{51}$ ergs with a peak luminosity of $\sim 10^9$ $L_\odot$. The ejected matter has velocities of $\sim 10^4$ km s^{-1}. Supernovae are classified

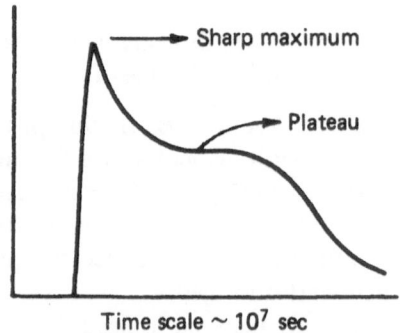

Time scale $\sim 10^7$ sec

Figure 11.2. Type II supernova.

into two major types, type I and type II. Type I are observed in elliptical glaxies as well as in spirals and irregular galaxies where there are population II stars. Type II supernovae appear only in the arms of spiral galaxies where younger stars are located. There is a characteristic difference in the light curves of the two types.

The light curve of type I supernovae shows an exponential tail showing an additional energy source at late times. Radioactive decay $Ni^{56} \rightarrow Co^{56} \rightarrow Fe^{56}$ can apparently explain the observed light curve: the light curve of type II supernovae show a sharp maximum and then a plateau region. The spectra for type II show normal abundances, in particular, the hydrogen lines are quite strong whereas they are absent in the case of type I.

As shown in the schematic diagram of light curves, the high luminosity persists for about a year ($\sim 10^7$ s) but the remnants of type II may remain active for over 10^4 yr. The type I supernovae are believed to originate from low mass stars ($1-2\ M_\odot$) like white dwarfs or helium stars, and an accretion triggers the explosion which completely disrupts the star leaving behind no remnant. Type II supernovae, on the other hand, are believed to be exploding massive stars with an extended hydrogen envelope and then leave behind a neutron star (or a black hole if the mass is sufficiently large).

Since the binding energy of a neutron star is 10^{53} ergs, an emission of 10^{51} ergs means that only about 1% of the gravitational energy needs to be released in the supernovae explosion.

11.6. Supernova 1987 A

This supernova II explosion occurred in our neighboring irregular galaxy, the large Magellanic cloud. Quite a large amount of observational data have been obtained in this event, and they have posed some problems for the usually believed theoretical picture. Thus the progenitor of the supernova was apparently a blue supergiant of mass between 15 and 20 $M_\odot$, whereas the prevalent idea about the progenitor of such events is a red giant. Again the luminosity of this supernova was much below average.

Apparently, two bursts of neutrinos were observed with a time gap of about 9 h. While the latter burst was recorded simultaneously in Japan and the United States, the earlier burst was recorded only in Italy. The explanation is not quite clear and there is a tendency to regard the Italian observation as spurious. There has been a claim that gravitational radiation has also been recorded in Rome and in Maryland (in the United States).

The observations are still being analyzed and computer simulations are being attempted to understand the processes that have occurred. Hopefully, we shall have, finally, a clear understanding of the physical conditions in collapsed and exploding stars but, maybe, some of our cherished beliefs and ideas will crumble.

12. Pulsars

12.1. Introduction

Pulsars, as we shall see, have been identified with neutron stars. It may therefore seem appropriate that we should first have a discussion on neutron stars; however, many of the observed characteristics of pulsars can be understood without going deeply into the structure of neutron stars. We therefore introduce the discussion of pulsar characteristics before the study of neutron stars. The main observational findings are listed below.

1. Pulsars are emitters of electromagnetic radiation over a wide wavelength range, most commonly in the radio wavelength region. The frequency bandwidths are usually greater than 100 MHz. Sometimes they are associated with optical and X radiation which are also of broad band type (including, sometimes, even hard γ-rays).

2. The radiation intensity varies periodically with time, and periods range all the way from a few milliseconds to a few seconds. In recent years, quite a number of millisecond pulsars have been discovered. The pulsar with the shortest time period discovered so far is PSR 1937 + 214 with a period about 1.56 ms. A number of pulsars are known to be in binaries. The period of any particular pulsar is remarkably constant, there being usually only a very slow steady increase; thus, the rate of change of period P, $\dot{P}$ is often $\approx 10^{-15}$ s s^{-1}. Recently, a 110 ms pulsar has been discovered in the Globular cluster M15, 2 arc seconds away from its center, whose period is gradually decreasing instead of increasing with $\dot{P} = -2 \times 10^{-17}$ s s^{-1} (Wolszczan et al., 1989). This has, however, been explained as due to a changing Doppler shift caused by its acceleration toward us at the rate of 5×10^{-6} cm s^{-2}, probably under the gravitational attraction of neighboring matter.

3. Occasionally, however, the periods show a sudden abrupt decrease, these are referred to as "glitches." Thus, for the pulsar in the Crab nebula, the fractional change $\Delta P/P$ due to a glitch has been found to be $\sim 10^{-8}$, while for the Vela pulsar, the corresponding fraction is $\sim 10^{-6}$. However, glitch is not a universal phenomenon in the sense that some pulsars (e.g., those named after Hulse and Taylor) have not shown any glitch over the several years during which they have been observed. Besides "glitches," the period of pulsars occasionally shows fluctuations commonly ascribed to noise. These fluctuations are, however, of a magnitude much much smaller than in glitches.

4. The *duty cycle*, i.e., the fraction of the period with measurable emission, is usually only 1–5%.

5. *Distribution in space.* Pulsars are concentrated in and near the galactic plane and have *some correlation with supernova remnants.* However, this correlation is more a theoretical belief rather than an observational finding, for only five pulsars have an association with SNRs. Thus, the peculiar situation has arisen that we "assume" that there should be an association of pulsars with SNRs, and then proceed to "explain" why such associations are not usually observed (Manchester and Taylor, 1977; Blandford et al., 1983). (No pulsar has, so far, been observed in the remnant of supernova 1987 A.) This last observation is of significance so far as the birth of neutron stars is concerned. Regarding the distribution in depth, the parallax for most pulsars is so small that we have to depend in most cases on what is called "dispersion measure" for the estimation of distances. In this way, pulsar distances are found to range from about 50 pc to 20 kpc. From all this the conclusion has been drawn that the observed pulsars are mostly within our galaxy. Quite a number of millisecond pulsars and low mass X-ray binaries have been detected in globular clusters, the reason for which is not clear.

12.2. Distance from Dispersion Measure

The interstellar space is not altogether empty but there is a small number of free electrons. Consequently, the dielectric constant of the interstallar space deviates from unity. From Sellmeier's formula we have the refractive index μ as a function of the wavelength λ

$$\mu = 1 - \frac{ne^2\lambda^2}{2\pi mc^2},$$

where n is the number density of free electrons of charge e and mass m. It should be noticed that the departure of μ from unity is very small even in the radio wave region. The phase velocity U is

$$U = \frac{c}{\mu} \approx c + \frac{ne^2\lambda^2}{2\pi mc}.$$

The group velocity with which energy propagates is V

$$V = U - \lambda\frac{dU}{d\lambda},$$

or

$$\frac{dV}{d\lambda} = -\lambda\frac{d^2U}{d\lambda^2} = -\frac{ne^2\lambda}{\pi mc}.$$

This dispersion in the group velocity causes a time difference in the arrival of signals of different wavelengths, which have started at the same instant and covered the same distance. The time t being L/V, we have

$$\frac{\Delta t}{\Delta \lambda} = -\frac{L}{V^2}\frac{\Delta V}{\Delta \lambda} = +\frac{L}{V^2}\frac{ne^2\lambda}{\pi mc} \approx \frac{L}{c^2}\frac{ne^2\lambda}{\pi m},$$

or, expressed in terms of the angular frequency $\omega = 2\pi c/\lambda$,

$$\frac{\Delta t}{\Delta \omega} = -\frac{4\pi e^2 nL}{mc\omega^3} = -\frac{4\pi e^2}{mc\omega^3}D, \qquad (12.1)$$

where $D \equiv nL$ or, more rigorously, $D \equiv \int_0^L n\, dl$, when we take into consideration the possible variation of the electron number density n along the path of the ray. D is called the dispersion measure of the astronomical object. Observationally, from the time difference in the arrival of pulses of different wavelengths, we determine $\Delta t/\Delta \omega$ and thus from (12.1) we find D. If now the average value of electron concentration $\langle n \rangle$ is known, the distance L is found. The main point of uncertainty in this determination of the distance is the value of $\langle n \rangle$. The average over large distances within our galaxy is believed to be between 0.02 and 0.03 cm^{-3}.

The "Braking Index"

The braking index n is defined as $n \equiv -\Omega\ddot{\Omega}/\dot{\Omega}^2$, where Ω is the angular velocity of the pulsar. It is rather difficult to determine this index accurately from observation and, formerly, the value for the crab pulsar ($n = 2.5$) was alone considered reliable. However, more recent measurements have provided the value of n for a fair number of pulsars which we present in the following table:

PSR	n
0329 + 54	$(4.81 \pm 0.18) \times 10^3$
0531 + 21 (Crab)	2.515 ± 0.005
0540 + 23	$(2.5 \pm 0.05) \times 10$
0611 + 22	3.5×10^2
0823 + 26	-1×10^4
0833 − 45 (Vela)	$(4.2 \pm 1.3) \times 10$
0950 + 08	$(-5.2 \pm 0.4) \times 10^4$
1508 + 55	3.25×10^3
1541 + 09	$(-2.5 \pm 0.2) \times 10^5$
1604 − 00	$(1.8 \pm 0.4) \times 10^4$
1859 + 03	$(5.5 \pm .05) \times 10^3$
1900 + 01	$(5.0 \pm 1.2) \times 10^3$
1907 + 00	$(-8.7 \pm 0.7) \times 10^3$
1907 + 02	$(-9.5 \pm 3.1) \times 10^2$
1907 + 10	$(-5.5 \pm 0.2) \times 10^3$
1915 + 13	$(-4.2 \pm 0.3) \times 10$
1929 + 10	$(-2.8 \pm 0.1) \times 10^3$
2002 + 31	$(1.2 \pm 0.05) \times 10^2$
2020 + 28	$(1.2 \pm 0.1) \times 10^3$

The data are mostly given by Gullahorn and Rankin (1982) and collected in Ghosh (1984). It should be noted that n varies widely and can be of either sign.

12.3. Identification of Pulsars as Neutron Stars

When regular pulsations were first observed, it was suspected that they were signals produced by some intelligent living beings on some planetary object. However, such an idea was soon discredited, as the estimation of the distance of these pulsars showed that the rate of emission of energy was $\sim 10^{30}$ erg s^{-1}, only a few orders of magnitude less than that of the Sun. Such a powerful steady source was clearly beyond the capability of anyone tied down to a planet. Originally, the absence of any periodic change in the pulsar frequency led Hewish to conclude that the source was not tied to a planet.

Again, the magnitude of the period and its remarkable constancy indicated that the linear dimensions of the source must not be greater than Pc where P is the period. Thus it had to be a compact object. (If we consider a millisecond pulsar, Pc is just a few hundred kilometres.) The tendency, therefore, at one time, was to identify pulsars with white dwarfs, the only compact objects previously observed. However, the following table is revealing:

Phenomenon	Time period P (theoretical)	Rate of change of $P = \dot{P}$ (theoretical)	Remark
1. Spin of white dwarf	>1 s	Positive	The observed period ranging from milliseconds
2. Vibration of white dwarf	>2 s	Negative	to seconds, and a small but definitely positive value of P
3. Orbital motion of a white dwarf in a binary	>4 s	Negative	settles the spinning neutron stars as the only viable model for pulsars.
4. Spin of a neutron star	$\geq 10^{-3}$ s	Positive	
5. Orbital motion of a neutron star in a binary	$>10^{-3}$ s	Negative and large (due to gravitational radiation)	

The pulsars were thus identified with spinning neutron stars; however, the millisecond pulsars may pose a problem. To overcome the centrifugal repulsion in these cases, we must have a very stiff equation of state. (For different equations of state and their role in stellar models, see the next chapter.) Thus it has been suggested that the millisecond pulsars may be pulsating (rather than rotating) neutron stars (Wang et al., 1989). But then the mechanism of conversion of radial pulsations into ratio wave emission is not clear, and a picture which proposes different models for different pulsation periods will not be very appealing. (For a model of electromagnetic radiation from a vibrating compact star, see Hoyle et al., 1964.)

12.4. The Energetics of Pulsar Emission

Nonaligned Vacuum Model

In this model, the pulsar is assumed to have a dipole moment μ in a direction inclined at α to the rotation axis (which is also a symmetry axis). Outside the pulsar, there is a vacuum without any free charges. (Note that if the dipole moment coincides with the rotation symmetry axis, the magnetic field will not vary with time and there would be no electromagnetic radiation.) The standard formula for a magnetic dipole radiation then gives, for the rate of radiation,

$$\frac{dE}{dt} = \frac{B_0^2 R^6 \Omega^4 \sin^2 \alpha}{6c^3},$$

where B_0 is the magnetic field at the pole on the surface due to a dipole of moment μ at the center of the star $= 2\mu/R^3$, R being the radius of the star (assumed nearly spherical), and Ω is the angular velocity. The source of energy is assumed to be the rotational kinetic energy, and hence the radiation would lead to a decrease of the angular velocity Ω given by

$$I\Omega\dot{\Omega} = -\frac{B_0^2 R^6 \Omega^4 \sin^2 \alpha}{6c^3}, \tag{12.2}$$

where I, the moment of inertia of the star, is assumed to remain strictly constant.

Before proceeding further, it is interesting to consider some order of magnitudes regarding the emission of energy. Consider the pulsar in Crab nebula. Observationally, Ω and $\dot{\Omega}$ are known

$$\Omega = 1.9 \times 10^2 \text{ radians/s},$$

$$\dot{\Omega} = -2.4 \times 10^{-10} \text{ radians/s}^{+2}.$$

We have no precise knowledge of the moment of inertia. However, theory indicates that neutron star masses would be $\sim 1\ M_\odot$, the radii $R \sim 10^6$ cm, so that $I \sim 10^{45}$ g cm^2. Putting in these values we find, for the Crab pulsar, $I\Omega\dot{\Omega} \sim 5 \times 10^{38}$ erg s^{-1}. However, a calculation based on the electromagnetic energy, as received by us in the pulses, shows that the electromagnetic energy coming out per second is of the order of $\sim 10^{31}$ erg s^{-1}. The discrepancy is too great to be brushed aside, and it has been suggested that most of the energy emitted is absorbed within the plasma in the Crab nebula and only a tiny fraction escapes. There is a significant fact supporting this idea, the energy emitted by the Crab nebula as a whole is of the same order of $\sim 5 \times 10^{38}$ erg s^{-1}. While this resolves a long-standing difficulty regarding the source of energy of the Crab nebula, it shows that the assumption of a charge-free space surrounding the star is seriously in error. An alternative explanation of the energy discrepancy is a particle radiation or a particle wind from the neutron star.

12.5. The Magnetic Field at the Pulsar Surface

Going back to equation (12.2) and assuming $\sin^2 \alpha \sim 1$, we get for B_0 a value of $\sim 10^{12}$ G. This figure is considered plausible as a contraction of linear dimensions by a factor of $\sim 10^5$ (which is the value of $R_\odot/R$) and is expected to lead to an increase in B by a factor of $\sim 10^{10}$ according to the flux conservation theorem, and hence magnetic fields of $\sim 10^2$ G in ordinary stars may be enhanced to $\sim 10^{12}$ G in pulsars. (Some support for this idea has come from the discovery of magnetic fields $\sim 10^8$ G in some white dwarf stars.)

Another, and perhaps a more reliable, method of estimating the magnetic fields of pulsars has come from the discovery of line features in the X-ray spectrum of some pulsars. Thus Trümper et al. (1978) found a feature at 58 keV in the Her X-1 spectrum. Depending on whether it is an emission or an absorption line, the field strength B comes out as $(4-5) \times 10^{12}$ G. Similarly, a 20 keV line from 4 U 0115 − 63 gives $B \sim 2 \times 10^{12}$ G.

An estimate of B_0 based on (12.2) with typical values of $I \sim 10^{45}$ g cm^2 and $R \sim 10^6$ cm shows that over 90% of pulsars have B_0 lying between 10^{12}–$10^{13.5}$ G. However, the magnetic fields for millisecond pulsars are found to have much lower values (10^8–10^9 G). It has been suggested that millisecond pulsars may be old neutron stars (age $\sim 10^9$ yr instead of $\sim 10^7$ yr for ordinary pulsars), in which the magnetic field has decayed to small values but the stars have experienced a spin-up due to mass transfer from a close binary (Van den Heuvel, 1984).

Recently, a 4.5 ms pulsar has been discovered in the globular cluster 47 Tucanae which is in a binary orbit with eccentricity 0.32 (Ables et al., 1989). If the millisecond pulsar were spin-up by mass transfer from its companion, eccentricity would have decreased as a result. Further a few millisecond pulsars have no companions. This is explained by the assumption that γ-rays from the millisecond pulsar ionized and ablated the companion, gradually reducing its size until the latter disappeared (Kluzniak et al., 1988). This theory was confirmed by the discovery of a nebulosity around the millisecond pulsar PSR 1957 + 20 and a companion of very low mass $\sim 0.02\ M_\odot$ (Bailyn, 1989). Such pulsars have been nicknamed "black widow pulsars," as the event reminds us about a species of female spiders which kills the male counterpart.

Quite a different idea about pulsar magnetic fields has sometimes been put forward; namely, that the pulsar does not inherit the field but the magnetic field is generated in the neutron star after their birth, the time scale required to build up magnetic fields of the requisite order being $\sim 10^5$ yr. This may explain the remarkable paucity of association of pulsars with supernova remnants (Blandford et al., 1983; Woodward, 1984).

12.6. The Age of Pulsars

Assuming the constancy of I, R, B_0, and α, (12.2) may be integrated to give

$$\frac{1}{\Omega^2} - \frac{1}{\Omega_0^2} = \frac{B_0^2 R^6 \sin^2 \alpha}{3c^3 I} t,$$

where we have chosen $t = 0$ to be the birth instant of the pulsar when the rotational velocity was Ω_0. If the present value of Ω is much less than Ω_0 we get

$$t = \frac{3c^3 I}{B_0^2 R^6 \Omega^2 \sin^2 \alpha} \equiv \frac{T}{2} \qquad \begin{array}{l} \text{(the age will be significantly less} \\ \text{if } \Omega_0 \text{ is not very different from } \Omega), \end{array}$$

where

$$T \equiv -\frac{\Omega}{\dot{\Omega}} = \frac{6c^3 I}{B_0^2 R^6 \Omega^2 \sin^2 \alpha}.$$

The age of the Crab pulsar from the observed value of Ω and $\dot{\Omega}$ comes out as ~ 1260 yr (to be precise 1243 yr was the value given by Groth and Gullaborn from an observation in 1972), whereas the age obtained from the Chinese records of a supernova outburst is ~ 940 yr. The agreement, considering the crudeness of the assumptions, may be regarded as excellent. For the Vela, both the estimated age of the supernova explosion and the pulsar slow-down age are $\sim 10^4$ yr. Another way of estimating the age of pulsars comes from the observation that many pulsars have a high proper motion directed away from the galactic plane (Sieber and Wielebinski, 1981). Assuming that the pulsars are born close to the galactic plane, we can then calculate a "kinematic age" τ_{kin} of any pulsar. For $\tau_{kin} \lesssim 5 \times 10^6$ yr, the spin-down age $-\Omega/2\dot{\Omega}$ agrees with τ_{kin}.

12.7. Calculation of the Braking Index

From (12.2) $\dot{\Omega} \propto \Omega^3$; hence, the braking index n is 3. This is not far from the value observed for the Crab pulsar ($n = 2.5$), although the agreement cannot be said to be good. Some tried to investigate the value of the braking index by supposing that the change in Ω was due to the emission of gravitational radiation. For an object which has strict axial symmetry about the rotation axis, conditions are stationary and no gravitational radiation occurs. If, however, the equatorial plane sections are elliptic, instead of circular with eccentricity ε, then assuming the density to be uniform, the rate of the loss of energy by gravitational radiation is $(32G/5c^3)I^2\varepsilon^2\Omega^6$, where I is the moment of inertia about the rotation axis and the ellipticity is small. Thus $\dot{\Omega} \propto \Omega^5$ and the braking index is 5, showing no improvement as far as the Crab pulsar is concerned.

We have already presented a table showing that n varies widely and may be of either sign. While fairly complicated formulas for the braking index have been proposed, taking into account the change of the inclination between the rotation axis and the dipole axis as well as the decay of the magnetic field, they are inadequate to explain the large range of the observed braking index, and the significance of the data of Gullahorn and Rankin has been in doubt. Thus, Blandford et al. remark, "The nominal braking indices represented by Gullahorn and Rankin, ranging up to 10^5 and of both signs are evidently spurious" and opine "timing noise makes braking index determination very uncertain" (Blandford et al., 1984; Ghosh, 1984).

12.8. The Nonvacuum Model

We have already seen that the idea of a charge-free vacuum surrounding the pulsar suffers from an inconsistency in that we have to suppose that the energy emitted by the pulsar is almost completely absorbed in its neighborhood, where there is no one to do the job. A more direct demonstration of the inconsistency came from Goldreich and Julian (1969), who pointed out that the intense magnetic field for the pulsars has to be associated with a strong electric field, as the conductivity σ in the star is quite high (a simplifying assumption, $\sigma \to \infty$, is considered not inappropriate). These electric fields are strong enough to cause an emission of charges from the surface of the star which build up a plasma outside the star.

Assuming $\sigma \to \infty$, we get, for the electric field just inside the star,

$$\mathbf{E}^{(i)} = -\frac{\mathbf{v}}{c} \times \mathbf{B}.$$

Taking $\mathbf{B}$ to be the field due to a dipole with the rotation axis

$$\mathbf{B} = \frac{2\mu}{r^3}(\cos \theta, \tfrac{1}{2} \sin \theta, 0),$$

and hence

$$\mathbf{E}^{(i)} = \frac{2\mu\Omega}{cr^2}(\tfrac{1}{2} \sin^2 \theta, -\sin \theta \cos \theta, 0).$$

Assuming, for the moment, that the outside is charge-free, the outside field will be static as well (because of the alignment of the magnetic dipole with the rotation axis). Hence, for the external field,

$$\mathbf{E}(e) = -\nabla \phi, \qquad \nabla^2 \phi = 0.$$

Breaking up the exterior field into multipole components, we find that the boundary conditions (the continuity of E^θ, E^ϕ) are satisfied by the quadrupole field

$$\mathbf{E}^{(e)} = -\frac{\mu R^2 \Omega}{2cr^4}[3 \cos^2 \theta - 1, 2 \sin \theta \cos \theta, 0].$$

As the normal component of the displacement vector $\mathbf{D}$ is discontinuous (although the dielectric constant throughout is unity), there is a surface charge of density ρ_s

$$\rho_s = \frac{1}{4\pi}[E_r^{(e)} - E_r^{(i)}]_{r=R} = -\frac{\mu\Omega\cos^2\theta}{2\pi cR^2} = \frac{B_0 R\Omega}{4\pi c}\cos^2\theta,$$

with the surface magnetic field $B_0 = 2\mu/R^3 \sim 10^{12}$ G this gives, for ρ_s and the electric intensity, a value of $\sim 10^8$ e.s.u.

Such a strong field causes an emission of the charges from the surface, and it is easy to see that gravitation is totally inadequate to stop this escape, for the ratio of the electrostatic force to the gravitational force is

$$\frac{eE}{GMm_e/R^2} \sim \frac{10^{-2}}{10^{-13}} = 10^{11}$$

for an electron and $\sim 10^8$ for a proton. The problem thus reduces to a study of the motion of the plasma and the electromagnetic fields. It turns out that in a region where the magnetic field lines form closed loops, the charged particles corotate with the pulsar. This situation extends up to a distance where the tangential velocity of the particles tends to the light velocity, i.e., up to $R_0 = c/\Omega \approx 0.5 \times 10^9$ P cm. Up to these distances the field is $\sim B_0(R/r)^3$. At greater distances, we have a wave zone with $\mathbf{E}$ normal to $\mathbf{B}$. The amount of energy radiated may be estimated by considering a continuity of the wave field with the near zone field at $r = R_c$. Thus the energy radiated per second is

$$\sim 4\pi R_c^2 c B_0^2 \left(\frac{R}{R_c}\right)^6 \approx \frac{B_0^2 R^6 \Omega^4}{c^3}.$$

Thus the rate of energy loss is of the same form as in the vacuum model, and so the braking index comes out the same from both models.

However, it has not been possible, so far, to construct a self-consistent model for the currents and fields surrounding the star, and hence the picture remains disturbingly incomplete. Two models of pulsed emission have been suggested. In the polar cap model, the emission is from near the magnetic polar region. As the magnetic axis is, in general, not aligned with the rotation axis, the source region rotates with the star and hence the cone of the emitted radiation also rotates. During the time that the beam sweeps the Earth, we receive the radiation and that constitutes the duty cycle (the pencil beam). There is also another model, in which it is supposed that the beam is tangential to the light cylinder and perpendicular to the rotation axis (the knifelike beam).

Theoretical work on pulsars received a boost after a number of pulsating X-ray sources, called X-ray pulsars, were discovered by the astronomical satellite UHURU in 1971. These are believed to be neutron stars in binary systems which accrete mass from their companions that are usually normal stars. The accreted gas may form hot spots on the neutron stars (probably at the magnetic poles), which send out searchlight-like beams towards the Earth as the neutron stars rotate. The first X-ray pulsars detected were Cen X-3 and Her X-1. When a binary system is a source of radiation in both optical and X-ray wavelengths we may easily determine the masses of the components. We shall return to the case of X-ray pulsars in Chapter 15.

12.9. Observational Determination of Pulsar Masses

Observational determination of pulsar masses has, so far, been possible only for those who are members of binary systems. The case of the Hulse–Taylor pulsar which we have already discussed is in a way exceptional; that in this case, all the characteristics of orbital motion including the periastronic motion has been accurately determined. For any binary system, however, we can determine the mass function by observation of the first-order Doppler change of the pulsation period (see (14.10))

$$f(m_1, m_2, i) = \frac{(m_1 \sin i)^3}{(m_1 + m_2)^2} = \frac{(a_1 \sin i)^3}{G} \left(\frac{2\pi}{T}\right)^2. \tag{12.3}$$

For these X-ray binaries, which have an optical member and a pulsating X-ray source, we obtain two mass functions, i.e., besides (12.3),

$$f(m_2, m_1, i) \equiv \frac{(m_2 \sin i)^3}{(m_1 + m_2)^2} = \frac{(a_2 \sin i)^3}{G} \left(\frac{2\pi}{T}\right)^2. \tag{12.4}$$

These provide two equations for the three unknowns m_1, m_2, and i. Different types of observation-like X-ray eclipse duration, variation in the optical light curve, or scintillation studies (Bahcall, 1979; Lyne, 1984) may help up assign the value of i. We then have a determination of the pulsar mass.

The results show that all the pulsar masses so far determined may have a value near 1.4 $M_\odot$ within the estimated errors. However, it should be emphasized that, other than the Hulse–Taylor binary, the observations are limited to pulsars showing pulsed emission in the X-ray region. Lyne's observation of PSR 0655 + 64 leads to the result

$$\frac{m_c^3}{(m_p + m_c)^2} = 0.103 \text{ or } 0.072 \, M_\odot.$$

This is consistent with a pulsar mass $m_p \sim 1.4 \, M_\odot$ but the companion mass m_c is then less than 0.8 $M_\odot$. The companion star in this case is invisible and the above estimate would mean that the companion is also a compact object.

12.10. Cooling of Neutron Stars—Theory and Observation

The supernova explosions, which are believed to lead to the birth of neutron stars, leave them at very high temperatures $T > 10^{11}$ K. At this temperature, even for nuclear densities, the neutrons are nondegenerate and the β-decay of neutrons and the inverse process proceed quite fast (recall that the free neutron half-life is only a few minutes). Neutrinos, thus produced, take away the energy at a rapid rate (this is the so-called URCA process) and the star cools down to temperatures of $\sim 10^9$ K in about 1 day. However, the neutrons and protons now become degenerate (but nonrelativistic) while the electrons are relativistically degenerate. The simple URCA process, i.e., the reactions,

$$n \rightarrow p + e^- + \bar{\nu}_e,$$

$$p + e^- \rightarrow n + \nu_e,$$

can no longer take place as the following consideration shows.

The overall charge neutrality means that the number densities of protons and electrons are equal. In the degenerate state, this means that the occupied phase space volumes, which are proportional to the cubes of the Fermi momentum p_f, are the same for electrons and protons, i.e.,

$$(p_f)_{\text{electron}} = (p_f)_{\text{proton}}. \tag{12.5}$$

However, the condition of β equilibrium requires the chemical potentials to be equal for either side of the reaction. Effectively, this means, for the Fermi energies E_f,

$$(E_f)_{\text{neutron}} = (E_f)_{\text{electron}} + (E_f)_{\text{proton}}. \tag{12.6}$$

The nonrelativistic protons and neutrons and ultrarelativistic electrons (12.5) and (12.6) give, for the kinetic energies ($E_{\text{kin}} = E_f - mc^2$) near the Fermi surface,

$$(E_{\text{kin}})_{\text{proton}} = \frac{(p_f)^2_{\text{proton}}}{2m} \ll (p_f)_{\text{proton}} c = (E_{\text{kin}})_{\text{electron}} \approx (E_{\text{kin}})_{\text{neutron}}$$

and

$$(p_f)_{\text{electron}} = (p_f)_{\text{proton}} \ll (p_f)_{\text{neutron}}.$$

Thus, a decay of the neutron near the Fermi surface releases only a small energy but a large momentum. The neutrino cannot simultaneously take care of both, as for the neutrino $p = E/c$; thus, the decay cannot conserve both energy and momentum. Hence neutrino emission can take place only in the presence of an additional particle

$$n + n \rightarrow n + p + e^- + \bar{\nu}_e,$$
$$n + p + e^- \rightarrow n + n + \nu_e, \tag{12.7}$$

where the additional neutron serves to ensure the conservation of energy and momentum. However, in the degenerate situation the reaction rates become strongly temperature-dependent, as only the Fermi "tail" region $\sim KT/E_f$ effectively contributes. Thus, while in the normal URCA process the reaction rates are independent of temperature, it now becomes proportional to T^8. There is also another change, while the simple neutron decay is dominated by the weak interaction, the presence of the additional neutron partaking in reaction (12.7) brings in the strong interaction as well. Taking all this into account we find for a star of mass M and uniform density ρ, the luminosity due to reaction (12.7) the following formula (Friman and Maxwell, 1979)

$$L = 5 \times 10^{39} \frac{M}{M_\odot} \left(\frac{\rho_{\text{nucl}}}{\rho} \right)^{1/3} T_9^8, \tag{12.8}$$

where L is the rate of energy loss for the star in erg s^{-1}, ρ_{nucl} is the typical nuclear density, and $T_9 = T/10^9$, T being the temperature in degrees Kelvin. While this modified URCA process is believed to be the usual mechanism of cooling, the presence of neutron superfluidity, pion condensation, or free quarks may modify the situation considerably.

12.11. The Influence of Superfluidity

The neutrino emission considered above is decelerated when the neutrons are paired together to form a superfluid, for in that case we must first break the pairs. If this requires an energy Δ, the processes are slowed down by a factor of $e^{-\Delta/KT}$. Cooling is also affected by a reduction in the heat capacity in the superfluid state.

However, a new process of neutrino emission may then assume importance. In the star crust where there are nuclei, the electrons may give rise to a neutrino–antineutrino pair (neutrino bremsstrahlung)

$$e^- + (\text{nucleus}) \rightarrow e^- + \nu_e + \bar{\nu}_e + (\text{nucleus}). \tag{12.9}$$

For this reaction the luminosity comes out as (Maxwell, 1979)

$$L \sim 5 \times 10^{39} \frac{M_c}{M_\odot} T_9^6, \tag{12.10}$$

where M_c is the crust mass. Note the occurrence of T^6 in place of T^8 in (12.8).

12.12. The Influence of Pion Condensation

Pions may give rise to neutrinos, neutrinos acting as stand-by particles,

$$\pi^- + n \rightarrow n + e^- + \bar{\nu}_e,$$

$$\pi^- + n \rightarrow n + \mu^- + \bar{\nu}_\mu,$$

$$n + e^- \rightarrow n + \pi^- + \nu_e,$$

$$n + \mu^- \rightarrow n + \pi^- + \nu_\mu.$$

The calculation of luminosity based on the above equations gives (Maxwell et al., 1977)

$$L \sim 2 \times 10^{45} \frac{M}{M_\odot} \frac{\rho_{\text{nucl}}}{\rho} T_9^6. \tag{12.11}$$

Note that the coefficient ($\sim 10^{45}$) exceeds that in (12.8) by a factor $> 10^5$, so that if a pion condensate is present (12.11) will dominate and give rise to very fast cooling.

12.13. The Influence of Quarks

Quarks, if present, may give rise to neutrinos by β-decay

$$d \rightarrow u + e^- + \bar{\nu}_e,$$

$$u + e^- \rightarrow d + \nu_e,$$

where u and d indicate the up and down quark, respectively. Making some additional assumptions, Iwamoto (1980) gives, for the corresponding luminosity,

$$L \sim 1.3 \times 10^{44} \frac{M}{M_\odot} T_9^6, \tag{12.12}$$

which is close to the luminosity due to pion decay (12.11).

There is also a loss of energy, due to optical radiation, given by

$$L \sim 4\pi R^2 \sigma T_e^4 = 7 \times 10^{36} \left(\frac{R}{10}\right)^2 T_{e,7}^4, \tag{12.13}$$

where $T_{e,7} = T_e/10^7$, T_e being an "effective" surface temperature and R is in kilometers, L, as before, is in erg s^{-1}. While in the main body of the neutron star, the temperature is maintained uniform due to the high thermal conductivity of the degenerate electrons, the surface temperature T may be taken $\sim (10\,T)^{2/3}$, T being the internal temperature. Thus, for $T \sim 10^8$ K, $T_e \sim 10^6$ K. (Actually high conductivity makes $T\sqrt{g_{00}}$ rather than T uniform.)

The effect of a magnetic field is to reduce the photon opacity so that the photons may come from lower depths where the temperature is higher, hence the photon luminosity increases and cooling becomes a little faster. To translate the energy loss formulas to the rate of fall of temperature, we have to know the thermal capacity of the steller material. For the degenerate condition, the heat capacity per particle is

$$C_v = \pi^2 \frac{(x^2 + 1)^{1/2}}{x^2} K \left(\frac{KT}{m_0 c^2}\right),$$

where $x = p_f/m_0 c^2$.

We can now put the fall of the temperature formula to the test if we can determine the age as well as the temperature. The age will be known from the supernova explosion which gave birth to the pulsar (if records of that exist, as in the case of the Crab and Vela) and the temperature can be inferred from X-ray observation. For a surface temperature of $\sim 10^6$ K, the black body radiation has a peak at a wavelength of $\sim 10^{-7}$ cm, i.e., in the soft X-ray region.

The satellite observation (from the Einstein observatory) could detect soft X-rays from a number of supernova remnant regions, and it is hypothesized that the source of this X-ray may be neutron stars within the SNRs. However, three considerations cast doubt on the reliability of the interpretation of these observational data:

(1) Only in very few cases, namely the Crab, the Vela, and that in MSH 15–52, is there conclusive evidence of the existence of pulsars in association with SNRs.
(2) The X radiation may have an origin other than thermal, so that the temperatures determined from X-ray observations are to be regarded as upper bounds.
(3) The soft X-rays suffer absorption in the interstellar space to an extent which is uncertain.

For a detailed report on the comparison of theory and observation the reader is referred to the paper by Nomoto and Tsuruta (1981).

Apparently, the temperatures for SN 1006, and perhaps also for Cas A and Tycho, are lower than the theoretically expected value and we may be tempted to conclude the presence of pion condensation and/or free quarks in some neutron stars. However, for the reasons already stated, such conclusions are to be treated with caution. For the Crab, in which the existence of a neutron star is known, there is no necessity for postulating pion condensation, etc. It is doubtful whether Cas A and SN 1006 contain any stellar remnant (Harnden, Jr. and Seward, 1984).

Note. Several points in connection with the above discussion should be noted.

(1) We have presented the formulas for neutrino luminosity assuming uniform density. In reality, we must take a suitable model using an appropriate equation of state.
(2) There are some general relativity effects besides that taken account of in the TOV equation; first, the equation of thermal equilibrium is modified to $T\sqrt{g_{00}} = $ constant, and second, the observed luminosity is also altered by a factor g_{00}.
(3) It can be shown that under the condition prevailing in the neutron stars, the neutrinos emerge freely through them.

Problems

1. Estimate the travel time difference from a pulsar at a distance of 2.5 kpc for the radiations of wavelengths 1 m and 1.05 m. Take the average free-electron number density in the intervening space to be 0.02 cm^{-3}.

2. Estimate the energy loss due to magnetic dipole radiation for a pulsar using the data for the Hulse–Taylor pulsar given at the end of Chapter 9.

3. Considering a neutron star to be a uniform sphere of density 10^{15} g cm^{-3}, calculate the minimum period with which it can rotate without being disrupted by centrifugal forces.

4. Show that the Fermi energy of neutrons at nuclear density correspond to a temperature of $\sim 10^{11}$ K. Hence justify the assumption that for neutron stars older than a few second, the thermal contributions to pressure and density are negligible.

5. The dipole field given in Section 12.8 may be obtained from the potential $A_\mu = (0, 0, 0, \mu(\sin^2 \theta/r))$ in spherical polar coordinates and flat Minkowski space. Show that if the background geometry is given by the Schwarzschild metric, then A_ϕ is changed to

$$A_\phi = -\frac{3\mu \sin^2 \theta}{r}\left[\frac{1}{x^3}\ln(1 - x) + \frac{1}{x^2}\left(1 + \frac{x}{2}\right)\right],$$

where $x = 2m/r$. Determine the percentage change in A_ϕ (from the flat space value) at the surface of a pulsar where $x = 0.1$ (Ginzburg and Ozernoi, 1965).

13. Spherically Symmetric Star Models

13.1. Introduction

In the older astrophysical theory, as well as in the discussions of Fowler and Chandrasekhar on white dwarf stars, the Newtonian equations of gravitation and mechanics were used. As we have already seen, this was apparently justified, for in these cases the crucial factor GM/Rc^2 was really small compared to unity. When, however, this factor would approach unity, the use of the general theory of relativity seems imperative and in the following we derive an equation of hydrostatic equilibrium in general relativity. This equation is commonly referred to as the Oppenheimer–Volkoff (1939) equation. The basic assumptions are:

(a) Matter constitutes a perfect fluid with the energy–stress tensor

$$T_\nu^\mu = (p + \rho)v^\mu v_\nu - p\delta_\nu^\mu, \tag{13.1}$$

where p and ρ are the pressure and mass-density of the fluid, respectively, and v^μ is the velocity vector of the fluid.

(b) The system is spherically symmetric; as we have already seen, it allows us to write the metric in the form

$$ds^2 = e^\nu \, dt^2 - e^\lambda \, dr^2 - r^2(d\theta^2 + \sin^2 \theta \, d\phi^2). \tag{13.2}$$

(c) The system is in equilibrium and this leads to two conditions. First, the metric coefficients are independent of t, so that ν and λ are functions of r alone. Second, the velocity has no space component in the coordinate system of (13.2), i.e.,

$$v^0 = e^{-\nu/2}, \qquad v^1 = v^2 = v^3 = 0,$$

where we write 0 for the time coordinate, and 1, 2, 3 for the coordinates r, θ, ϕ, respectively. Putting these values of v^μ in (13.1), T_ν^μ has only diagonal components nonvanishing

$$T_\nu^\mu = \begin{vmatrix} \rho & 0 & 0 & 0 \\ 0 & -p & 0 & 0 \\ 0 & 0 & -p & 0 \\ 0 & 0 & 0 & -p \end{vmatrix}. \tag{13.3}$$

(d) The metric field $g_{\mu\nu}$ is determined by the equation

$$R_\nu^\mu - \tfrac{1}{2}R\delta_\nu^\mu = -8\pi T_\nu^\mu, \tag{13.4}$$

where the units have been chosen so that $G = c = 1$. Substituting from (13.2) and (13.3) in (13.4) we obtain three nontrivial equations

$$8\pi p = e^{-\lambda}\left(\frac{\nu'}{r} + \frac{1}{r^2}\right) - \frac{1}{r^2}, \tag{13.5}$$

$$8\pi p = e^{-\lambda}\left(\frac{\nu''}{2} - \frac{\lambda'\nu'}{4} + \frac{\nu'^2}{4} + \frac{\nu' - \lambda'}{2r}\right), \tag{13.6}$$

$$8\pi\rho = e^{-\lambda}\left(\frac{\lambda'}{r} - \frac{1}{r^2}\right) + \frac{1}{r^2}. \tag{13.7}$$

However, it is convenient to replace (13.6) by the relation $T_{\nu;\mu}^\mu = 0$, which holds identically in view of (13.4). This gives

$$p' = -(p + \rho)\left(\frac{\nu'}{2}\right). \tag{13.8}$$

Primes in the above equations indicate differentiation with respect to r.

13.2. The Tolman, Oppenheimer–Volkoff Equation

Equation (13.7) can be formally integrated to give

$$e^{-\lambda} = 1 - \frac{2m}{r}, \tag{13.9}$$

where

$$m(r) \equiv 4\pi \int_0^r \rho/r^2\, dr. \tag{13.10}$$

Also from (13.5) and (13.7)

$$8\pi(p + \rho) = \frac{e^{-\lambda}}{r}(\lambda' + \nu'). \tag{13.11}$$

Eliminating λ from (13.11) with the help of (13.9) we get

$$8\pi(p + \rho) = \left(1 - \frac{2m}{r}\right)\frac{\nu'}{r} + \frac{1}{r}\left(8\pi\rho r - \frac{2m}{r^2}\right).$$

Lastly, substituting from (13.8) for ν', we finally have the Oppenheimer–Volkoff equation

$$\frac{dp}{dr} = -\frac{(p + \rho)(4\pi pr^3 + m)}{r(r - 2m)}. \tag{13.12}$$

Equation (13.12) differs from the corresponding Newtonian equation by adding terms involving pressure to density terms in the numerator (e.g., ρ is replaced by $p + \rho$ and m by $\frac{4}{3}\pi r^3(\bar{\rho} + 3p)$, where $\bar{\rho}$ is an "average" density), and in the denominator $r(r - 2m)$ occurs in place of r^2. Both these lead to a steeper pressure gradient.

We have to integrate (13.12) to build up a stellar model but that is only possible if we have an additional equation connecting p and ρ. In ordinary stars, where thermonuclear reactions leading to energy generation occur, there are modification to the energy–stress tensor T_ν^μ as well, to take care of the flux of radiation or heat. However, in these cases, the departure from Newtonian physics can well be neglected. At present, we are primarily interested in compact bodies where the pressure arises essentially from degeneracy, i.e., we consider cold catalyzed matter. Here we have a relation between p and ρ. Thus to integrate (13.12) we have to append the appropriate equation of state.

Before studying the equation of state, let us see exactly how the integration is performed if $p = p(\rho)$ is known. As an input we take the value of the central density ρ_c. This determines p_c while p' and m vanish at $r = 0$. Equation (13.10) then determines m for an infinitesimal increase in r, and plugging in this value of m and ρ_c and p_c into (13.12) determines the value of p' allowing a determination of p at the next step. For the p thus found the equation of state determines ρ, and we go over the whole process once again to determine the values of the variables at the next step. In this way, the computation of p, ρ, m for increasing values of r goes on until we arrive at $p = 0$. We identify this as the boundary of the star. The process is repeated for different values of ρ_c.

Problem

1. Integrate the Oppenheimer–Volkoff equation for constant ρ (Schwarzschild interior solution) and obtain the maximum value of m/r.

13.3. The Equation of State for Cold Catalyzed Matter

The word "cold" signifies that the conditions are such that the pressure comes principally from a degenerate system, but the temperature may not be low in the usual laboratory sense. Indeed, white dwarf stars and pulsars may have temperatures of $\sim 10^7$ K, but the electrons (in white dwarfs) and neutrons (in pulsars) are degenerate and are the principal contributors to the pressure. Thus these are, in our nomenclature, cold bodies and in the equation of state the temperature has hardly any appreciable effect. "Catalyzed" signifies that all possible thermonuclear reactions have proceeded to completion, so that the composition has attained the state corresponding to minimum energy consistent with the constraints of the system.

For densities lying between 10^5–10^7 g cm^{-3} we have seen that the pressure is given by that due to degenerate free electrons with a correction for the

Coulomb interaction. However, when the density becomes greater, at a certain stage the Fermi energy of the electrons exceeds ~ 1.8 MeV, which is the threshold for the reaction

$$p^+ + e^- \rightarrow n + \nu,$$

(i.e., a proton and electron unite to form a neutron and a neutrino). Electrons then begin to be absorbed by the nuclei, whereby protons are converted into neutrons and we have nuclei progressively richer in neutrons. The neutrinos escape and will be left out of our consideration. The opposite process of β disintegration is, however, impeded, as an electron with an energy less than the Fermi energy cannot be liberated. This neutronization begins at a density of $\rho \sim 1.2 \times 10^7$ g cm^{-3} (the reader may calculate the corresponding electron Fermi energy to convince himself) and goes up to $\rho \sim 3.4 \times 10^{11}$ g cm^{-3}. For ρ in this region the equation of state has been given by Baym et al. (1971) (hereafter referred to as BPS).

In obtaining the BPS equation, consider the total energy of the system as the sum of the masses of nuclei, the energy of degenerate free electrons ε_e, and the lattice energy e_L due to Coulomb interaction. Thus, the energy density ε is

$$\varepsilon = \frac{n}{A} M(A, Z) + \varepsilon_e + e_L,$$

where n is the nucleon number density, A is the mass number, and Z is the atomic number of the nucleus. The BPS procedure is to start with a value of n (i.e., density ρ because $\rho \sim 1.66 \times 10^{-24} n$ g cm^{-3}), then find out the values of A and Z which lead to the minimum of ε. BPS used the empirical tabulated values of $M(A, Z)$, the number density of the electrons is nZ/A, hence these three suffice to determine ε_e. The lattice energy was also given in terms of these three variables

$$\varepsilon_L = -1.44 \, Z^{2/3} e^2 \left(\frac{nZ}{A}\right)^{4/3}.$$

With ε_{min} thus obtained as a function of n, the pressure p is given by the usual thermodynamical formula

$$p = n^2 \left[\frac{\partial(\varepsilon/n)}{\partial n}\right]_{A,Z}.$$

The influence of neutronization by β capture was shown by a shift in the stable nuclear species from $^{56}_{26}$Fe to $^{118}_{36}$Kr as ρ changed from 10^7 to 4.3×10^{11} g cm^{-3}, i.e., Z/A changed from 0.46 to 0.31.

At still higher densities (i.e., $\rho > 4 \times 10^{11}$ g cm^{-3}) the neutron–proton ratio becomes so high that a more stable state is attained by an escape of neutrons from the nuclei, commonly referred to as neutron drip. Thus a new phase of free neutron gas now appears and the energy density is a function of five variables (instead of three in the lower density regime), these are:

(1) the baryon number density n (or, equivalently, the density ρ);
(2) the neutron number density n_n outside the nuclei;
(3) the mass number of the nucleus A;
(4) the charge number of the nucleus Z; and
(5) the volume V_n of the nucleus (A, Z).

The last factor is introduced because the volume of the nuclei is affected by the pressure of the neutron gas and the nuclear mass in turn depends on its volume. As before, the idea is to obtain the equilibrium value of the energy density for a given n, by minimizing the energy density considering the variation of the other four factors ((2)–(5) in the above list). The mass of the nucleus (A, Z, V_n) was calculated by Baym et al. (1971) (hereafter referred to as BBP) by taking into account the Coulomb energy, the surface energy, and also the energy due to nucleon–nucleon interaction. The last factor was again taken over from many body calculations. BBP carried the calculations up to $\rho \sim 5 \times 10^{14}$ g cm^{-3} as, for still higher densities, the condition of nuclear matter required further scrutiny. The BBP equation of state showed a sharp drop in the adiabatic index Γ, as neutron drip sets in having a minimum of about 0.37 and then rises to the value $\frac{4}{3}$ at about $\rho \sim 7 \times 10^{12}$ (see Fig. 13.1). Stability considerations show that the equilibrium of a star is stable only if Γ is $\geq \frac{4}{3}$ at the center. Hence stars with a central density lying between 4×10^{11} g cm^{-3} and 7×10^{12} g cm^{-3} are in an unstable equilibrium.

At still higher densities of $\rho > 5 \times 10^{14}$ g cm^{-3} the picture of a lattice built up of distinct nuclei breaks down and, further, muons appear with appreciable concentration and have to be taken into account. We shall not go into the form of the equation of state in these regions but consider an extremely

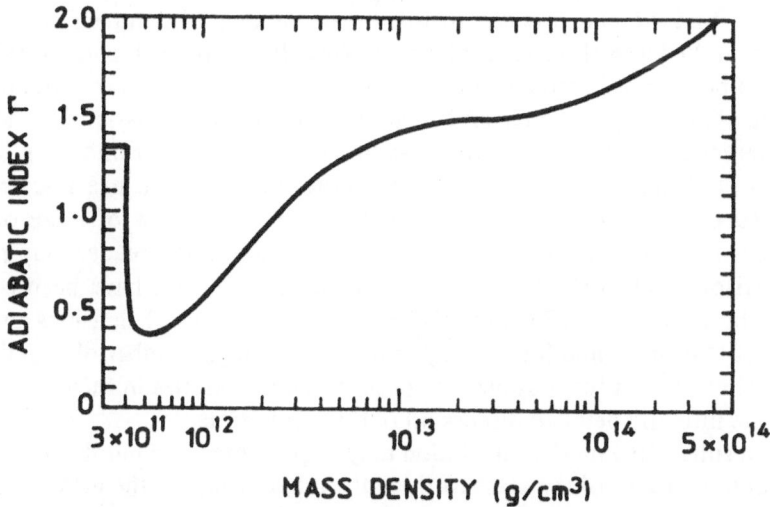

Figure 13.1. The adiabatic index $\Gamma = 2 \ln \rho/2 \ln n_b$ as a function of mass density. (After G. Baym, H.A. Bethe, and C.J. Pethick, *Nucl. Phys.* A175, 255 (1971).

idealized (and so unrealistic) situation in which there is an assembly of electrons, protons, and neutrons in equilibrium. Charge neutrality requires the number of electrons and protons to be same, and in the ultrarelativistic regime when all the particles are degenerate it would mean that the Fermi energy E_e of the electrons is the same as that of the protons E_p. Now consider the reaction which maintains the equilibrium between these particles

$$p^+ + e^- \rightleftarrows n + \tilde{\nu}.$$

Obviously, if E_n is the Fermi energy of neutrons

$$E_n = E_p + E_e = 2E_e \qquad \text{or} \qquad p_n = 2p_e;$$

hence, the number of neutrons would be eight times the number of electrons (or protons). Thus the idealized consideration shows how a star in such circumstances would be predominantly made up of neutrons; hence the name "neutron stars" (Canuto, 1974, 1975; Baym and Pethick, 1975, 1979).

We have not taken into consideration the possible presence of pions, these would apparently appear at a density somewhat higher than nuclear densities. The pions, being bosons there, would be the typical Bose–Einstein condensation to the zero momentum state and thus would not contribute to pressure. Hence in the region $\rho > \rho_{nucl}$ there is a softening of the equation of state. However, we will not go into a discussion of quantitative estimates as they are not only complicated but also rather uncertain.

13.4. A Model of a Neutron Star and the Mass Limits

We have seen that a large literature has grown on the equation of state at high densities, and we can proceed to investigate models of compact spherical objects on the basis of the Oppenheimer–Volkoff equations. However, calculation of neutron star models predated these investigations on the equations of state. Thus Oppenheimer and Volkoff worked on the basis of an ideal noninteracting degenerate neutron gas in 1939, and Wheeler and his school used in their model calculations an equation of state which has since been superseded by the BBS and BBP equations. (For a review of these older works, which are now principally of historical interest, the reader may refer to Harrison et al. (1965).) More recent model calculations have been commonly based on the BPS and BBP equations for central densities up to $\sim 5 \times 10^{14}$ g cm^{-3}, and for still higher densities quite a number of equations of state have been used, but because of several uncertainties in them we have not gone into any detailed discussion of these equations.

The results of the model calculation may be presented in a number of ways. We can have a plot of the mass M (here the mass M means the gravitational mass which occurs in the exterior Schwarzschild field) against the central density ρ_c or against the radius R (Figs. 13.2 and 13.3). The general feature of these curves is the occurrence of a maximum mass at a finite value of the

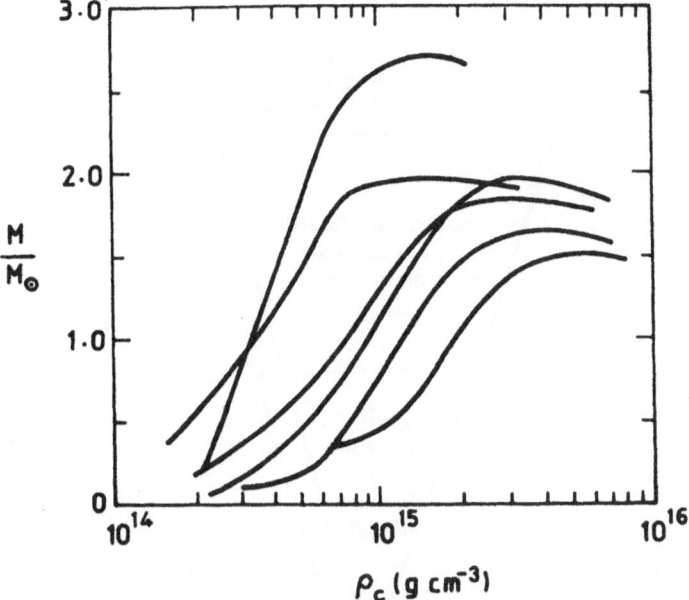

Figure 13.2. Gravitational mass central density for various equations of state. (After G. Baym and C.J. Pethick (1979). Reproduced with permission from the *Annual Review of Astronomy and Astrophysics*, Vol. 17, © 1979 by Annual Reviews, Inc.). We omit details about the equations of state.

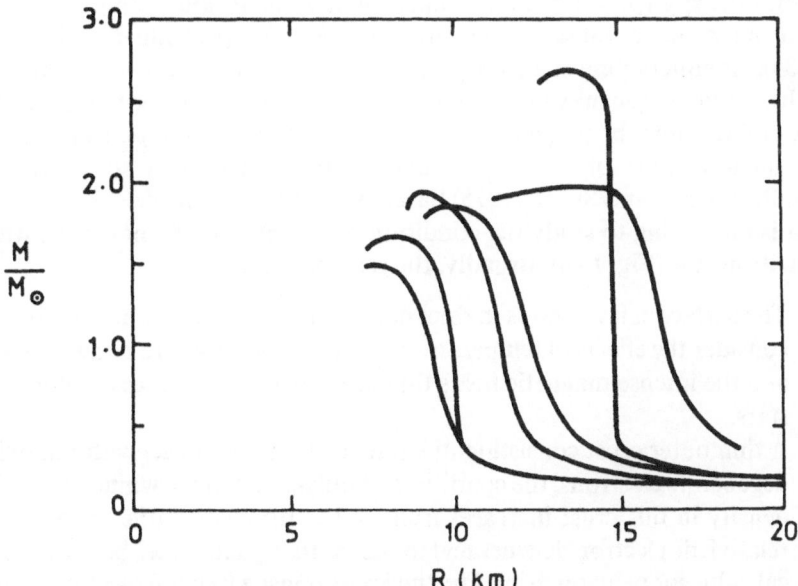

Figure 13.3. Gravitational mass versus radius for the same equations of state depicted in Fig. 13.2. (After G. Baym and C.J. Pethick (1979). Reproduced with permission from the *Annual Review of Astronomy and Astrophysics*, Vol. 17, © 1979 by Annual Reviews, Inc.).

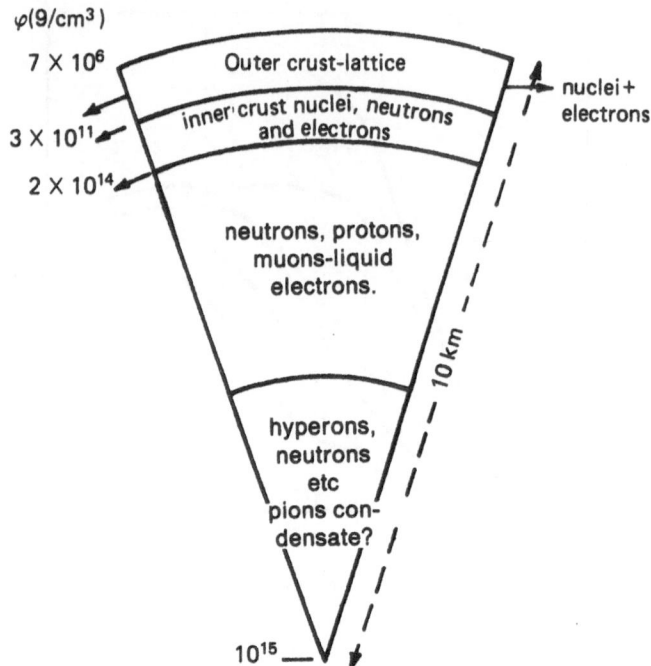

Figure 13.4. Schematic diagram of a neutron star showing conditions at different depths.

central density ($\rho_c \sim 10^{15}$) and approximately of 10 km radius. The maximum mass has a value ranging from 0.7 $M_\odot$ (Oppenheimer and Volkoff (1939), noninteracting neutron gas) to a value of little above 2.5 $M_\odot$, the higher values of the maximum being obtained for the stiffer equations of state (the stiffness may be judged by the value of $\Gamma \equiv d \ln p/d \ln \rho$). In particular, the highest value of 2.7 $M_\odot$ occurs for the equation of state given by Pandharipande and Smith (1975) based on mean scalar fields.

It is interesting to study the condition of matter at different depths within a neutron star (Fig. 13.4). Broadly, the regions are:

(a) The surface, a few metres in thickness, is an atmosphere where we have to consider the effects of temperature (i.e., the conditions are nondegenerate) and the intense magnetic fields that are believed to be present in neutron stars.

(b) A thin outer crust consisting of a lattice of ionized nuclei with relativistic degenerate electrons, the conditions simulate that in the white dwarfs. The density in this crust increases from $\sim 7 \times 10^6$ g cm^{-3} (the condition for relativistic electron degeneracy) to $\sim 3 \times 10^{11}$ g cm^{-3} (i.e., below the critical value for neutron drip). The thickness is just a fraction of a kilometer.

(c) The inner crust is somewhat thicker, the density (which is above 3×10^{11} g cm^{-3}) leads to a degenerate neutron gas in addition to neutron-rich nuclei. The neutron gas may be a superfluid due to formation of

"Cooper pairs" of neutrons. The density at the bottom of the crust is of the order of nuclear density $\sim 2 \times 10^{14}$ g cm^{-3}.

(d) Below the inner crust the density is the same as the nuclear density. Here we have a dominantly neutron-rich regime with protons and electrons with equilibrium between them. The neutrons and protons may be in a superfluid condition so that we have superconductivity as well.

(e) The central core region, extending up to a few kilometers from the center. Our knowledge and understanding of this region is rather uncertain; there may be hyperons, pions which have undergone Bose–Einstein condensation, and in this case the density is high enough to have a quark phase. Indeed, it has been speculated that in the high density regime, there may be a third branch of quark stars in the $M - \rho_c$ diagram.

There is a general belief, based on the "core convergence" in the evolution of stars (see Chapter 11), that most neutron stars would have masses of $\sim 1.4\ M_\odot$, and this has received some support from the observational determination of pulsar masses, but we must note that so far only a few pulsar masses have been determined.

All the proposed equations of state lead to a maximum mass limit. The question now arises, How is this existence of a limiting mass dependent on the equation of state that we have used? Is it possible that, at sufficiently high densities (where the validity of the currently used equations of state is rather questionable), the real equation of state is such that there is no mass limit? Further, if even then there is a maximum possible mass, what is its value? We shall see how these questions have been answered.

13.5. The Problems of the Upper Mass Limit of Neutron Stars

As the equation of state for densities above ρ_{nucl} ($\sim 5 \times 10^{14}$ g cm^{-3}) is not definitely known, we cannot integrate the Oppenheimer–Volkoff equation when such high densities occur. However, we can arrive at an upper mass limit basing our discussion on some very plausible assumptions:

(1) As an approximation, we neglect rotation and magnetic fields and the system is static and spherically symmetric.

(2) The material of the star can be regarded as a perfect fluid, i.e., the tangential stresses vanish (this may not be true, as for some densities the material may solidify).

(3) The density is nonnegative.

(4) The condition $dp/d\rho \geq 0$ is satisfied everywhere; otherwise, any decrease in volume locally, however small, would lead to a decrease in pressure of the material there and hence to further compression of the region. This $dp/d\rho \geq 0$ is just the condition for local stability. Further, as p is positive for ordinary densities, p will be positive for all densities under consideration.

For the equilibrium of a fluid we obtain, from the Raychaudhuri equation (see Chapter 17),

$$\dot{v}^{\mu}_{;\mu} = 4\pi(\rho + 3p), \tag{13.13}$$

where v^{μ} is the velocity vector and $\dot{v}^{\mu} = v^{\mu}_{;\alpha}v^{\alpha}$. Equation (13.13) is true quite generally, for the case of spherical symmetry we have

$$ds^2 = e^{\nu}\,dt^2 - e^{\lambda}\,dr^2 - r^2(d\theta^2 + \sin^2\theta\,d\phi^2), \tag{13.14}$$

$$v^{\mu} = e^{-\nu/2}\delta^{\mu}_0, \tag{13.15}$$

$$e^{-\lambda} = 1 - \frac{2m}{r}, \tag{13.16}$$

with

$$m(r) = \int_0^r 4\pi\rho r^2\,dr \equiv \tfrac{4}{3}\pi\bar{\rho}r^3. \tag{13.17}$$

Equation (13.17) defines the "mass" m and the "mean" density $\bar{\rho}$. Substituting from (13.15) and (13.13), we obtain

$$\frac{e^{-(\nu+\lambda)/2}}{r^2}\frac{d}{dr}\left[e^{-\lambda/2}r^2\frac{d}{dr}(e^{\nu/2})\right] = 4\pi(\rho + 3p). \tag{13.18}$$

Eliminating p from (13.18) with the help of the following equation, obtained from (13.5), (13.9), and (13.17),

$$e^{-\lambda}\frac{\nu'}{r} = 8\pi\left(p + \frac{\bar{\rho}}{3}\right), \tag{13.19}$$

we obtain

$$e^{-(\nu+\lambda)/2}r\frac{d}{dr}\left[\frac{e^{-\lambda/2}}{r}\frac{d}{dr}(e^{\nu/2})\right] = 4\pi(\rho - \bar{\rho}). \tag{13.20}$$

From the Oppenheimer–Volkoff equation it follows that dp/dr is negative unless $(1 - 2m/r)$ becomes negative. However, such a situation which makes r a timelike coordinate is not consistent with the static condition. Hence $dp/dr < 0$ and, consequently, in view of the assumption (13.16), ρ is also a nonincreasing function of r. Hence $\bar{\rho} \geq \rho$, the equality occurring in the case of uniform density (the Schwarzschild interior solution).

Hence from (13.20)

$$\frac{d^2}{dx^2}(e^{\nu/2}) = \frac{4\pi e^{\nu/2}}{r^2}(\rho - \bar{\rho}) \leq 0, \tag{13.21}$$

where

$$dx = r\,dr\,e^{\lambda/2} \quad \text{or} \quad x = \int_0^r r\,dr\,e^{\lambda/2}. \tag{13.22}$$

Inequality (13.21) indicates that $Y \equiv (d/dx)(e^{\nu/2})$, is in general a decreasing function of x, and is a constant only in the case of uniform density. Hence the total change of $e^{\nu/2}$ from $x = 0$ to $x = x$ will be underestimated by taking the values of Y at x and multiplying by the interval x, i.e.,

$$(e^{v/2})_{x>0} - (e^{v/2})_{x=0} \geq Y_x.x,$$

or as

$$(e^{v/2})_{x=0} > 0,$$

$$Y_x \equiv \frac{d}{dx}(e^{v/2}) \leq \frac{(e^{v/2})_x}{x}.$$

Restoring the variable r, the above inequality reads

$$\frac{e^{-\lambda/2}}{2r}\frac{dv}{dr} \leq \frac{1}{x}. \tag{13.23}$$

Again, we can set a lower bound to x. Note that for any $a > r$,

$$\frac{m(r)}{r} = \tfrac{4}{3}\pi\bar{\rho}_r r^2 \geq \tfrac{4}{3}\pi\bar{\rho}_a r^2 = \frac{m(a)}{a^3}r^2.$$

Hence

$$x_{r=a} \equiv \int_0^a dr\, r\left(1 - \frac{2m}{r}\right)^{-1/2} \geq \int_0^a dr\, r\left[1 - \frac{2m_a}{a^3}r^2\right]^{-1/2}$$

$$= \frac{a^3}{2m_a}\left[1 - \left(1 - \frac{2m_a}{a}\right)^{1/2}\right]. \tag{13.24}$$

Recall now the Oppenheimer–Volkoff equation

$$\frac{dp}{dr} = -(p + \rho)\frac{v'}{2} = -\frac{(p + \rho)(4\pi pr^3 + m)}{r(r - 2m)}. \tag{13.25}$$

Using (13.25) to eliminate v' from (13.23), and the bound of x as given in (13.24), we get

$$\frac{m + 4\pi r^3 p}{r^3(1 - 2m/r)^{1/2}} \leq \frac{2m}{r^3}\left[1 - \left(1 - \frac{2m}{r}\right)^{1/2}\right]^{-1}.$$

Simplifying the above inequality, we get a quadratic expression in m/r

$$9\left(\frac{m}{r}\right)^2 + \frac{4m}{r}(6\pi r^2 p - 1) + 4\pi r^2 p(4\pi r^2 p - 2) \leq 0,$$

or

$$\frac{m}{r} \leq \tfrac{2}{9}[1 - 6\pi r^2 p + (1 + 6\pi r^2 p)^{1/2}]. \tag{13.26}$$

Quite generally, therefore, we get a bound on M/R where R is the boundary value of r at which $p = 0$ and $M \equiv m(R)$

$$\frac{M}{R} \leq \frac{4}{9}.$$

This bound on the value of M/R was first obtained by Buchdahl (1959), it showed that the high redshift of quasars could not be gravitational. Subsequently, Chandrasekhar (1964) showed that stability considerations may further reduce the bound for M/R.

The results above will be used only for $\rho_s > \rho_{nucl} \sim 5 \times 10^{14}$ g cm^{-3}. Below that value we use some standard equations of state. Thus we speak of the star as consisting of a central core ($\rho \geq \rho_{nuc}$) surrounded by an envelope of $\rho < \rho_{nucl}$. Suppose the core extends up to r_0 where the density and pressure are, respectively, ρ_0 and p_0. Then from (13.26)

$$\frac{M_0}{r_0} \leq \tfrac{2}{9}[1 - 6\pi r_0^2 p_0 + (1 + 6\pi r^2 p_0)^{1/2}],$$

where M_0 stands for the core mass $4\pi \int_0^{r_0} \rho r^2 \, dr$. However, as $p_0 \ll \rho_0$ for densities somewhat beyond the nuclear densities, the above inequality reads

$$\frac{M_0}{r_0} \leq \frac{4}{9}. \tag{13.27}$$

In all the above inequalities, the equality sign holds good for the uniform density case. Thus these relations are called "optimal" meaning, thereby, that we can actually construct solutions satisfying the equality relation as well. The problem now is to perform the integration of the Oppenheimer–Volkoff equations for the envelope (i.e., $r \geq r_0$) with different values of r_0 and M_0.

Besides the inequality (13.27) there is another restriction as well, coming from the nonincreasing nature of ρ:

$$M_0 \geq \tfrac{4}{3}\pi \rho_0 r_0^3. \tag{13.28}$$

Results of the calculation with $\rho_0 = 5 \times 10^{14}$ g cm^{-3}, and assuming the Baym, Bethe, Pethick, and Sutherland equation of state (i.e., a combination of BBP and BPS equations in the envelope) are given in the literature (Hartle, 1978). It is found that the contribution of the envelope to the limiting mass is almost insignificant (less than 1%), and the limiting mass is thus nearly the same as the core mass which from (13.27) and (13.28) is

$$M_{max} = \frac{4}{9}\left(\frac{1}{3\pi\rho_0}\right)^{1/2},$$

or using

$$\rho_{nucl} = 2.8 \times 10^{14} \text{ g cm}^{-3},$$

$$M_{max} = 6.8\left(\frac{\rho_{nucl}}{\rho_0}\right)^{1/2} M_\odot \simeq 5\ M_\odot.$$

This maximum mass occurs for uniform density in the core, with $dp/d\rho \to \infty$ at the center. If we introduce an additional restriction $dp/d\rho \leq 1$ (frequently referred to as the causality condition, implying that violation of this condition would lead to the velocity of compressional waves exceeding the velocity of light and thus leading to a breakdown of causality), the value of the limiting mass comes down to about 3 $M_\odot$. This is the value usually taken as the theoretical upper bound of neutron star masses. We have seen in the last section that if we calculate on the basis of the currently proposed equations of state, the maximum mass ranges up to 2.7 $M_\odot$ (Baym and Pethick, 1979; Arnett and Bowers, 1977).

There remain some questions which have not yet been clearly answered. What is the influence of rotation and magnetic fields in this mass limit? Is the fluidity assumption valid at all densities and if it is not, how is the limiting mass affected thereby? Lastly, there is the condition of stability, does that consideration reduce the mass limit further?

13.6. The Influence of Rotation, etc., on the Mass Limit

For white dwarfs, the maximum mass, i.e., the Chandrasekhar limit, is affected only slightly by rigid rotation. The rotational velocity is subject to the condition that the centrifugal force must be less than gravity at the equator, and Anand (1968) shows that in this case the limiting mass is

$$M_{\text{lim}} = \frac{5.92}{\mu_e^2} M_\odot,$$

i.e., there is only an increase of about 8% over the Chandrasekhar value.

For nonrigid rotation, however, the effect on the mass limit is considerable. Indeed, Hoyle (1947) showed that in this case there is no mass limit. The same conclusion was arrived at by Ostriker et al. (1966). Detailed models of very massive white dwarfs in fast nonuniform motion were constructed by Ostriker and Bodenheimer (1968). However, no observational evidence for the existence of such massive white dwarfs seems available (Greenstein et al. (1977)).

In the case of neutron stars, nonrigid rotation seems unlikely. Bardeen and Wagonar (1971) found that arbitrary large masses can be held in equilibrium (in the form of disks) if we take the rotation to be arbitrarily large. However, such equilibrium for large rotation velocity is not stable.

For slow and rigid rotation, model calculations were done by Butterworth and Ipser (1975, 1976) and extended by Friedman et al. (1984) for the fast-moving pulsar (period, 1.6 ms). For the uniform rotation of this period the maximum mass is increased by $\sim 14\%$ for the softest equations of state and by $\sim 30\%$ for the stiffest equations of state.

Solidification of the neutron star core, a possibility at densities exceeding the nuclear density, may influence the mass limit. Indeed, we have hardly any justification for assuming spherical symmetry in that case, and there have been no investigations of nonspherical distributions. However, assuming spherical symmetry and also that the density is nonincreasing outwards, the maximum possible mass in equilibrium is given by

$$M_{\text{max}} = 13.6 \left(\frac{10^{14} \text{ g cm}^{-3}}{\rho_c} \right)^{1/2} M_\odot,$$

if we allow for anisotropy in pressure. Thus there is an increase of about 20% over the value found for fluid spheres.

13.7. Note on the Stability of Compact Objects

Before concluding this discussion on the masses of neutron stars we give, in outline, some important results on the stability of the equilibrium of compact objects. We begin with spherically symmetric fluid distributions (i.e., non-rotating and nonmagnetic stars). We consider an infinitesimal perturbation $\xi(\mathbf{r}, t)$ in the position of the fluid element at time t ($\mathbf{r}$ is the unperturbed position vector at equilibrium). Applying this perturbation to the hydro-dynamic and thermodynamic equations obeyed by the system, we obtain an equation of the form

$$\frac{\partial^2 \xi^i}{\partial t^2} = Q_k^i \xi^k, \tag{13.29}$$

where all the terms, nonlinear in ξ^k, have been thrown out because of the assumed infinitesimal nature of ξ^k. The absence of a term linear in $\partial \xi^i / \partial t$ (i.e., a term irreversible in time) is due to neglect of the dissipative forces. The coefficients Q_k^i involve the fluid properties p and ρ and the adiabatic index Γ.

If we now try for a solution (a normal mode of oscillation in the form

$$\xi^i(\mathbf{r}, t) = \eta^i(\mathbf{r}) e^{i\omega t},$$

where ω is a constant, (13.29) reduces to an eigenvalue equation for ω^2. If ω^2 comes out positive (i.e., ω is real), $\exp(i\omega t)$ indicates an oscillation about the equilibrium configuration, and as any dissipation (which is nevertheless pre-sent, although we have neglected it) damps out such an oscillation, we con-clude that the system will return to its unperturbed equilibrium state, i.e., the system is stable. If, on the other hand, ω^2 turns out to be negative there are some exponentially increasing perturbations, i.e., the equilibrium is unstable. In the critical case, $\omega^2 = 0$, the system after perturbation continues to be in equilibrium, i.e., there are two infinitesimally neighboring equilibrium states and the equilibrium is neutral (Fig. 13.5).

To be more specific, consider the case of radial perturbation (η^i being a function of r alone, the origin being at the center). Equation (13.29), written out explicitly, is then (we omit the derivation)

$$\frac{d}{dr}\left[\Gamma p \frac{1}{r^2} \frac{d}{dr}(r^2\eta)\right] - \frac{4\,dp}{r\,dr}\eta + \omega^2 p\eta = 0. \tag{13.30}$$

It then turns out that the condition for stability (i.e., for ω^2 to be positive) is

$$\bar{\Gamma} > \tfrac{4}{3}, \tag{13.31}$$

where

$$\bar{\Gamma} = \frac{\int_0^R \Gamma p r^2 \, dr}{\int_0^R p r^2 \, dr},$$

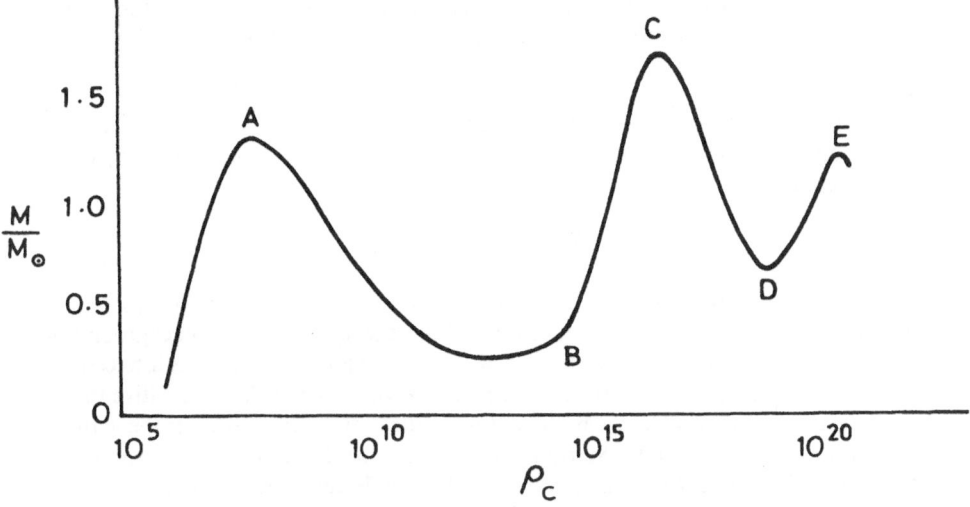

Figure 13.5. Schematic plot of mass against central density.

R being the boundary value of r. The above condition holds for the Newtonian case: if general relativity considerations are introduced the stability condition is modified to

$$\bar{\Gamma} - \tfrac{4}{3} > K \frac{2GM}{Rc^2},\tag{13.32}$$

where K is a constant depending on the density distribution and is usually of the order unity (see, e.g., Chandrasekhar, 1964). Thus, for compact objects, general relativity makes the condition for stability somewhat more stringent.

Besides the above condition on $\bar{\Gamma}$, there are other criteria for the transition from stability to instability. One which is of special interest to us is the following theorem.

If, with any particular equation of state, M (the equilibrium mass, as a function of the central density ρ_c) has a critical point, then one radial mode changes its stability property at this point.

Note that the theorem as stated has a degree of indefiniteness, e.g., it may happen that an unstable mode changes to stability and the system becomes stable or it may be that there were more than one unstable modes and, although one of them has changed to stability, the equilibrium continues to be unstable. There is also the possibility that the change is from, say, one mode instability to two mode instability (see Misner and Zapolsky, 1964).

We can understand the above theorem by the following argument. If the $M - \rho_c$ curve has a critical point, then a change from the critical ρ_c to $\rho_c + \delta\rho_c$ would give an equilibrium configuration at the same M and the same total baryon number (see the problem below). We may consider the change from ρ_c to $\rho_c + \delta\rho_c$ as an infinitesimal perturbation and the above shows that the system is in neutral equilibrium with respect to this perturbation. Hence, for this perturbation, stability will undergo a change at this stage.

Problem

Show that for cold gas spheres in equilibrium, if M and $M + \delta M$ are the masses corresponding to the baryon numbers N and $N + \delta N$ ($\delta M \ll M$, $\delta N \ll N$), then, correct up to first-order quantities,

$$\delta M = \frac{p + \rho}{n} e^{\nu/2} \, \delta N,$$

where n is the number density of the baryons. (Use the relation $dn = n/(p + \rho) \, d\rho$ and note that $((p + \rho)/n)e^{\nu/2}$ is a constant throughout the sphere.) In the background of the above theorem, a look at the typical $M - \rho_c$ curve for cold spheres is interesting.

While the exact position of the critical points A, B, C, D, and E are sensitive to the particular equation of state, the broad nature of the curve is tthe same for the various plausible equations of state. Starting from the left we have stability up to A, the Chandrasekhar limit. At A, the lowest radial mode becomes unstable and thus the region A to B represents unstable equilibrium. After B stability is recovered and C represents the maximum mass of neutron stars. The unstable region CD is followed by DE where we may have, as yet, an undiscovered regime of very dense stars (quark stars?). Note that B indicates a lower limit to the neutron star mass at about $\sim 1 \; M_\odot$ and $\rho_c \sim 10^{14}$ g cm^{-3}. Recall, in this connection, the low value of Γ for density $\rho > 10^{11}$ g cm^{-3} owing to the onset of neutron drip.

Problems

1. Obtain the metric inside a sphere of constant density ρ in equilibrium, which fits continuously with the metric of the Schwarzschild vacuum metric using:
   (a) equations (13.5)–(13.8); and
   (b) the Tolman, Oppenheimer–Volkoff equation.
   (The continuity conditions in the present case are the continuity of pressure, λ, ν, and ν'.)

2. Show that the propositions of Problem 1, Chapter 10, hold good in general relativity as well, using the Tolman, Oppenheimer–Volkoff equation in place of the equation of hydrostatic equilibrium given in Chapter 10.

3. Prove, among all spherically symmetric equilibrium configurations of perfect fluid which contains a specified number of baryons, that the configuration which extremizes the Schwarzschild mass M ($\equiv \int 4\pi \rho r^2 \, dr$) satisfies the Tolman, Oppenheimer–Volkoff equation. The baryon number A is given by

$$A = \int n(r) 4\pi r^2 \left(1 - \frac{2m}{r}\right)^{1/2} dr,$$

where $n(r)$ is the baryon number per unit proper volume at r (see the book by Harrison, Thorne, Wakano, and Wheeler (1965) cited in the Bibliography).

4. If the equation of state is of the form

$$\rho = \text{constant } n^\gamma, \qquad p = (\gamma - 1)\rho,$$

where n is the baryon number density and $1 \le \gamma \le 2$, prove that the total baryon number, the mass, and the radius of equilibrium configuration are all finite even if the central density is infinite (see Harrison et al. (1965) cited in Problem 3).

14. Black Holes

14.1. Introduction

We have had some discussions on black holes in connection with the Schwarzschild and Kerr metrics. We now propose to take up some advanced topics as listed below:

(i) The no-hair theorem and the uniqueness of the Kerr–Newmann metric for black holes.
(ii) The cosmic censorship hypothesis, the area increase theorem, and black hole thermodynamics. The possibility of black hole evaporation.
(iii) The observational evidence about the existence of black holes; in particular, the identification of Cygnus X-1 as a black hole.

Most of the theoretical topics require discussion at a level above what is appropriate in a book such as the present one. Consequently, we shall merely indicate the general trend of development and omit the detailed mathematical arguments.

The black hole is characterized by a horizon of the topology of a sphere which, at least classically, acts as a one-way membrane preventing any communication from the "inside" region going out. True, by the reckoning of an outside observer, a collapsing body never shrinks beyond the horizon or, in other words, the horizon is never formed; in reality, after a finite time (which is quite short once gravitational collapse sets in) the collapsing body can affect the outside universe only through its gravitational and electrostatic fields (we do not consider the existence of magnetic monopoles).

14.2. The No-Hair Theorem

The no-hair theorem states that the metric for a black hole in matter-free space will be uniquely determined by three parameters which may be identified with the mass, the electric charge (and/or the magnetic monopole charge), and the angular momentum of the source. (Identification can be done by means of the motion of test particles in the weak field region.) The three quantities are conserved and can be represented by surface integrals over the sphere at infinity.

This theorem is based on the following results obtained by different workers (for a detailed review, which is somewhat backdated, see Carter (1979)):

(1) If there is a field of integral spin s in the Schwarzschild background, then only those multipole components 2^l, having $l < s$, survive, higher multipole components being radiated away. This result is due to Price (1972a, b). Thus, for the electromagnetic field ($s = 1$), only the monopole survives (i.e., the electric charge or the magnetic charge if magnetic monopoles exist) and, for the gravitational field ($s = 2$), only the mass and angular momentum survive. However, we must qualify these assertions by stating that Price's result has been proved only for weak fields, i.e., fields which do not alter the background Schwarzschild field.
(2) A black hole cannot exert any weak or strong interaction forces caused by leptons or hadrons once they go inside the black hole (Hartle, 1971; Bekenstein, 1972a, b; Teitelboim, 1972).
(3) Two black holes which have the same mass, angular momentum, and charge are indistinguishable to an external observer, even if one is composed of matter and the other of antimatter.

In view of results (2) and (3), it is not meaningful to talk of the conservation of leptons or baryons in a black hole.

A closely related theorem is that of the uniqueness of the Kerr–Newmann black hole. The Kerr, Reissner–Nordström, and Schwarzschild black holes are special cases of this. The Kerr and Kerr–Newmann black holes have axial symmetry while the other two are spherically symmetric. One way of proving this theorem would be to start with an arbitrary isolated distribution of matter, and then to show that as the system evolves all departures from the Kerr–Newmann situation progressively decrease, tending to disappear as the collapse develops a horizon. Such a proof has not been given in perfect generality. An alternative method which has been adopted is to assume that a horizon is already there and then prove that the Kerr–Newmann metric is the unique solution of the field equations. This has indeed been done by Mazur (1982), who proved that the only possible exterior solution for a stationary rotating black hole with a nondegenerate event horizon is the Kerr–Newmann solution with $m^2 - a^2 - Q^2 > 0$, where m is the mass, a is the angular momentum per unit mass, and Q is the electric charge. Mazur considers magnetic charge as well, that merely replaces Q^2 by $Q^2 + p^2$, where p is the magnetic charge.

It has been conjectured that no naked singularities (i.e., singularities which are not enclosed by an event horizon) can occur as a result of the collapse of a physically realistic distribution of matter. In other words, an observer cannot receive any communication from a singularity. Otherwise, particles created near a singularity can be detected by the observer. This is known as the cosmic censorship hypothesis.

14.3. The Laws of Black Hole Physics

We give a sketch of the proof of the area nondecrease theorem. We have already seen that the horizon in the Kerr metric is given by $r_h = m + \sqrt{m^2 - a^2}$ and is generated by null lines. This result can be generalized, i.e., even when the conditions are not stationary, or when the outside space is nonempty, any horizon that is formed is generated by null geodesics.

We introduce, in a Lorentz invariant manner, an idea of the cross section of a pencil of null geodesics in the following manner. Consider a unit timelike vector (i.e., the world-line of a possible observer) and this defines a 3-space. The area of the two-dimensional space, orthogonal to the null rays in the ordinary three-dimensional sense, cut off by the pencil, is the cross section of the pencil. This area may be shown to be independent of the choice of the unit timelike vector (i.e., it is Lorentz invariant).

It can now be show that the change in this cross section α, as we proceed along the null geodesic, is given by

$$\frac{1}{\sqrt{\alpha}} \frac{d^2 \sqrt{\alpha}}{d\lambda^2} = -\sigma^2 + \tfrac{1}{2} R_{\alpha\beta} K^\alpha K^\beta, \tag{14.1}$$

where K^α is the future directed null vector tangential to the generators and $d/d\lambda = K^\mu(\partial/\partial x^\mu)$. Further

$$\sigma^2 = \tfrac{1}{2} g^{\alpha\mu} g^{\beta\nu} K_{\mu;\nu} K_{\alpha;\beta} - \tfrac{1}{4}(K^\alpha_{;\alpha})^2. \tag{14.2}$$

The scalar σ is called the shear and is a measure of the change of shape of the cross section of the pencil. Consider now the Einstein equations

$$R_{\alpha\beta} K^\alpha K^\beta = -8\pi [T_{\alpha\beta} - \tfrac{1}{2} T g_{\alpha\beta}] K^\alpha K^\beta$$
$$= -8\pi T_{\alpha\beta} K^\alpha K^\beta. \tag{14.3}$$

If the energy density as observed by observers moving along the null lines is nonnegative, then (14.1) gives, in view of (14.2) and (14.3),

$$\frac{d^2 \sqrt{\alpha}}{d\lambda^2} \leq 0,$$

where the sign of equality holds for nonshearing geodesics in empty space. It is now clear that if the cross section α is once contracting (i.e., $d\sqrt{\alpha}/d\lambda < 0$), then the contraction will be accelerated and, in a finite value of the affine parameter λ, the cross section α would collapse to zero, i.e., the pencil will converge to a point.

As the horizon is generated by these null geodesics, this convergence would correspond to the collapse of the horizon to a naked singularity. We now introduce, following Hawking, the cosmic censorship hypothesis which denies the possibility of naked singularities, as that would make the future unpredictable. Hence, the assumption that at a certain stage the cross section α is contracting is wrong. Hence the theorem of the nondecreasing area of the horizon. This theorem is referred to as the second law of black hole physics.

There is also a first law, it states a relationship between the differences of mass δm, the area δA, and the angular momentum δJ, for a black hole which is embedded in vacuum both initially and finally

$$\delta m = \frac{K}{8\pi} \delta A + \Omega \, \delta J. \tag{14.4}$$

Here Ω is "the angular velocity" $\equiv a/m$ and J is the angular momentum $= ma$. The area of the horizon A is the area of the two space spanned by the coordinates θ and ϕ at r_h. For a Kerr black hole

$$A = 4\pi(r_h^2 + a^2),$$

while K is known as the surface gravity of the black hole. For the Kerr black hole

$$k = \frac{(r_h - m)}{r_h^2 + a^2}.$$

14.4. Black Hole Thermodynamics

Equation (14.4) is quite analogous in form to the first law of thermodynamics

$$dU = T \, dS - p \, dv, \tag{14.5}$$

and, further, the nondecreasing property of the horizon area cannot but bring to mind the nondecreasing property of entropy. Bekenstein (1973) further found that consideration of the information lost to the world outside the horizon, when something enters the horizon, is closely correlated to the area increase, and he suggested the identification of the entropy of the black hole with a constant multiple of the area. However, there are two shortcomings in Bekenstein's proposal; first, the constant of proportionality between the entropy S and the area A could not be specifically determined, and, second, if we are to assign a temperature to the black hole, then the requirement of thermal equilibrium demands a flux of energy out from the black hole when the surroundings are at a lower temperature. However, this is expressly barred in classical general relativity.

The way out was found by Hawking. Considering a black hole formed by collapse in which initially there are no particles at infinity, and considering the condition that the renormalized vacuum stress tensor $\langle T^{\mu\nu} \rangle$ should be smooth on the horizon, Hawking showed that in the final state at infinity there is an outgoing stream of particles at temperature T (for the Schwarzschild black hole)

$$T = \frac{\hbar c^3}{8\pi K G m}, \tag{14.6}$$

where K is the Boltzmann constant. (To have an idea of the order of magnitude note that this gives, in degrees Kelvin, $T = 6.2 \times 10^{-8}(M_\odot/m)$, so that at a black hole whose mass is of the same order as the Sun, the temperature is extremely low.)

From (14.4) we now have (again for the Schwarzschild black hole)

$$\delta(mc^2) = \frac{c^6}{G^2} \frac{1}{32\pi m} dA, \tag{14.7}$$

where we have restored c and G. Comparing (14.5), (14.6), and (14.7)

$$S = \frac{Kc^3}{G\hbar} \frac{A}{4}. \tag{14.8}$$

Equation (14.8) holds for the Kerr black hole as well, but in that case the outgoing radiation is not purely thermal due to the superradiance associated with the ergosphere.

A black hole is thus an object of negative thermal capacity; any loss of energy leads to an increase in the temperature because of (14.6), and hence by spontaneous emission of radiation is disappears altogether in a finite time. We can work out the time for complete evaporation quite easily. The rate of the decrease in mass due to radiation is

$$\frac{dE}{dt} = \frac{d}{dt}(mc^2) = -\sigma A T^4 = -4\pi\sigma\left(\frac{2Gm}{c^2}\right)^2\left(\frac{c^3}{8\pi GmK}\right)^4$$

$$= -\frac{16\pi\sigma G^2}{c^4}\left(\frac{c^3}{8\pi GK}\right)^4 \frac{1}{m^2},$$

where σ is the Stefan constant $= 5.7 \times 10^{-5}$ erg cm^{-2} s^{-1} K^{-4}. Integrating, we get the time τ for complete evaporation in years

$$\tau = \frac{256}{3} \frac{\pi^3 G^2 K^4}{\sigma e^6 \hbar^4} m^3 \sim 10^{10}\left(\frac{m}{10^{15}\,\text{g}}\right)^3 \text{yr}$$

Thus only for "mini" black holes of mass $m \leq 10^{15}$ g is there the possibility of a complete evaporation within the time scale since the big bang. It is to be noted that the rate of energy emission diverges as $m \to 0$, hence if mini black holes were formed in the early universe as a result of density fluctuations, some of them might be emitting strong radiation right now. The possibility of detecting such radiation has been discussed, but no observational confirmation seems available so far.

Problems

1. Consider particles moving in circles around a Kerr black hole with velocity vector $U^\alpha = U^0(\delta_0^\alpha - \delta_\phi^\alpha(g_{\phi t}/g_{\phi\phi}))$ (i.e., locally nonrotating). Show that U^α/U tends to a finite limit independent of θ as the radius of the orbit tends to the horizon. (This limit is called the surface gravity and its constancy over the horizon is referred to as the zeroth law of black hole physics.)

2. The area nondecrease theorem allows us to define an irreducible mass as

$$m_{ir} = \sqrt{\frac{A}{16\pi}}.$$

Show that for a Kerr black hole

$$m_{ir}^2 = \tfrac{1}{2}[m^2 + (m^4 - J^2)^{1/2}] \qquad (J = am).$$

3. Calculate the increase of entropy that would occur if the Sun is converted into a Schwarzschild black hole (take the Sun to be a sphere of protons and electrons at a temperature of 2×10^6 K and radius of 7×10^{10} cm and $M = 2 \times 10^{33}$ g). Calculate also the entropy of a sphere of black body radiation of the same mass and temperature as the black hole.

4. Suppose a Schwarzschild black hole decreases its radius by dr due to Hawking radiation. Assuming the radiation to be of the same temperature as the black hole, calculate the increase in entropy and explain it.

5. Using the area theorem, show that for two equal Schwarzschild black holes coalescing to form a single hole, the maximum efficiency (energy released/total initial energy) is less than 30%.

14.5. The Identification of a Black Hole—Cygnus X-1

When we raise the question, "Do black holes exist in nature?" we are faced with a logical difficulty. Do we mean that it was present from the very beginning, i.e., the big bang, and are we not to ponder how it came into being? That seems to be not very appealing, but if we consider alternatively that the black hole was the result of gravitational collapse, then we have to face the problem that the black hole comes into being only after infinite time in the reckoning of the outside observer, whereas the universe itself has only a finite span of life. Be that as it may, an extreme condition when the luminosity of the collapsed object is greatly reduced, due to the combined gravitational and Doppler shifts, is attained in a relatively short time and that is what matters observationally.

The criteria, which would lead us to consider that an object is a black hole, are the following:

(a) it must be a compact object with an intense gravitational field;
(b) it must be invisible directly, i.e., no radiation or particle of any type should come to us directly from the object (except for a very small mass, Hawking radiation is extremely weak); and
(c) its mass should be greater than three times the solar mass.

The third condition may seem uncalled for as there is nothing, in principle, against a black hole of smaller mass (especially, if formed in the early universe); however, it is hardly possible to distinguish such a black hole from a white dwarf which has cooled down to become nonradiating, or a neutron star which does not pulsate, for they would also satisfy the first two criteria. The third condition eliminates the two possibilities. (Recall the limiting mass theorem.)

There are several candidates like Cyg X-1, A0620-00, LMX-X1, and X-3 which appear to be black holes, and the most eligible object which seems to pass all the above tests is the Cygnus X-1—an object in the Cygnus constellation associated with X-ray emission. X-rays may arise typically in the following manner. Suppose that a compact object is a member of a binary system, the companion being an ordinary star. Owing to its intense gravitational field, the compact object will draw out gas from the companion. However, the gas will not fall directly towards the compact object; owing to angular momentum, the gas will rotate and drift towards the compact object in a spiral path. The accumulating gas will form an accretion disk in which there will be a radial gradient in the velocity. Viscous forces will then come into play and the gas would consequently be heated to a very high temperature. X-rays may then be generated either by thermal bremsstrahlung or by synchrotron radiation. The latter may occur when the compact object has a magnetic field, as in pulsars, but such magnetic fields are not associated with black holes if there is no magnetic monopole (recall the no-hair theorem). Further, with a pulsar we expect a pulsed X-ray beam, but in the case of Cygnus X-1, the radiation does not undergo any pulsation. That there is a compact object is evident from the fact that there are occasional X-ray bursts lasting only for about 1 ms, setting an upper bound to the linear dimension of the source region of 300 km.

The question now arises about the mass of the object. It has been possible to identify the companion star from the following observations:

(i) The onset of radio emissions from the companion in March, 1971, coincided with a change in X-ray emission.
(ii) The period of the star in its orbit is 5.6 days (as determined from the periodicity of the Doppler shift) and there is a modulation of the X-ray intensity with the same period.

From Newtonian dynamics, we get

$$\frac{M_x^3}{(M_x + M_0)^2} = \frac{4\pi^2 a_2^3}{GP^2}, \tag{14.9}$$

where P is the orbital time period (in the present case, 5.6 days), a_2 is the semimajor axis of the companion orbit, M_x and M_0 are, respectively, the masses of the compact object and the optical companion.

Doppler shift observation of the light from the companion star determines its velocity along the line of sight, and from an analysis of the velocity curve we can determine the orbit elements; in particular, the eccentricity e, the product of $a_2 \sin i$ where i is the angle between the line of sight and the normal to the orbit plane. In the present case, it is found that $e \leq 0.02$ so that we shall henceforth consider the orbit to be circular ($e = 0$). From (14.9) we get, for the mass function, f

$$f \equiv \frac{M_x^3 \sin^3 i}{(M_x + M_0)^2} = \frac{4\pi^2 (a_2 \sin i)^3}{GP^2} = 0.25 \, M_\odot. \qquad (14.10)$$

The last figure has been obtained from observation. Hence

$$M_x = \frac{(1 + q)^2}{\sin^3 i} \times 0.25 \, M_\odot, \qquad (14.11)$$

where we have written

$$q \equiv \frac{M_\odot}{M_x}.$$

From the observed spectrum of the companion star, we can estimate an effective surface temperature T_e so that, from Stefan's law, the observed luminosity L_{obs} at a distance D is

$$L_{obs} = \frac{\sigma R^2 T_e^4}{D^2},$$

where R is the radius of the companion star. Putting in the observed values we get, in the present case,

$$R \approx 10 \, R_\odot D,$$

where D is now in kiloparsecs. It is usual to introduce the nondimensional variable

$$x = \frac{R}{a_2 \sin i},$$

so that

$$x = \frac{10}{a_2 \sin i} R_\odot D \approx 1.2D. \qquad (14.12)$$

Another bound on R is obtained from the absence of any eclipse of the X-rays (Fig. 14.1)

$$R \leq a \cos i,$$

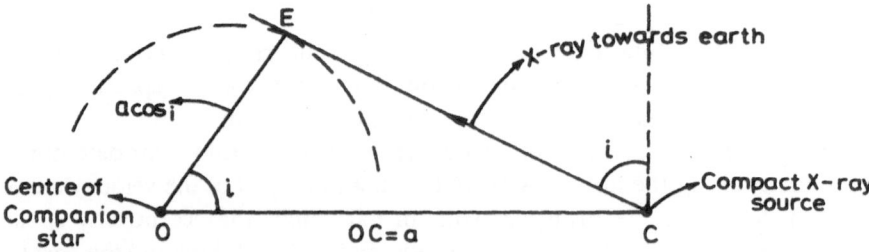

Figure 14.1. No X-ray eclipse requires the companion star not to obstruct the ray CE, i.e., its radius R must be less than OE.

where a is the distance between the centers of the compact object and the companion. The above inequality is equivalent to

$$x \leq (1 + q) \cot i,$$

or

$$\sin^2 i \leq \frac{(1 + q)^2}{x^2 + (1 + q)^2}, \tag{14.13}$$

substituting (14.13) into (14.11)

$$M_x \geq \frac{[x^2 + (1 + q)^2]^{3/2}}{(1 + q)} \times 0.25 \; M_\odot.$$

Taking q as a variable parameter, we find that the minimum value of M_x is

$$(M_x)_{min} = \frac{3\sqrt{3x^2}}{2} \times 0.25 \; M_\odot = 0.85 \; M_\odot D^2,$$

where in the extreme right step we have used (14.12). It is estimated that D is greater than 2 kpc, so that

$$M_x > 3.4 \; M_\odot.$$

This happens to be the most modest lower bound to M_x. We may introduce the additional assumption that the companion star must lie within its Roche lobe (i.e., the first equipotential encompassing both the objects). Such an assumption is justified as, otherwise, gas from the companion star would be drawn out very fast and form a dense column cutting off the X-rays so that we would not observe a steady X-ray emission. This assumption means

$$R \leq a[0.38 + 0.2 \log q],$$

and the analysis then leads to $M_x > 6.5 \; M_\odot$ for $D = 2$ kpc.

Again a careful analysis by Bahcall (1979) has given $M_x = (9\text{–}15) \; M_\odot$. Considering all these we can state with fair confidence that the compact object is neither a white dwarf nor a neutron star but a black hole.

It is true that alternative models have been proposed which consider the compact object to be not a black hole and they cannot be proved straight-away to be wrong, but the general consensus is that these models are highly artificial and do not command confidence unless some additional observational data appear to support them.

14.6. The Possible Locale of the Occurrence of Black Holes

Speculation about the possible origin of black holes have centered round the following:

(a) the gravitational collapse of single stars following the completion of nuclear burning, e.g., Cygnus X-1;

(b) the gravitational collapse of star clusters or very massive stars leading to massive black holes, imagined to be present in globular clusters, active nuclei, and quasars; and

(c) the density fluctuation in the early universe leading to mini black holes.

Globular clusters are spherical regions of radius ranging from 50 to 100 pc and have a large number of stars ranging from a few thousands to even millions, there being an increase in the number density toward the center of the clusters, as in the case of galaxies. The age of the globular clusters in our galaxy is estimated to be $\sim 10^{10}$ yr, so that we may expect that many clusters have been dead by now and quite a number are in the process of dying.

It is therefore likely that black holes are there in the clusters; further, a black hole once formed in such a dense collection of stars would increase its mass enormously by progressively sucking in matter and finally drifting towards the center of the cluster. Thus, we expect that black holes of mass even as high as $\sim 10^8\ M_\odot$ may be present in the centers of globular clusters as in also some galaxies.

The very high energy emission from quasars and active galactic nuclei suggests gravitational energy as the possible source, and models have been worked out in which they contain super massive black holes which have accretion disks around them (Salpeter, 1961; Zel'dovich, 1964).

In the very early universe, as a result of fluctuations, black holes might have been formed. A relation between the mass M of such black holes and the epoch of their formation t was given by Hawking (1971)

$$M \sim 10(t/10^{-4}\ \text{s})\ M_\odot.$$

Such black holes were termed primordial black holes, in short PBH, and their small masses require them to evaporate within the age of the present universe. However, as we have already noted, no such evaporation radiation has as yet been confirmed.

14.7. The Quasi-Stellar Objects (Quasars)

In 1960, few objects were discovered which were characterized by broad emission spectra, high redshifts, and sharp starlike point source appearance.* These objects have been named quasi-stellar objects (QSOs) or, in short, quasars. To date, about four thousand such objects have been discovered and the highest redshift observed so far (mid-1990) is 4.73. If the redshifts of the quasars are cosmological, they are at very great distances (the redshift 4.73 corresponds to a distance of 8.6 billion light years) and their luminosities have to be $\sim 10^{44}$–10^{47} ergs s^{-1}. At first it seemed puzzling to associate such high luminosities with objects which appeared pointlike, and attempts were made to find explanations of the high redshift in causes other than cosmological.

---

* The precise determination of redshifts was done first in 1963.

However, such models have not obtained general acceptance and the present trend of thinking is that the quasars are active galactic nuclei whose luminosity arises from accretion disks formed around supermassive black holes. As we shall see in the next chapter, such luminosities are possible only if the black hole mass is in the range 10^6–$10^9\ M_\odot$.

The observational data on quasars are quite interesting. Many quasars vary in intensity, the period of variation ranging from a fraction of a day to several months. A variation period of 1 day would indicate a linear dimension of the source $\sim 2.6 \times 10^{10}$ km. If the luminosity is indeed due to a stationary accretion disk, then the disk particles must be in orbits around the black hole with a diameter of this order. However, for the Schwarzschild black hole, the minimum stable orbit diameter is $12GM/C^2 \sim 18\ M/M_\odot$ km. Hence we have

$$18\,\frac{M}{M_\odot} \lesssim 2.6 \times 10^{10} \qquad \text{or} \qquad M \lesssim 1.4 \times 10^9\ M_\odot,$$

which is consistent with our estimated mass of the black hole on the basis of luminosity. Note that we have taken the black hole field to be the Schwarzschild field, this seems permissible as the angular momentum of the black hole is presumably small.

Besides some discrete frequencies from which the redshift is determined, the quasars have a continuous emission spectrum as well. The continuous spectrum extends over a wide range and the intensity follows a power law v^{-n} with $0.5 \le n \le 1$ showing that the emission is nonthermal. There are often absorption lines in the quasar spectra and these perhaps arise from the absorption from intervening galaxies or gas clouds.

The study of quasars has led to some observations which have been interpreted in terms of what is called the gravitational lens effect. In the next section we present a short report on this effect.

14.8. Gravitational Lens

We have seen that the bending of light in the gravitational field of the Sun constitutes one of the crucial tests of the general theory of relativity. This bending is, in a way, similar to the refraction of light at the interfaces of two media, and thus the phenomena associated with the effect of gravitational fields on light are broadly termed as gravitational lens phenomena.

However, there are some basic differences between optical lensing and gravitational lensing. In the former, light rays from a point source are brought to a focus (real or virtual) and we say that an image has been formed. In gravitational lensing, the null geodesics which connect the source and the observer are split up in such a manner that the observer receives light from a number (finite or infinite) of different directions and hypothesizes sources (or images) in those directions. Thus the problem in gravitational lensing is basically to find out the null geodesics connecting the source and the

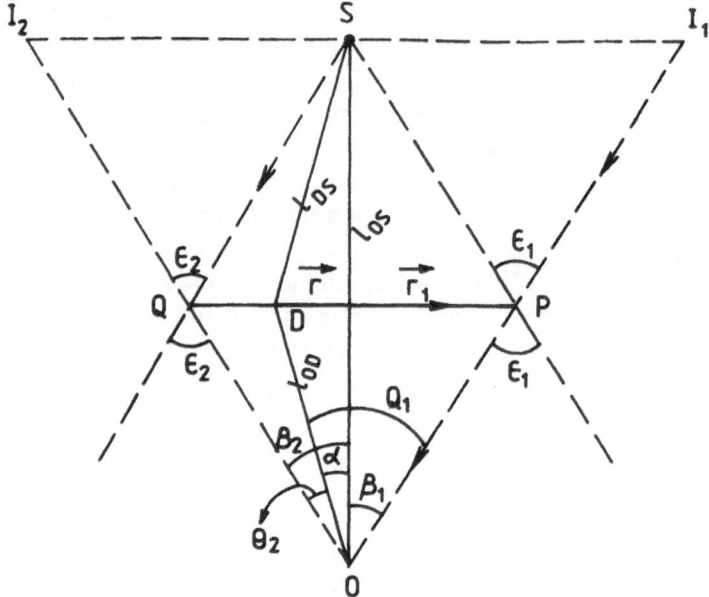

Figure 14.2

observer. In this sense, we may say that the image production is objective in optical lensing and subjective in gravitational lensing.

Consider the simple case of a point source S, a deflector D, and an observer O, as in Fig. 14.2 (see Saslaw et al., 1985). The alignment is such that O receives light from all sides of the deflector, the figure shows the light from the left and right of SO. Corresponding to the two rays SPO and SQO, the observer sees two images in the direction OI_1 and OI_2. Assuming the deflections are small, we have

$$SI_1 = \varepsilon_1 l_{DS} = \beta_1 l_{OS}, \tag{14.14}$$

$$\theta_1 = \frac{r_1}{l_{OD}}. \tag{14.15}$$

Again as r_1 is the impact parameter (see (4.38))

$$\varepsilon_1 = \frac{4GM}{c^2 r_1}, \tag{14.16}$$

where we have assumed the deflector to be a point mass with a Schwarzschild field. From (14.14)–(14.16) we get

$$\beta_1 \theta_1 = \frac{\mu}{l_{OD}^2} = \theta_0^2 \quad \text{(say)}, \tag{14.17}$$

where

$$\mu = \frac{4GM}{c^2} \frac{l_{OD} l_{DS}}{l_{OS}}. \tag{14.18}$$

Note that θ_0 does not involve the index 1 and hence we shall also have

$$\beta_2 \theta_2 = \theta_0^2,$$

or   (14.19)

$$\beta_1 \theta_1 = \beta_2 \theta_2.$$

But from Fig. 14.2

$$\theta_1 + \theta_2 = \beta_1 + \beta_2.$$ (14.20)

Hence from (14.19) and (14.20), $\beta_1 = \theta_2$, $\beta_2 = \theta_1$, and so $\theta_1 \theta_2 = \theta_0^2$. The angle θ_0 has been called the cone of inversion (Liebes, 1964). For the exceptional case of perfect alignment (i.e., SDO lying in a straight line) $\theta_1 = \theta_2$, and the observer will see a ring of angular radius θ_0. Such a ring is sometimes referred to as the Einstein ring, as this result was first noted by Einstein (1936).

It may appear that the Einstein ring may act as the detecton of black holes and also serve to determine the black hole mass. However, the exact alignment required for the Einstein ring is highly improbable and so far no such ring has been observed. If the alignment is only slightly broken there should be two crescents of different angular radii facing each other. In the case of the crucial test described in Chapter 4, the Sun is an obstacle of finite size such that the ray on only one side is received by the observer on the Earth and only a deflected source is seen.

What resembles part of an Einstein ring was observed by Soucail et al. (1987). It was a narrow strip of light in the form of an arc extending over $\gtrsim 20''$ in the galactic cluster Abell 370. The arc has a complicated structure, the longest arc has a measured redshift of 0.724 and smaller arcs have redshifts of $\gtrsim 1.2$, whereas the redshift of the Abell cluster is much less, namely, 0.374. Thus, following a suggestion by Paczynski (1987), the arc has been interpreted as the image of more distant galaxies by the Abell cluster (Grossman and Narayan, 1989; for an earlier attempt see Narasimha and Chitre, 1988).

The first observations regarding gravitational lensing came from quasars. Walsh et al. (1979) discovered two quasars with an angular separation of 6.1 arc seconds and identical redshift $z = 1.41$, and the spectra of the two quasars were also identical. While we cannot rule out that there are two exactly similar quasars very close together, such a situation appears highly improbable, and the generally accepted idea is that they are just images of a single quasar formed by gravitational lensing. Support for this idea came from the discovery of a dim but giant elliptical galaxy with a redshift of 0.39 lying slightly off center from a line between the two images. We could attribute the lensing action to this galaxy.

There are several other cases of multiple quasar observations and we give in Table 14.1, following Chitre (1988), the number of images and angular separations in some cases. It may be noted that with one exception, in all the other cases noted in Table 14.1 the number of images is even. However, a theorem due to Burke (1981) states that the number of images formed by a transparent gravitational lens should be odd. Thus it has been suggested that

Table 14.1

Object	z	Number of images	Angular separation in arc seconds
0957 + 561	1.41	2	6.1
1115 + 080	1.72	4	2.3
2016 + 112	3.27	3	3.4
2345 + 007	2.15	2	7.3
2237 + 031	1.70	2	1.2
1635 + 267	1.96	2	3.8
0023 + 171	0.95	2	5.9

there may be some unobserved images of low luminosity. However, Burke's theorem is only true for a transparent deflector, i.e., it must be such that all the rays hitting a plane near the deflector also hit a plane near the receiver. It may be that this transparency condition is not satisfied in many cases.

Quite a novel type of situation may arise in the presence of cosmic strings. Cosmic strings are almost one dimensional and may be in the form of closed loops or endless lines. The simplest idealized cosmic string is straight, infinitely long, and of negligible cross section. The energy–momentum tensor for the string is of the δ function form, the mass per unit length is exactly equal to the tension, and other components of the energy tensor vanish. Under these circumstances the geometry outside the string is given by

$$ds^2 = dt^2 - dz^2 - dr^2 - r^2(1 - 4\mu)^2 \, d\phi^2, \tag{14.21}$$

where μ is the mass per unit length of the string and $0 \le \phi \le 2\pi$. The string lies along the z-axis. If we make the transformation $\phi' = (1 - 4\mu)\phi$, the line element (14.21) goes over to the Minkowski form

$$ds^2 = dt^2 - dz^2 - dr^2 - r^2 \, d\phi'^2. \tag{14.22}$$

However, the domain of ϕ' is $0 \le \phi' \le 2\pi(1 - 4\mu)$. Consider the 2-space spanned by the r and ϕ coordinates. Following Gott (1985) we may represent it by the surface of a cone of semiangle. $\alpha = \sin^{-1}(1 - 4\mu)$ (Fig. 14.3).

However, in the transformed coordinates r and ϕ', the cone is transformed to a plane with a wedge of semiangle $4\pi\mu$ (Fig. 14.4). Note that the points

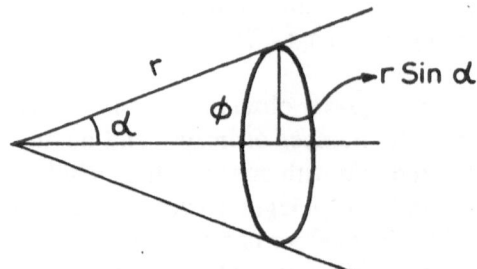

Figure 14.3. Note that on the surface of the cone $dl^2 = dr^2 + r^2 \sin^2 \alpha \, d\phi^2$, $\sin \alpha = 1 - 4\mu$.

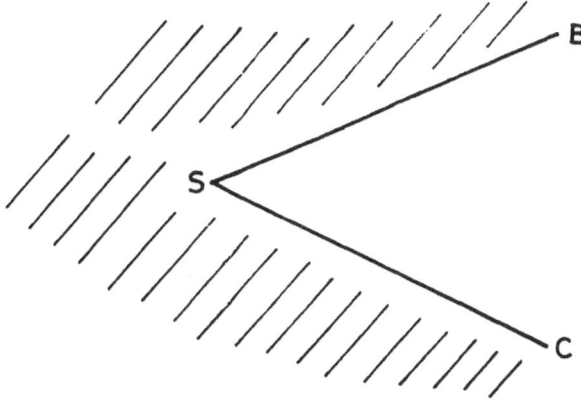

Figure 14.4

$\phi = 0$ and $\phi = 2\pi$ are identical on the cone, but are mapped to distinct points on the plane. Thus points on SB and SC are to be identified while there is no region in the cone corresponding to the unshaded region BSC. Suppose now that a source Q and observer O are on opposite sides of S. Their angular separation is π; so that the difference of their ϕ' coordinates is $\pi (1 - 4\mu)$.

Figure 14.5 represents the position of the source Q and the observer O, both O_1 and O_2 correspond to O. Hence two distinct rays QO_1 and QO_2 will now go from Q to the observer O. Figure 14.5 is the situation from the point of view of Q. The same phenomenon as seen from the observer O is depicted in Fig. 14.6. Observer O sees two images in the directions OQ_1 and OQ_2. From the triangle OSQ_1, assuming that the angle $4\pi\mu$ is small, we have, for the angular separation ψ of the images,

$$\frac{\psi/2}{l} = \frac{4\pi\mu - \psi/2}{d} = \frac{4\pi\mu}{l + d},$$

or

$$\psi = 8\pi\mu l(l + d)^{-1}. \tag{14.23}$$

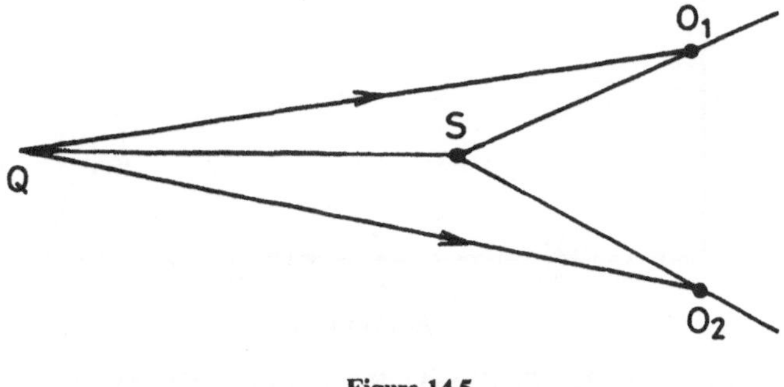

Figure 14.5

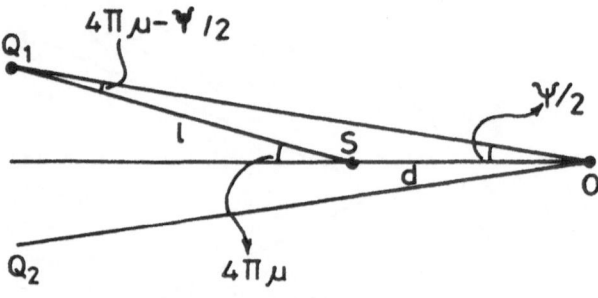

Figure 14.6

In the above, l is the distance of the source from the string and d is that of the observer from the string. The formula requires modification when the plane of observation is not orthogonal to the string and also when there is a transverse velocity of the string. Thus, the general formula for the angular separation between the image is

$$\psi = 8\pi\mu \cdot l(l + d)^{-1} \sin \theta \gamma^{-1}(1 - \mathbf{n} \cdot \mathbf{v})^{-1},$$

where $\mathbf{v}$ is the transverse velocity of the string (in units, $c = 1$), $\gamma = (1 - v^2)^{-1/2}$,

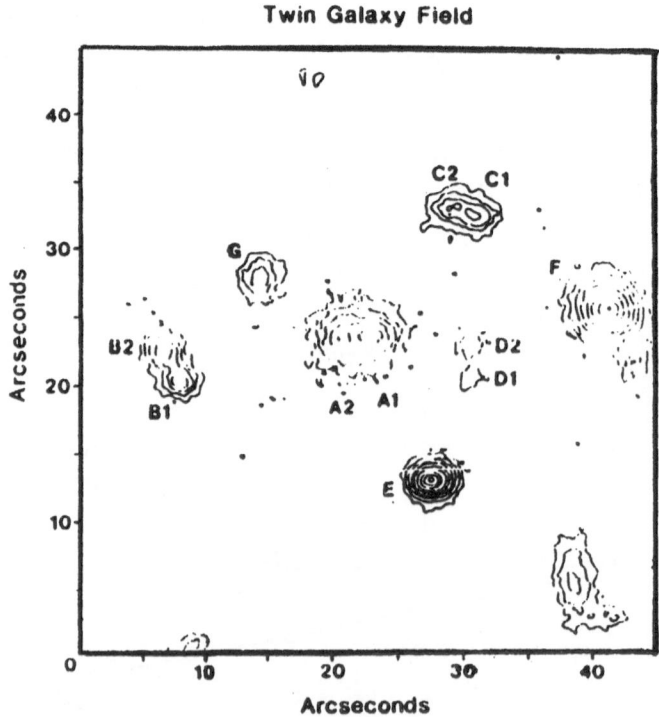

Figure 14.7 (After L.L. Cowie and E.H. Hu, *Astrophys. J*, **318**, L33, 1987).

and **n** is the unit vector along the line of sight with $\mu \sim 10^{-6}$, the expected separation ψ from (14.23) is of the order of a few seconds of arc.

Figure 14.7 shows four galaxy pairs A, B, C, and D within a small region of the sky (40 arc seconds on each side). The observation was made by Cowie and Hu (1987) who found that each pair has a separation of about 2 arc seconds, and the members of each pair have the same redshift but different pairs differ in their redshifts. All this is consistent with the idea that the pairings are due to the action of a cosmic string lying between us and the four single galaxies at different redshifts. However, the cosmic string is a child of GUT and so the average astronomer feels some reluctance in accepting this explanation in the absence of further corroborative evidence.

When observations reveal what is apparently a lensing phenomenon, attempts are made to build up theoretical models to reproduce the observed characteristics. However, there are a large number of parameters to be taken into consideration, involving the source and the lensing system. In fact, the lensing system may even consist of a number of deflecting objects at different redshifts, the so-called thick lens of Penrose. In view of all this, the theoretical models are by no means uniqe and, consequently, conclusions drawn from any particular model remain doubtful. Nevertheless, we may hope to have an idea of the amount and distribution of dark matter, which although directly not observable, would take part in the deflection light. It has also been suggested that in the case of sources of varying intensity, the lens phenomenon may lead to an alternative (and may be a more correct) determination of the Hubble constant. It may also be that some cases of anomalous redshifts will find an explanation as lensing phenomena.

15. Accretion onto Compact Objects

15.1. Introduction—Spherically Symmetric Accretion

The accretion of gas onto compact objects plays an important part in the emission of X-rays. We give an elementary discussion of some aspects of accretion as a preliminary to the account of X-ray observations and their interpretation.

The simplest case of accretion occurs when a compact object, like a neutron star or a black hole, is in an atmosphere of accreting gas which extends a large distance from the object. To simplify matters we introduce the following assumptions:

(a) The situation can be considered stationary and spherically symmetric. This means, on the one hand, that there is no other massive object in the neighborhood and that in the accreting atmosphere all the physical variables, e.g., pressure, density, and fluid velocity, are independent of time.

(b) The influence of the accreting atmosphere on the gravitational field is negligible and the increase in the mass of the black hole or the neutron star due to accretion may be neglected in comparison to its mass, i.e., $\dot{m} \cdot \Delta t \ll m$ where Δt is some characteristic time for the accretion considered. In view of this assumption, the field in which the accreting fluid moves is always given by the Schwarzschild metric.

(c) The velocity of the fluid at large distances from the object tends to vanish.

(d) The field is regular everywhere.

(e) The flow is adiabatic so that we have an equation of state $p = f(\rho)$ connecting the pressure and energy density of the accreting fluid. In particular, we assume a polytropic equation of state.

The basic equations of our problems are:

(i) Energy–momentum conservation for the accreting gas

$$T^{\mu\nu}_{;\nu} = 0, \tag{15.1}$$

with

$$T^{\mu\nu} = (p + \rho)v^{\mu}v^{\nu} - pg^{\mu\nu}.$$

Note that as we are not considering the accreting fluid as a source of the Einstein gravitational field (15.1) does not follow from the field equation, but is an independent assumption corresponding to Euler's equations of hydrodynamics and the equation of continuity. However, the use of the covariant derivative and the Schwarzschild metric takes care of the gravitational influence of the compact object at the center. The units we adopt are such as to make $c = G = 1$.

(ii) The baryon number conservation

$$(nv^{\mu})_{;\mu} = 0, \tag{15.2}$$

n being the baryon number density and v^{μ} is the velocity vector of the accreting gas. From (15.1) and (15.2) we obtain, with the Schwarzschild metric,

$$\frac{\rho'}{p + \rho} = \frac{-(Ur^2)'}{Ur^2}, \tag{15.3}$$

$$UU' + \frac{m}{r^2} = -\frac{p'}{p + \rho}\left(1 - \frac{2m}{r} + U^2\right), \tag{15.4}$$

$$\frac{n'}{n} = -\frac{(Ur^2)'}{Ur^2}. \tag{15.5}$$

From (15.3) and (15.5) we get

$$\frac{d\rho}{dn} = \frac{(p + \rho)}{n}, \tag{15.6}$$

where m is the mass of the black hole and we have written U for v^r and the primes denote differentiation with respect to r. Writing a for the velocity of sound $(dp/d\rho)^{1/2}$, (15.3) and (15.4) are two simultaneous equations for $\rho'/(p + \rho)$ and U'/U. Separating them we get

$$\frac{U'}{U}\left[U^2 - a^2\left(1 - \frac{2m}{r} + U^2\right)\right] + \frac{m}{r^2} - \frac{2a^2}{r}\left(1 - \frac{2m}{r} + U^2\right) = 0, \tag{15.7}$$

$$\frac{\rho'}{p + \rho}\left[U^2 - a^2\left(1 - \frac{2m}{r} + U^2\right)\right] + \frac{2U^2}{r} - \frac{m}{r^2} = 0. \tag{15.8}$$

Equation (15.5) gives, on integration,

$$4\pi\mu n U r^2 = \text{constant} = \dot{m}, \tag{15.9}$$

where μ is the baryonic rest mass and $\dot{m}$ is identified with the accretion rate (in reality, it is the rest mass of the baryons which accrete to the compact object per unit time and is therefore somewhat less than the actual mass accreting). Again from (15.6) we have

$$\frac{p'}{p + \rho} = \left[\ln\left(\frac{p + \rho}{n}\right)\right]'.$$

Using this in (15.4), we get the integral

$$\left(\frac{p + \rho}{n}\right)^2\left(1 - \frac{2m}{r} + U^2\right) = \text{constant}. \tag{15.10}$$

Let us look back to (15.7) and (15.8). The expression

$$\left[U^2 - a^2\left(1 - \frac{2m}{r} + U^2 \right) \right]$$

tends to $-a^2$ for $r \to \infty$ because of assumption (c), while for $r = 2m$ it is equal to $U^2(1 - a^2)$ which would be positive as the sound velocity a cannot exceed the light velocity unity. Hence the expression has a zero for some value of $r > 2m$. Condition (d) requires that at this r (say r_c) the following relations must hold (see (15.7) and (15.8)):

$$U_c^2 - a_c^2\left(1 - \frac{2m}{r_c} + U_c^2 \right) = 0, \qquad (15.11a)$$

$$\frac{m}{r_0} - 2a_c^2\left(1 - \frac{2m}{r_0} + U_c^2 \right) = 0, \qquad (15.11b)$$

$$2U_c^2 - \frac{m}{r_0} = 0. \qquad (15.11c)$$

Obviously, only two of the three relations (15.11a–c) are independent. Eliminating m, we get

$$U_c^2 = \frac{a_c^2}{(1 + 3a_c^2)}. \qquad (15.11d)$$

In view of (15.9) we may have the impression that the accretion will be determined by two arbitrary parameters, the values of n and Ur^2 as $r \to \infty$. However, the constraints (15.11) do not allow us to choose them independently. To see this, suppose the gas obeys the equation of state

$$p = Kn^\Gamma, \qquad (15.12)$$

where K and Γ are constants with $\Gamma > 1$. Equation (15.6) then gives

$$\frac{d\rho}{dn} - \frac{\rho}{n} = Kn^{(\Gamma-1)},$$

which on integration yields

$$\rho = n\mu + \frac{Kn^\Gamma}{\Gamma - 1}, \qquad (15.13)$$

where we have fixed up the integration constant by the condition that, for $n \to 0$, the energy density equals the rest mass density of baryons.

The velocity of sound a can now be expressed in terms of n

$$a^2 = \left(\frac{dp}{d\rho} \right) = \left(\frac{dp}{dn} \right) \Big/ \left(\frac{d\rho}{dn} \right) = (\Gamma - 1) - \frac{n\mu(\Gamma - 1)}{p + \rho},$$

or

$$\left(\frac{p + \rho}{n\mu} \right) = \left(1 - \frac{a^2}{\Gamma - 1} \right)^{-1}. \qquad (15.14)$$

Using the above relation, (15.10) becomes

$$\left(1 - \frac{a_\infty^2}{\Gamma - 1}\right)^2 \left(1 - \frac{2m}{r} + U^2\right) = \left(1 - \frac{a^2}{\Gamma - 1}\right)^2. \tag{15.15}$$

Here a_∞ indicates the sound velocity at $r \to \infty$ and we have used the assumption $U \to 0$ as $r \to \infty$. Equation (15.15) along with (15.11) provides three independent relations involving the unknowns u_c, r_c, and a_c. We can solve these equations. The exact relation is

$$\left(1 - \frac{a_\infty^2}{\Gamma - 1}\right) = \left(1 - \frac{a_c^2}{\Gamma - 1}\right)(1 + 3a_c^2).$$

In the case $a_c \ll 1$, we get

$$a_c^2 \approx \frac{2a_\infty^2}{5 - 3\Gamma} \qquad \text{for} \quad \Gamma < \tfrac{5}{3},$$

$$\approx \frac{2a_\infty}{3} \qquad \text{for} \quad \Gamma = \tfrac{5}{3}, \tag{15.16}$$

$$U_c \approx a_c, \tag{15.17}$$

$$r_c \approx \frac{m}{2a_c^2} \approx \left(\frac{5 - 3\Gamma}{4a_\infty^2}\right)m \qquad \text{for} \quad \Gamma < \tfrac{5}{3},$$

$$\approx \frac{3m}{4a_\infty} \qquad \text{for} \quad \Gamma = \tfrac{5}{3}. \tag{15.18}$$

The value of $\dot{m}$ can now be evaluated from (15.9) in terms of a_∞ if n_c is expressed in terms of a_∞. This can be done by using (15.12), (15.13), and (15.14)

$$1 + \left(\frac{K\Gamma}{\Gamma - 1}\right)\frac{n^{\Gamma - 1}}{\mu} = \left(1 - \frac{a^2}{\Gamma - 1}\right)^{-1} \approx 1 + \frac{a^2}{\Gamma - 1}, \tag{15.19}$$

so that

$$\frac{n_c}{n_\infty} = \left(\frac{a_c}{a_\infty}\right)^{2/\Gamma - 1} \tag{15.20}$$

Hence from (15.9), (15.16), (15.17), (15.18), and (15.20)

$$\dot{m} = 4\pi\mu m^2 n_\infty a_\infty^{-3}\lambda, \tag{15.21}$$

which is formally correct for $\Gamma \le \tfrac{5}{3}$ but λ has the values

$$\lambda = 2^{(9 - 7\Gamma)/2(\Gamma - 1)} \cdot (5 - 3\Gamma)^{(3\Gamma - 5)/2(\Gamma - 1)} \qquad \text{for} \quad \Gamma < \tfrac{5}{3},$$

$$= \tfrac{1}{4} \qquad \text{for} \quad \Gamma = \tfrac{5}{3}. \tag{15.22}$$

For numerical computation we have to introduce G and c, and then (15.21) gives, for $\Gamma = \frac{5}{3}$,

$$\dot{m} \sim 10^{-15} \, (m/M_\odot)^2 \left(\frac{\rho_\infty}{10^{-24} \text{ g cm}^{-3}} \right) \left(\frac{v_\infty}{10^6 \text{ cm s}^{-1}} \right)^{-3} M_\odot \text{ yr}^{-1},$$

(15.23)

where the velocity of sound at infinity v_∞ is in cm s^{-1} and $\rho_\infty = \mu \cdot n_\infty$ is in g cm^{-3}.

To find the radiation emitted by the hot accreting gas, as a result of electron–ion and electron–electron bremsstrahlung, we have to calculate the temperature and the density near the horizon. We first give a model calculation assuming the perfect gas equation and then indicate how the results are modified in more realistic situations.

Neglecting a_∞^2 (which is, say, $\sim 10^{-8}$) in (15.15) we get at the horizon

$$U_h^2 \left(1 - \frac{a_h^2}{\Gamma - 1} \right)^{-2} = 1,$$

where the subscript h indicates values at the horizon. Now, using (15.19), we have

$$U_h^2 \left[1 - \frac{a_\infty^2}{\Gamma - 1} \left(\frac{n_h}{n_\infty} \right)^{\Gamma - 1} \right]^{-2} = 1.$$

(15.24)

Again from (15.9) and (15.21)

$$\frac{n_h}{n_\infty} = \frac{a_\infty^{-3} \lambda}{4 U_h},$$

(15.25)

so that (15.24) and (15.25) together give

$$U_h^2 \left[1 - \frac{1}{(\Gamma - 1)} \left(\frac{\lambda}{4 U_h} \right)^{\Gamma - 1} a_\infty^{5 - 3\Gamma} \right]^{-2} = 1.$$

We thus have

$$U_h \approx 1 \qquad \text{for} \quad 1 < \Gamma < \tfrac{5}{3},$$

and

$$U_h^2 \left[1 - \frac{3}{2^{11/3} U_h^{2/3}} \right]^{-2} = 1 \qquad \text{for} \quad \Gamma = \tfrac{5}{3},$$

or

$$U_h + \frac{3 U_h^{1/3}}{2^{11/3}} \approx 1 \quad \rightarrow \quad U_h \approx 0.78.$$

(15.26)

Thus the velocity U_h is independent of the mass of the black hole or the conditions at infinity.

We introduce the temperature T by the perfect gas equation

$$p = nkT.$$

(15.27)

Also, as $kT_\infty \ll \mu c^2$,

$$a_\infty^2 = \frac{kT_\infty}{\mu}. \tag{15.28}$$

From (15.27) and (15.12)

$$kT = Kn^{\Gamma-1}.$$

Using the relations (15.9) and (15.28) in (15.25), we finally have

$$\frac{T_h}{T_\infty} = \left[\frac{\lambda}{4u_h} \left(\frac{kT_\infty \Gamma}{\mu} \right)^{-3/2} \right]^{(\Gamma-1)},$$

or, with $\Gamma = \frac{5}{3}$, T_h becomes independent of T_∞ and is given by

$$T_h = \left(\frac{1}{16 \times 0.78} \right)^{2/3} \frac{3\mu c^2}{5k} \approx 1.2 \times 10^{12} \text{ K}, \tag{15.29}$$

where we have used for μ, the mass of the hydrogen atom. However, at such high temperatures, hydrogen would be completely ionized and the electrons would become relativistic. This makes the use of (15.27) and (15.12) inappropriate in the present case. Taking these into consideration, Γ is no longer a constant and the temperature at the horizon comes out somewhat lower, namely $T_h \sim 8 \times 10^{10}$ K, practically independent of T_∞ (provided T_∞ is not so high as to make the electrons relativistic, i.e., $kT_\infty \ll m_e c^2$). The material density at the horizon varies roughly as T_∞^{-1} and we have

$$\frac{\rho_h}{\rho_\infty} = 3.88 \times 10^{11} \left(\frac{T_\infty}{10^4 \text{ K}} \right)^{-1}.$$

The accreting material is an electrically neutral plasma of electrons and protons (Brinkman, 1980).

The emissivities in the extreme relativistic case are given by the formulas

$$\varepsilon_{ep} = 12\alpha r_0^2 n_e n_p c K T \left[\frac{3}{2} + \ln \left(\frac{2KT}{m_e c^2} \right) - \gamma \right],$$

$$\varepsilon_{ee} = 24\alpha r_0^2 n_e^2 c K T \left[\frac{5}{4} + \ln \left(\frac{2KT}{m_e c^2} \right) - \gamma \right],$$

(for electron–proton and electron–electron bremsstrahlung, respectively), where α is the fine structure constant $= e^2/c_\hbar = 1/137$, $r_0 = e^2/m_e c^2$, n_e and n_p are the concentrations of electrons and protons, and $\gamma = 0.577$ (Maxon, 1972); the formulas can be obtained fairly easily by using the formulas given in Landau and Lifshitz (1982) or in Jackson (1975). If we now take the accreting plasma to be optically thin, the luminosity comes out as

$$L \approx 10^{21} \left(\frac{n_\infty}{l e m^{-3}} \right)^2 \left(\frac{T_\infty}{10^4 \text{ K}} \right)^{-2} \left(\frac{M}{M_\odot} \right)^{+3} \text{ erg s}^{-1}, \tag{15.30}$$

which gives a very low efficiency for radiation except for massive black holes, i.e., for

$$M \sim 10^9 \, M_\odot.$$

So far we have considered only the integrated luminosity. So far as the spectral distribution is concerned, the dependence of emissivity at a frequency v involves the exponential function $e^{-hv/kT}$. Thus the intensity falls sharply for $hv > kT_h$, which in our case is in the ~ 10 MeV region. We can thus expect considerable emission of even hard X-rays and γ-rays.

The above calculation considers the accreting matter to be a fluid; however, suppose a black hole as within a cluster of stars. The accreting stars would then constitute a system of collisionless particles, and the rate of accretion is then somewhat larger as the pressure gradient in the case of accreting fluids opposes the inward motion. The other assumption that we have used is that the system is optically thin, i.e., the radiation passes out without any significant absorption in the accreting atmosphere and that the whole process is adiabatic. This implies that the energy less in the form of radiation is small compared to the heat developed in the compression. Both these assumptions may be justified for stellar mass black holes. In case the accretion is onto a neutron star rather than onto a black hole, the motion of the accreting matter stops at the hard surface of the star and almost the entire kinetic energy is converted into radiation. Thus the luminosity would be enhanced.

Theoretically, accretion is opposed by the force due to outward flux of the radiation. We may consider the force to arise from the Thomson scattering by the electrons, and then communicated to the protons via electrostatic interaction. This consideration leads to a limit to the luminosity known as the Eddington limit. The average radially outward force per electron due to radiation flux may be written as $\sigma L/4\pi r^2 c$ where L is the luminosity and σ is the electron scattering cross section for Thompson scattering ($\sigma = 0.66 \times 10^{-24}$ cm^2). The expression can be arrived at in two slightly different ways, we may take the result from the classical electromagnetic theory that the directed flux L gives rise to a radial force of magnitude $L/4\pi r^2 c$ per unit area, and then take the Thomson scattering cross section as the effective classical area of the electron. Alternatively, we may note that in each scattering, the electron gains some momentum from the radiation and then averages out for all angles of scattering. Now the limit to the luminosity and accretion is determined by a balance between this radial outward force and the attraction due to gravitation on a hydrogen atom (our picture of the accreting matter is a proton–electron plasma with overall electrical neutrality). This gives

$$\frac{L\sigma}{4\pi r^2 c} = \frac{GMm_H}{r^2} \quad \text{or} \quad L_{lim} = \frac{4\pi GMm_H c}{\sigma},$$

where L_{lim}, the Eddington limit, comes out as

$$L_{lim} = 1.3 \times 10^{38} \, (M/M) \text{ erg s}^{-1}.$$

Indeed, in general, there seems to be a cut off in the observed luminosities at the values given above.

15.2. Disk Accretion

Spherically symmetric accretion occurs only if an isolated compact object with small rotation is in a gaseous atmosphere. A more complicated type of accretion takes place in a binary system. The members of a binary would both be revolving about their center of mass and the gas from the noncompact star would be drawn out. However, this gas would not fall radially towards the compact object but owing to its angular momentum would describe a curved path. Depending on the impact parameter, the particles of gas may either be absorbed by the compact object or fall back into the mother star or escape into interstellar space. The gas which remains bound to the compact object ultimately forms an accretion disk whose thickness, perpendicular to the plane of the binary, is small compared to the radial dimension. Any particular element is almost in a steady circular orbit about the compact object, but owing to the viscosity between different layers of the accretion disk the particles lose their energy and slowly spiral towards the compact object and ultimately accrete. For a simplified discussion we assume that steady state conditions obtain and make a nonrelativistic analysis.

The calculation of accretion and the consequent energy generation now proceed as follows:

(i) The gas particles in the accretion disk are in nearly circular orbits, hence their orbital velocity is given by

$$v^\phi = \left(\frac{GM_A}{r}\right)^{1/2}, \tag{15.31}$$

or the angular velocity Ω is

$$\Omega = \left(\frac{GM_A}{r^3}\right)^{1/2}. \tag{10.32}$$

(ii) The conservation of mass gives, for the accretion rate,

$$\dot{M} = 4\pi rhv^r\rho = 2\pi r\Sigma v^r = \text{constant}, \tag{15.33}$$

where v^r is the radial velocity, $2h$ is the thickness of the disk, and ρ is the density at r. Thus $\Sigma = 2\rho h$ is the mass per unit area of the disk. The constancy of M is a consequence of the steady state assumption.

(iii) The conservation of angular momentum gives

$$\dot{M}v^\phi r = W(\eta) + \dot{J}, \tag{15.34}$$

where the left-hand side represents the rate of angular momentum loss from the disk due to accretion, $W(\eta)$ is the viscous torque, and $\dot{J}$ represents the rate of change of angular momentum of the accreting object.

The viscous stress T is given by the phenomenological expression

$$T = \eta_{\text{shear}} = \eta r \frac{d\Omega}{dr} = \tfrac{3}{2}\eta \left(\frac{GM_A}{r^3}\right)^{1/2},$$

so that

$$W(\eta) = \text{stress} \times \text{area} \times r,$$
$$= 6\pi\eta h (GM_A r)^{1/2}. \tag{15.35}$$

Regarding $\dot{J}$, we note that it cannot exceed the angular momentum carried by a mass M at the inner edge of the disk $(r = r_i)$ and hence we write

$$\dot{J} = \beta \dot{M} r_i v_i^\phi = \beta \dot{M}(GM_A r_i)^{1/2} \qquad (\beta < 1). \tag{15.36}$$

Hence from (15.34), using (15.31), (15.35), and (15.36)

$$\dot{M}[1 - \beta(r_i/r)^{1/2}] = 6\pi\eta h. \tag{15.37}$$

Equation (15.37) shows that ηh is not a constant throughout the disk.

(iv) The rate of generation of the energy per unit volume due to viscosity is viscous stress × shear which comes out to be $\tfrac{9}{4}\eta(GM_A/r^3)$. Hence the total energy generated in the disk per unit time is

$$\dot{Q} = \int_{r_i}^{\infty} \tfrac{9}{4}\eta \frac{GM_A}{r^3} 2\pi r \, dr \, 2h, \tag{15.38}$$

where the upper limit of integration has been put at infinity as the radius of the outer edge far exceeds the inner edge radius r_i. Finally, using (15.37) to eliminate η in (15.38) and integrating, we get

$$\dot{Q} = \frac{GM_A \dot{M}}{r_i}(\tfrac{3}{2} - \beta). \tag{15.39}$$

Assuming that all the heat generated is radiated out, we may take $\dot{Q}$ to give the total luminosity of the disk. The situation here is completely different from that in spherical accretion, there the process is assumed to be adiabatic, i.e., the heat generated goes to increase the temperature of the gas and only a very small fraction of the energy is radiated out.

It may appear rather strange that although the heat generated is due to viscosity, η does not occur explicitly in (15.38). The reason is that the rate of accretion $\dot{M}$ is itself determined by the viscosity equation (15.37), and thus the viscosity appears implicitly in (15.39).

We think about the spectrum of radiation in the following manner. Assuming that the disk is optically thin, the radiation will come out freely from the surfaces on either side. Hence, radiation coming from any one of the surfaces from the annular region r to $r + dr$ is (see (15.38))

$$\tfrac{9}{8}\eta \frac{GM_A}{r^3} 2\pi r \, dr \, 2h.$$

Introducing the Stefan–Boltzmann σT^4 law for the radiation, we define an effective temperature

$$T_{\text{eff}}^4 = \frac{9}{8\sigma} \eta \frac{GM_A}{r^3} 2h$$

$$= \frac{3}{8\sigma} \frac{GM_A}{r^3} \dot{M}[1 - \beta(r_i/r)^{1/2}]. \tag{15.40}$$

In the last step we have used (15.37). Note that the effective temperature has a maximum at a value of $r = (49/36\beta^2)r_i$. Assuming that each region of the accretion disk radiates as a black body it is clear that, owing to the superposition of black body radiation at different temperatures, the maximum in the energy density frequency curve will be very flat, unlike the peak in the spectrum of a black body at a single definite temperature. However, the exact temperature can be estimated only if we have further information about the accreting system such as the mass of the accretion body, the viscous property, and the optical absorption of the matter in the disk.

It is important to note that efficiency in the case of disk accretion is quite high. Thus, from (15.38), we have

$$\frac{\text{luminosity}}{\text{mass accreted}} = \frac{\dot{Q}}{\dot{M}} = \frac{GM_A}{r_i}(\tfrac{3}{2} - \beta) \geq \frac{GM_A}{2r_i}.$$

For the case of white dwarfs and neutron stars, the inner radius r_i may be closed to their surface radii, and the efficiency comes out to be $\sim 10^{-4}$ for white dwarfs and $\sim 10^{-1}$ for neutron stars and even somewhat higher for black holes. Thus, for neutron stars or black holes, the luminosity $L \sim 10^{-1} \cdot \dot{M}c^2 \sim 10^{37}$ erg s^{-1} for an accretion rate of $\sim 10^{-9}$ $M_\odot$ yr^{-1} ($\sim 10^{17}$ g s^{-1}). This luminosity is about four orders of magnitude higher than that of the Sun.

Also from (15.40), taking $\beta \ll 1$, we have for the temperature at r_i,

$$T_i \approx \left(\frac{3}{8\sigma} \frac{GM_A}{r_i^3} \dot{M}\right)^{1/4} = 1.4 \times 10^7 \text{ K} \left(\frac{3GM_A}{r_i c^2}\right)^{3/4} \left(\frac{M_A}{M_\odot}\right)^{-1/2} \times \left(\frac{\dot{M}}{10^{17} \text{ g s}^{-1}}\right)^{1/4}.$$

Thus with $M_A \approx M_\odot$ and $\dot{M} \approx 10^{17}$ g s^{-1}, the temperature is $\sim 10^7$ K and the peak of the radiation intensity would be in the X-ray region (~ 1 keV).

We have talked about gas ejected from one member of a binary system. We can go a little deeper into the possible mechanics of ejection, broadly they are:

(a) *Stellar wind.* The winds owe their origin to events occurring in the stellar interior (i.e., not due to the presence of the companion compact object). Such winds often have an intensity of $\sim 10^{-6}$ $M_\odot$ yr^{-1} and this can result in an accretion rate of $\sim 10^{-9}$ $M_\odot$ yr^{-1} ($\sim 10^{17}$ g s^{-1}). The compact object acts as an obstacle for the wind resulting in a shock front. Some material flowing through the shock front may be decelerated and captured by the compact object.

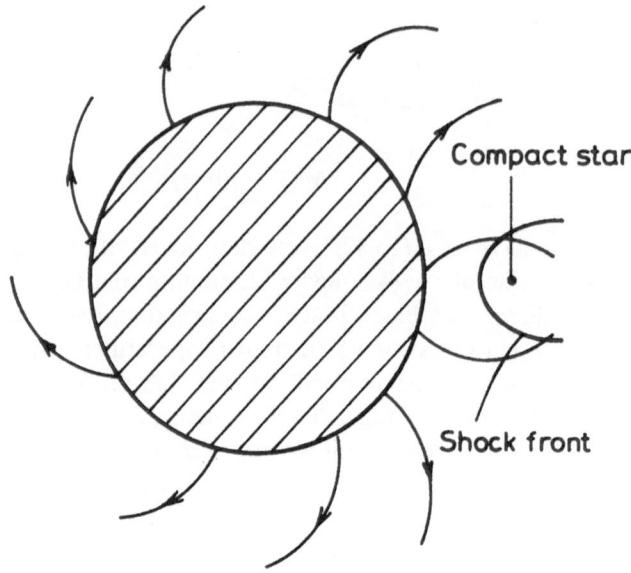

Figure 15.1. Accretion due to stellar winds for ordinary star and shock formation about the compact object.

(b) *Roche lobe overflow.* The normal companion of the binary in this case expands to fill up the space up to the Roche lobe (which is the first equipotential encompassing both members of the binary). The matter can now easily flow through the Lagrange point to form a disk about the compact star. Quite a large mass transfer can occur by this mechanism, but if the density of the accreting gas becomes too large, the gas may shut out the escaping radiation.

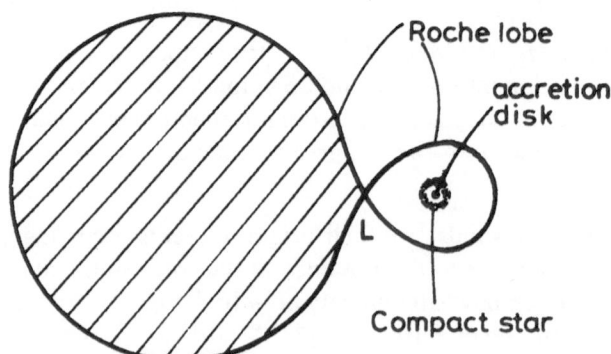

Figure 15.2. Accretion due to Roche lobe overflow through the Lagrange point L. Ordinary star extends up to the Roche lobe.

15.3. Compact X-Ray Sources

The observations on X-rays were the first to indicate that violent events are taking place in the universe and gave rise to what is called high-energy astrophysics. We give, in outline, the observational picture regarding X-rays from compact objects.

Number and Distributions

The fourth Uhuru Catalogue (Forman et al., 1978) listed 339 X-ray sources. Later satellite observations have detected more weak sources. Most of these are galactic, as evident from their situation at low latitude and distances ranging from a few hundred to one thousand parsecs. However, Her X-1 and Sco X-1 have high latitudes, though they are within our galaxy. More distant extragalactic sources are distributed fairly isotropically.

Time Variability

The time variability of X-ray intensity can be divided broadly into three types:

(a) The periodically varying sources known as X-ray pulsars. The period is usually of the order of a second. Unlike ordinary pulsars, which are detected by pulsed radio waves, X-ray pulsars usually show a spin-up, i.e., a decrease in the pulse period with time.

The X-ray pulsars are thought to be binary systems one of which is a spinning neutron star. (Black holes excluded for a periodic variation are not to be expected with them.) In many cases, the companion star is a visible star with a rather large mass ($\sim 10\text{--}20 \ M_\odot$). The theoretical picture envisages that gas from the visible companion is turned toward the polar regions of the neutron star, owing to the interaction between the electric charge of the accreting gas and the intense magnetic field associated with the neutron star. We thus have two hot spots emitting intense X radiation. The nonalignment of the magnetic axis with the rotation axis gives rise to the observed pulsations with a period equal to the rotation period of the neutron star. The spin-up is attributed to the angular momentum imparted by the accreting gas, which more than compensates for the usual loss of angular momentum due to any radio emission from the neutron star. The masses of the neutron stars in the case of a number of binary X-ray pulsars have been estimated, they all have values in the neighborhood of the theoretically probable value of $1.4 \ M_\odot$.

(b) The second type of temporal variation that is observed is a rather random variation in intensity, whose characteristic time scale may be anything from a fraction of a second (~ 1 ms in the case of Cyg X-1 which is thought to be a black hole) to a number of years. These peculiar variations may be due to changes in the accreting rate but the picture is far from clear.

(c) The third type are the so-called "bursters." In these cases there is a sharp rise in the intensity, reaching a peak value within a few seconds and then falling back to the original value in about a minute. Within this burst phase about 10^{39} ergs are emitted in the form of X radiation. The bursters occur mostly near the galactic center or in the globular clusters. In between two successive bursts there can be an interval of anything between a few hours to a few days but there is no regular periodicity whatsoever.

Amongst bursters, the Rapid Burster MXB 1730-335 is peculiar in that it gives bursts of energy at quick succession, so much so that sometimes there is a few thousand bursts in a single day. Sometimes, the more common type of burst is referred to as type I bursts and the rapid burster as type II bursts. The explanation of the bursts is not quite clear but the general belief is that the type I bursts are due to thermonuclear flashes. The accreting gas, primarily consisting of hydrogen and helium may form a dense cloud around a nonmagnetic neutron star. (Apparently, the absence of pulsation may be attributed to this or to an alignment of the rotation and magnetic axis.) If the temperature of the neutron star is sufficiently high, hydrogen can undergo fusion to helium and the helium in its turn may be fused to form carbon. These fusion reactions which are transient phenomena and in which almost all the combustible matter is, for the moment, exhausted gives rise to the X-ray bursts. It has been claimed that many of the observed characteristics of the bursts such as the maximum luminosity and the temporal behavior can be correctly obtained from a calculation based on the thermonuclear flash model. The type II bursts, however, apparently have a different origin and are commonly ascribed to some instability in the accretion disk.

Problems

1. Spherically symmetric accretion of a fluid on the basis of Newtonian physics was investigated by Bondi (1952). Using the adiabatic equation of state $p = k\rho^\Gamma$, the equation of continuity, and Euler's hydrodynamical equations, we obtain the following relations:

$$\frac{\rho'}{\rho} + \frac{u'}{u} + \frac{2}{r} = 0,$$

$$uv' + \frac{a^2\rho'}{\rho} + \frac{GM}{r^2} = 0.$$

2. Solve the above two simultaneous equations for u' and ρ' and show that for a smooth, singularity free flow with $u \to \infty$ as $r \to 0$, we have the constraints that at a certain radial distance r_c

$$u_c^2 = a_c^2 = \frac{1}{2}\frac{GM}{r_c}.$$

Compare this with (15.11).

3. Using the above results, show that the Newtonian rate of accretion is given by

$$\dot{M} = 4\pi\lambda(GM)^2\rho_\infty \cdot a_\infty^{-3},$$

with λ the same as in the general relativity case. How far is this formula identical with the general relativity formula?

4. Show that for a neutral plasma of electrons and protons, in which the electrons are ultrarelativistic while the protons are nonrelativistic, the adiabatic equation of state is $p = kn^{\Gamma}$ with $\Gamma = 13/q$. Use the relations $p = nKT$, $\rho = (n\mu + \varepsilon)$, and $T\,ds = d(\rho v) + p\,dv$.

Part III
Cosmology

16. The Standard Cosmological Model

16.1. Introduction to the Friedmann Metric

We now go to the application of the general theory of relativity to the universe as a whole. The relevance of the general theory of relativity arises from the fact that the only effective interaction between the constituents of the universe is gravitational attraction. However, the cosmological problem puts forward a rather novel situation in the following sense.

Application of the general theory of relativity aims basically to extract information from the field equations and, if possible, to solve these equations and thus obtain the geometry and dynamics of the universe. These would require a knowledge of the sources of the field, i.e., the energy–momentum tensor T_ν^μ as well as some boundary conditions. Unfortunately, we have rather little knowledge of T_ν^μ in the universe, i.e., of the matter and energy distributions, of the stresses, and of the velocities. While it is true that today observations by optical and radio telescopes have given us some data about these items, the distributions seem so irregular and discontinuous at first sight that a mathematically faithful description becomes hopelessly complicated. Nor do the present observations give us a complete picture, as our observations are limited to a light cone. Further, while we have come to learn of a considerable presence of dark matter (i.e., matter which has no apparent electromagnetic interaction), the precise amount of this matter outside the galaxies remains obscure. The situation regarding the boundary condition is also intriguing. In common problems of electrodynamics or gravitation, we usually deal with a bounded system situated in an outside vacuum, and then it is appropriate to assume that at large distances the field vanishes sufficiently rapidly. Obviously, it would be preposterous to assume that all matter and energy of the universe are situated in our neighborhood in an all extending void.

Faced with these difficulties, Einstein in 1917 introduced a novel procedure with some very powerful but, nevertheless, arbitrary assumptions. The lack of any well-defined boundary condition he compensated for by assuming that the universe at large is homogeneous and isotropic. Perhaps behind this assumption were some philosophical considerations. We shall have occasion to examine later the present observational position regarding these assumptions of homogeneity and isotropy, which are often referred to as the cosmological principle. Further, Einstein assumed the system to be static, thus disregarding any large scale temporal change.

It is difficult to see how Einstein could reconcile himself to the idea of a static universe even as an approximation. In those days, we had no idea of galaxies as such and the distribution of stars seemed quite irregular. True, there were no evidence of systematic large scale motions such as would indicate a nonstatic condition, but, in general, there was an excessive emission of radiation compared to absorption of radiation. Hence the partition of energy into matter and radiation was continually changing and as matter and radiation, even of the same density, have different gravitational effects the system should change with time. It cannot be argued that the change is slow enough to be neglected, for in building a model of the universe, we have to consider the entire past and future. True, an evolving universe would either be periodic in some sense or have some sort of a beginning. A periodic universe seems inconsistent with the second law of thermodynamics because a beginning comes into contradiction with our idea of an eternally existing time. Indeed, present-day cosmology is very much concerned with these problems. The steady state idea that was put forward by Bondi, Gold, and Hoyle in 1948 did reconcile change with eternity, but then we had to bring in a beginning in the career of every bit of matter in the form of continuous creation.

Even leaving aside these problems, Einstein's assumptions led to a trivial solution. To see this, write the general static spherically symmetry metric (spherical symmetry is a consequence of isotropy)

$$dS^2 = e^\gamma \, dt^2 - e^\lambda \, dr^2 - r^2(d\theta^2 + \sin^2 \theta \, d\phi^2), \tag{16.1}$$

where γ and λ are functions of r alone. The energy–stress tensor is of the perfect fluid form

$$T_\nu^\mu = (p + \rho)v^\mu v_\nu - p\delta_\nu^\mu,$$

and the field equations give

$$8\pi pG = e^{-\lambda}\left(\frac{\gamma'}{r} + \frac{1}{r^2}\right) - \frac{1}{r^2}, \tag{16.2}$$

$$8\pi\rho G = e^{-\lambda}\left(\frac{\lambda'}{r} - \frac{1}{r^2}\right) + \frac{1}{r^2}, \tag{16.3}$$

$$p' = -(p + \rho)\frac{\gamma'}{2}. \tag{16.4}$$

Homogeneity requires $p' = 0$, so that from (16.4), $\gamma' = 0$ (if $p + \rho \neq 0$). Again from (16.2) and (16.3), because of the constancy of ρ,

$$e^{-\lambda} = 1 + 8\pi Gpr^2, \tag{16.5}$$

$$e^{-\lambda} = 1 - \frac{8\pi G\rho r^2}{3}. \tag{16.6}$$

Equations (16.5) and (16.6) require $p = \rho = 0$ (if negative values of p and ρ are not allowed).

Instead of giving up the static assumption, Einstein chose to change the field equations by adding a divergence-free term

$$R^{\mu}_{\nu} - \tfrac{1}{2}R\delta^{\mu}_{\nu} = -8\pi G T^{\mu}_{\nu} + \Lambda\delta^{\mu}_{\nu}, \tag{16.7}$$

where Λ is a constant, named the cosmological constant. Equations (16.2) and (16.3) are now modified to

$$8\pi pG + \Lambda = e^{-\lambda}\left(\frac{\gamma'}{r} + \frac{1}{r^2}\right) - \frac{1}{r^2}, \tag{16.8}$$

$$8\pi pG - \Lambda = e^{-\lambda}\left(\frac{\lambda'}{r} - \frac{1}{r^2}\right) + \frac{1}{r^2}, \tag{16.9}$$

and the solution given by Einstein is

$$e^{\gamma} = 1, \qquad e^{-\lambda} = (1 - \Lambda r^2), \qquad \rho = \Lambda/4\pi, \qquad p = 0.$$

The cosmological term introduces, in Newtonian language, a repulsive force which effectively balances the gravitational attraction and thus gives an equilibrium.

Note on the Cosmological Term

An additional justification for the cosmological term was advanced by Einstein. While the equations without the Λ-term admitted a solution for a completely empty space, namely, the Minkowski metric of the special theory of relativity, Einstein had the impression that with the Λ-term there would be no solution for empty space, and this nonexistence was considered by Einstein to be consistent with the ideas associated with Mach's principle. We shall not go into a discussion of Mach's principle as it is not very clearly defined nor does it seem to have provided a significant advance to our theoretical understanding. However, contrary to Einstein's conjecture, solutions for empty space exist even with the Λ-term as was shown shortly after by de Sitter (1917). His metric is

$$dS^2 = dt^2\left(1 - \frac{\Lambda r^2}{3}\right) - dr^2\left(1 - \frac{\Lambda r^2}{3}\right)^{-1} - r^2(d\theta^2 + \sin^2\theta\, d\phi^2). \tag{16.10}$$

In this static form, there is a horizon at $r = \sqrt{3/\Lambda}$. However, the metric can be transformed to the form

$$dS^2 = dt^2 - e^{\sqrt{\Lambda/3}\,t}[dr^2 + r^2\, d\theta^2 + r^2\sin^2\theta\, d\phi^2], \tag{16.11}$$

which has a time dependence. At present, the cosmological term is looked upon as arising from a "false vacuum" and plays a crucial role in inflationary models as we shall see later.

A few years later, in 1922, Friedmann showed that there are nonstatic solutions to the original Einstein equations, which are consistent with the assumptions of homogeneity and isotropy but are not static. Quite generally, the metrics are of the form

$$dS^2 = dt^2 - R^2 \left(\frac{dr^2}{1 - kr^2} + r^2 \, d\theta^2 + r^2 \sin^2 \theta \, d\phi^2 \right), \qquad (16.12)$$

where R is a function of t alone and K, called the curvature parameter, may be 0, +1, or −1. The metric can be transformed to the "isotropic" form

$$dS^2 = dt^2 - \frac{R^2}{(1 + k\bar{r}^2/4)^2} \left[d\bar{r}^2 + \bar{r}^2 \, d\theta^2 + \bar{r}^2 \sin^2 \theta \, d\phi^2 \right], \qquad (16.13)$$

where $r = \bar{r}(1 + k\bar{r}^2/4)^{-1}$. In future, we shall use (16.13) without using the overbar over r. Somewhat later on, it was observed by Robertson and Walker that the form of the metric (16.13) could be obtained directly from the ideas of homogeneity and isotropy without taking recourse to the field equations of general relativity. We mention below the theorems and the line of reasoning used by them to obtain the metric of form (16.13).

(1) If a space of n dimensions admits a group of motions with r parameters, where $r \geq (n - 1)$ and the orbits of the group are $(n - 1)$-dimensional hypersurfaces, then the metric of the n-dimensional space can be written in the form

$$dS^2 = \pm dx^{1^2} + g_{ik} \, dx^i \, dx^k,$$

where i and k run over the indices 2, 3, ..., n and the corresponding coordinates span the invariant hypersurfaces (i.e., the Killing vectors are all tangential to these hypersurfaces). The plus or minus sign indicates the possibility of a timelike or spacelike nature of the x^1 coordinate.

(2) A space of m dimensions can at most admit $m(m + 1)/2$ Killing vectors and when it admits this maximal number of Killing vectors its metric can be reduced to the form

$$\frac{\pm dx^{1^2} \pm dx^{2^2} \pm dx^{3^2} \pm \cdots \pm dx^{m^2}}{\left[1 + \frac{k}{4}(\pm x^{1^2} \pm x^{2^2} \pm \cdots \pm x^{m^2}) \right]^2},$$

with $k = 0$, +1, or −1. Such a space is called a space of constant curvature.

(3) Combining the above theorems and recalling that the space of general relativity is $(3 + 1)$-dimensional and that the 3-spaces admit the maximal six-parameter group (three for homogeneity and three of the rotation group for isotropy) we get the standard form (16.13).

(The reader may consult Eisenhart (1933) for detailed proofs of the theorems.) Note that the nonstatic form (16.11) of the de Sitter metric is obtained from (16.13) by taking $k = 0$ and $R = \exp(\sqrt{\Lambda/3})t$. However, the metric in this case admits a ten-parameter group.

16.2. Elementary Discussion of Standard Cosmology

The so-called standard cosmology rests on two assumptions:

(a) the cosmological principle of homogeneity and isotropy leading to the Friedmann metric; and
(b) the field equations of general relativity along with the familiar properties of matter.

We may expect that general relativity would determine the curvature parameter k and the scale function R if the source characteristics are known. That is indeed so, but as the sources are not precisely known we shall simply have a general discussion. The Friedmann metric itself puts severe restriction on the energy–momentum tensor, namely, $T_1^1 = T_2^2 = T_3^3$ and $T_\beta^\alpha = 0$ for $\alpha \neq \beta$. These restrictions are consistent with a perfect fluid energy–momentum tensor with velocity $V^\mu = \delta_0^\mu$. The field equations of general relativity then reduce to (writing $T_0^0 = \rho$ and $T_1^1 = T_2^2 = T_3^3 = -p$)

$$\frac{8\pi\rho G}{3} = \frac{\dot{R}^2}{R^2} + \frac{k}{R^2}, \tag{16.14}$$

$$8\pi p G = -\frac{2\ddot{R}}{R} - \frac{\dot{R}^2}{R^2} - \frac{k}{R^2}. \tag{16.15}$$

Combining (16.14) and (16.15) it is easy to obtain

$$(\rho R^3)^{\cdot} + 3pR^2\dot{R} = 0, \tag{16.16}$$

and

$$\frac{\ddot{R}}{R} = -\frac{4\pi}{3}(\rho + 3p)G. \tag{16.17}$$

Equation (16.16) is the familiar energy conservation relation,

$$dU + p\, dV = 0,$$

as obtains in the case of adiabatic changes. Note that the proper volume element is αR^3. Equation (16.18) shows, in particular, that R cannot be a constant (i.e., there is no static model) nor can R have a minimum, provided $\rho + 3p > 0$. Thus a decreasing R can only end at the singular state $R = 0$. We introduce three definitions:

$$\text{Hubble parameter:} \quad H \equiv \dot{R}/R,$$

$$\text{Deceleration parameter:} \quad q \equiv -(\ddot{R}/R)H^{-2},$$

$$\text{Density parameter:} \quad \Omega \equiv (8\pi\rho G/3)H^{-2}.$$

Note that q and Ω are dimensionless pure numbers.

It is usually assumed that at the present epoch the pressure p may be neglected in comparison with the energy density ρ. With $p = 0$, (16.14) and

(16.15) yield

$$\frac{k}{R^2} = H^2(\Omega - 1) = H^2(2q - 1). \tag{16.18}$$

Thus $k = 0$ (i.e., the space sections will be flat) if $\Omega = 1$ or $q = \frac{1}{2}$. Note that, only in this case, Ω and q are constants independent of time. The corresponding density $\rho = 3H^2/8\pi$ is referred to as the critical density ρ_c. Also from (16.18) we have

$$k = -1 \quad \text{if} \quad \rho < \rho_c \quad \Leftrightarrow \quad \Omega < 1 \quad \Leftrightarrow \quad q < \tfrac{1}{2},$$

and

$$k = +1 \quad \text{if} \quad \rho > \rho_c \quad \Leftrightarrow \quad \Omega > 1 \quad \Leftrightarrow \quad q > \tfrac{1}{2}.$$

It is interesting to consider some geometrical peculiarities for different k. Using the metric (16.13), the magnitude of the linear dimension and the volume of the universe are, respectively, the infinite integrals

$$R \cdot \int_0^\infty \frac{dr}{(1 + kr^2/4)},$$

and

$$R^3 \int_{r=0}^{r\to\infty} \int_{\theta=0}^{\theta=\pi} \int_{\phi=0}^{\phi=2\pi} \frac{r^2 \sin\theta \, d\theta \, d\phi \, dr}{(1 + kr^2/4)^3}.$$

Although we have put the upper limit of r at infinity, in the case $k = -1$, the integral blows up at $r = 2$ and the upper limit is to be taken as 2. Both the above integrals diverge if $k = 0$ or -1 and the universe is then said to be open and infinite. If, however, $k = +1$, the integrals converge and we have a closed and finite universe although without any boundary.

The topological difference of space sections is related to an important difference in the termporal behavior of the cosmological models. To see this, assume simply that both p and ρ are nonnegative. For $k = 0$ or -1, (16.14) gives

$$\frac{\dot{R}^2}{R^2} \geq \frac{8\pi\rho G}{3} \geq 0.$$

Hence $\dot{R}$ has no zero except in the trivial case $\rho = p = 0$, and an increasing scale factor R would go on increasing for ever with $R \to \infty$ as $t \to \infty$. For $k = +1$, however,

$$\dot{R}^2 = \frac{8\pi\rho R^2 G}{3} - 1. \tag{16.19}$$

Again from (16.19)

$$(\rho R^2)' = -R\dot{R}(3p + \rho).$$

Hence with $\dot{R} > 0$, ρR^2 would decrease with time and, from (16.19), $\dot{R}$ will vanish at $R = (3/8\pi\rho)^{1/2}$. In view of (16.17), this corresponds to a maximum

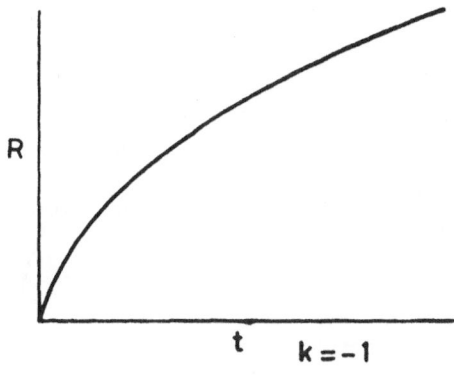

Figure 16.1

of R; R will now go on decreasing and distances go on shrinking, and as we have already noted R then ends at the singular state $R = 0$. However, the equations have time reversal symmetry, so while a decreasing R ends at the singularity, an increasing R must begin at a singularity. The original singularity is referred to as the big bang.

We can also estimate the time interval between the big bang and a state of given $H (= \dot{R}/R)$. As $\ddot{R}$ is always negative, a positive $\dot{R}$ at present was progressively greater in the past and hence R vanished at a time $T \leq H^{-1}$, the equality occurring in the limiting case $\ddot{R} = 0 \Leftrightarrow \rho = p = 0$. We present the behavior of R for different k, the arrows correspond to $\dot{R} > 0$ at present; for the case $\dot{R} < 0$, the arrow will be in the reverse direction (see Figs. 16.1, 16.2, and 16.3).

Without going into details right now, we only note that observations give the present value of H as lying between $1-2 \times 10^{10}$ yr^{-1}. Hence we may say

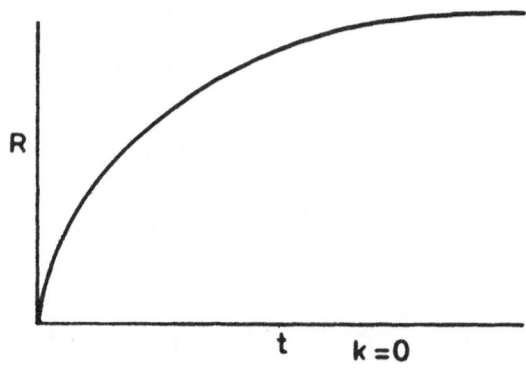

Figure 16.2

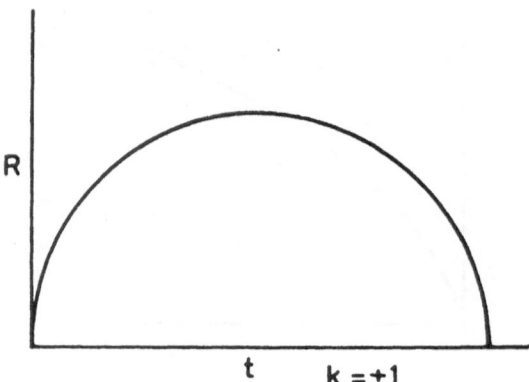

Figure 16.3

that the time that has lapsed since the big bang (called the age of the universe) is $\leq 2 \times 10^{10}$ yr.

Integration of (16.14) and (16.15)
Equations (16.14) and (16.15) are two equations involving three unknowns, ρ, p, and R, and thus form an underdetermined set. It is the usual practice to investigate the equation of state (i.e., the relation between p and ρ) at different stages and then integrate the equations. However, that necessitates computation. Here we consider two extreme conditions:

(a) $p = 0$—this is the situation obtaining at present, the case of an incoherent dust; and

(b) $p = \rho/3$—the equation of state for ultrarelativistic matter, a condition associated with the early hot universe.

With $p = 0$, (16.16) shows $\rho R^3 = \text{constant} = 3a^2/8\pi$ (say). Then (16.14) reads

$$dt = \pm \frac{dR}{(a^2/R - k)^{1/2}}, \tag{16.20}$$

where the positive and negative signs correspond to expanding and contracting situations, respectively.

(i) For $k = 0$ (flat open space sections), the integral of (16.20) is

$$R = (\tfrac{3}{2}at)^{2/3}, \qquad H \equiv \frac{\dot{R}}{R} = \frac{2}{3t}. \tag{16.21}$$

(ii) For $k = +1$ (closed space of positive curvature)

$$R = a^2 \sin^2 \psi, \qquad t = \frac{a^2}{2}[2\psi - \sin 2\psi],$$

$$H = \frac{\cot \psi \, \text{cosec}^2 \psi}{a^2}. \tag{16.22}$$

(iii) For $k = -1$ (open space of negative curvature)

$$R = a^2 \sinh^2 \psi, \qquad t = \frac{a^2}{2}(\sinh 2\psi - 2\psi),$$

$$(16.23)$$

$$H = \frac{1}{a^2} \coth \psi \ \mathrm{cosech}^2 \ \psi.$$

In all three cases, the time variable has been normalized to be zero at $R = 0$, i.e., the time is reckoned from the big bang singularity. The models of type (ii) involve circular functions and thus apparently R shows a cyclic behavior with time. However, it should be noted that the theory provides no way of passing through the singularities at $R = 0$, and hence it is erroneous to say that these are cyclic universe models. For $p = \rho/3$, (16.16) gives $\rho R^4 = $ constant and we get from (16.20)

$$dt = \pm \frac{dR}{(a^2 R^{-2} - k)^{1/2}},$$

where $a^2 = \frac{8}{3}\pi G\rho R^4$. Again we have three cases:

(i) $\qquad\qquad k = 0, \qquad R = (2at)^{1/2}, \qquad\qquad H = (2t)^{-1}, \qquad\qquad (16.24)$

(ii) $\qquad\qquad k = +1, \qquad R = (2at - t^2)^{1/2}, \qquad H = \dfrac{a - t}{t(2a - t)}, \qquad (16.25)$

(iii) $\qquad\qquad k = -1, \qquad R = (2at + t^2)^{1/2}, \qquad H = \dfrac{a + t}{t(2a + t)}. \qquad (16.26)$

Note that for $t \to 0$, in all three cases we get $R = (2at)^{1/2}$, as in that situation the energy density $\rho \ (\sim R^{-4})$ dominates over the curvature $(\sim R^{-2})$.

The Spectral Shift for the Friedmann Metric

So far we have concentrated on the mathematical consequences of the Friedmann metric. We shall now point out a significant observational aspect of the Friedmann metric.

Note first of all, that we have come across the continuous variables ρ and p, whereas the luminous matter is present in the form of discrete stars, galaxies, galactic clusters, etc. The situation is similar to that in fluid mechanics where we also consider a continuous fluid although it consists, in reality, of discrete molecules. There we justify consideration by the argument that the theory applies only at a macroscopic level where the linear dimensions involved are much greater than those of molecules. Here, too, we say that in applying the Friedmann metric we should consider it valid only on a large scale, the linear dimensions being much greater than those of galaxies and galactic clusters. Thus ρ is the density of the smoothed-out matter and the energy distributions and the pressure p originates from the incoherent motion of the galaxies, etc., as well as from any radiation (or, more generally, massless particles) that may be present on a large scale. The condition $V^\mu = \delta_0^\mu$ signifies that the overall distribution is at rest in the coordinate system but for random motions which give rise to pressure.

Consider now the light emitted from a galaxy and received by another galaxy. As the system is completely homogeneous and isotropic, we may consider the origin to be at the receiving galaxy and the coordinates of the emitting galaxy as $r_0, 0, 0$. As the light track is a null geodesic, we have

$$\int_{t_e}^{t_o} \frac{dt}{R} = \int_0^{r_0} \frac{dr}{(1 + kr^2/4)} \quad \text{(independent of time)},$$

where t_e and t_o are the instants of emission and observation, respectively. Hence if Δt_e is the time interval between the emission of two pulses and Δt_o is that between their receptions, we have

$$\frac{\Delta t_e}{R_e} = \frac{\Delta t_o}{R_o},$$

where the Δt's are considered sufficiently small so that the R's are effectively constant over them. Both the emitter and the observer assign the same velocity c to light (recall the special theory of relativity), and hence if λ_e and λ_o represent the wavelengths of light emitted and received, then

$$\frac{\lambda_e}{\lambda_o} = \frac{c\Delta t_e}{c\Delta t_o} = \frac{R_e}{R_o}.$$

Or, finally, for the spectral shift

$$z \equiv \frac{\Delta \lambda}{\lambda_e} = \frac{\lambda_o - \lambda_e}{\lambda_e} = \frac{R_o - R_e}{R_e} = \left(\frac{\dot{R}}{R}\right)_0 (t_o - t_e).$$

In the last step we have assumed that the time of travel $(t_o - t_e)$ is small enough to justify the neglect of the higher powers of $(t_o - t_e)$. Of course, this step will not be justified if $\dot{R}$ happens to vanish at $t_o \approx t_e$.

For small $t_o - t_e$, we can write $(t_o - t_e) = D/c$ where D is the distance between the emitter and observer, and thus we have

$$z = \left(\frac{\dot{R}}{R}\right)\frac{D}{C} = H_0 \frac{D}{C}. \tag{16.27}$$

Note that z will be positive (or negative) according to whether $\dot{R}$ is positive (or negative). Hence an apparent increase (or decrease) in wavelength corresponds to a temporarily increasing (or decreasing) R. As observations show a systematic redshift of light from galaxies, R is an increasing function of time signifying an increasing distance between the galaxies. This mutual recession of galaxies and the expansion of any region of space determined by a family of galaxies is referred to as the expansion of the universe. We may now use the usual formula for the Doppler shift, valid for relative velocities $v \ll c$,

$$\frac{\delta \lambda}{\lambda} = \frac{v}{c}.$$

Comparing with (16.27)

$$v = H_0 D.$$

This linear relation between the "velocity" and "distance" was discovered by Hubble from his observations and is known as Hubble's law. However, the reader should note that the words velocity and distance are not to be naively understood. We have already noted that H is called the Hubble parameter and its present value is $\sim (10^{10} \text{ yr})^{-1}$ up to an uncertainty factor of 2.

The Relation Between z and the Time of Emission

It is clear that z will increase with distance and hence a greater z corresponds to an earlier epoch of emission; z may thus be used as a measure of both time and distance. We can deduce a formula connecting the epoch of emission, the redshift z the deceleration parameter q_0, and the Hubble parameter H_0. In general, the formula is complicated and we shall consider only the special case of the dust universe of zero curvature, i.e., $p = 0$, $k = 0$, and $q_0 = \frac{1}{2}$. Such a universe is known as the Einstein–de Sitter universe and may be a faithful representation of the actual universe at present.

We have for this case (recall (16.21)),

$$R = (\tfrac{3}{2}at)^{2/3}, \qquad H = \frac{2}{3t}.$$

Hence

$$(1 + z) = \frac{R_0}{R_e} = \left(\frac{t_0}{t_e}\right)^{2/3},$$

or

$$t_e = \frac{t_0}{(1 + z)^{3/2}} = \tfrac{2}{3}H_0^{-1}(1 + z)^{-3/2}. \tag{16.28}$$

Problems

1. Prove that for a dust universe

$$H^2 = H_0^2(1 + z)^2(1 + 2q_0 z).$$

2. Replacing R by $R_0(1 + z)^{-1}$, show that the above relation gives

$$\frac{dz}{dt} = -H_0(1 + z)^2(1 + 2q_0 z)^{1/2}.$$

Integrate the above relation for different values of q_0 with the boundary condition $t = 0$ at $z \to \infty$.

3. Show that for the metric

$$ds^2 = e^\nu \, dt^2 - e^\mu (dr^2 + r^2 \, d\theta^2 + r^2 \sin^2 \theta \, d\phi^2),$$

with μ and ν functions of r and t, Einstein's equations are

$$8\pi T_r^r = e^{-\mu}\left(\frac{\mu'^2}{4} + \frac{\mu'\nu'}{2} + \frac{\mu' + \nu'}{r}\right) + e^{-\nu}\left(\ddot{\mu} + \tfrac{3}{4}\dot{\mu}^2 - \frac{\dot{\mu}\dot{\nu}}{2}\right),$$

$$8\pi T_\theta^\theta = 8\pi T_\phi^\phi = -e^{-\mu}\left(\frac{\mu'' + \nu''}{2} + \frac{\nu'^2}{4} + \frac{\mu' + \nu'}{2r}\right) + e^{-\nu}\left(\ddot\mu + \tfrac{3}{4}\dot\mu^2 - \frac{\dot\mu\dot\nu}{2}\right),$$

$$8\pi T_r^r = -e^{-\mu}\left(\mu'' + \frac{\mu'^2}{4} + \frac{2\mu'}{r}\right) + \tfrac{3}{4}e^{-\nu}\dot\mu^2,$$

$$8\pi T_t^r = -e^{(\nu-\mu)}8\pi T_r^t = e^{-\mu}\left(\dot\mu' - \frac{\dot\mu\nu'}{2}\right).$$

and hence deduce the Friedmann form of the metric.

Hint.

(a) Use the perfect fluid energy–momentum tensor.

(b) Note that isotropy requires the coordinate system to be comoving.

(c) Homogeneity requires p and ρ to be functions of t alone.

4. Using the metric

$$ds^2 = dt^2 + A^2\,dx^2 + B^2\,dy^2 + C^2\,dz^2,$$

where A, B, and C are functions of t alone, show that if the space is empty, then

$$p_1 + p_2 + p_3 = 1 \qquad \text{and} \qquad p_1^2 + p_2^2 + p_3^2 = 1,$$

with $A = t^{p_1}$, $B = t^{p_2}$, and $C = t^{p_3}$. (This is known as the Kasner metric (Kasner, 1927).) Discuss the restrictions set on p_1, p_2, and p_3.

5. Find the solution for a dust universe ($\rho \neq 0$, $p = 0$) with the metric of the form given in Problem 4.

6. For the metric

$$ds^2 = e^\nu\,dt^2 - e^\lambda\,dr^2 - e^\mu(d\theta^2 + \sin^2\theta\,d\phi^2),$$

show that the field equations are (for a perfect fluid)

$$8\pi p = \tfrac{1}{2}e^{-\lambda}\left(\frac{\mu'^2}{2} + \mu'\nu'\right) - e^{-\nu}(\ddot\mu - \tfrac{1}{2}\dot\mu\dot\nu + \tfrac{3}{4}\dot\mu^2) - e^{-\mu}$$

$$= \tfrac{1}{4}e^{-\lambda}(2\nu'' + \nu'^2 + 2\mu'' + \mu'^2 - \mu'\lambda' - \nu'\lambda' + \mu'\nu')$$

$$+ \tfrac{1}{4}e^{-\nu}(\dot\lambda\dot\nu + \dot\mu\dot\nu - \dot\lambda\dot\mu - 2\ddot\lambda - \dot\lambda^2 - 2\ddot\mu - \dot\mu^2),$$

$$8\pi\rho = -e^{-\lambda}\left(\mu'' + \tfrac{3}{4}\mu'^2 - \frac{\mu'\lambda'}{2}\right) + \tfrac{1}{2}e^{-\nu}(\dot\lambda\dot\mu + \dot\mu^2/2) + e^{-\mu},$$

$$8\pi T_t^r = 8\pi(p + \rho)v^r v_t = \tfrac{1}{2}e^{-\lambda}(2\dot\mu' + \dot\mu\mu' - \dot\lambda\mu' - \nu'\dot\mu).$$

7. Consider the metric given in Problem 3. Write $e^\mu r^2 = \xi^4$. Now show that, in vacuum, ξ is determined by the differential equation.

$$\frac{\partial\xi}{\partial x} = \tfrac{1}{4}\xi^2 + \tfrac{1}{16}\dot g^2\xi^6 + K,$$

where K is a constant, $x = \log r$, and g is a function of t alone, such that

$$e^\nu = \frac{16\dot\xi^2}{\dot g^2\xi^2}.$$

Show further that a transformation

$$\hat{t}^2 = \xi^4,$$

$$d\hat{t} = \frac{8\xi\dot{\xi}\xi_x}{\dot{g}(\xi^2 - 2m)}dt + \frac{\xi^6\dot{g}}{2(\xi^2 - 2m)}dx, \qquad \text{where} \quad \xi_x = \left(\frac{\partial\xi}{\partial x}\right) \quad \text{and} \quad m = -2K$$

reduces the metric to the Schwarzschild metric for a particle of mass m (Raychaudhuri, 1953).

8. Find out the transformation that reduces the metric (16.11) to (16.10).

9. Light from an object shows a redshift $z = 4$. Assuming that the shift is cosmological, calculate the epoch of emission using: (i) the Einstein–de Sitter model and (ii) the de Sitter metric.

16.3. The Observational Background of Cosmology

Is the Friedmann metric a faithful model of the universe? Broadly, this is the question that is to be settled by observations. To be more specific we pose the following questions:

(1) How far is the assumption of homogeneity and isotropy supported by observations?
(2) What is the value of the Hubble parameter and the consequent upper bound to the age of the universe? Is it consistent with the ages of different astronomical objects?
(3) Can we determine from observations the values of k, q_0, and ρ_0? Are the values consistent with the equations of the Friedmann model?
(4) Is there any evidence of the big bang origin?

Homogeneity and Isotropy

At first sight we may seek a verification of the assumptions of homogeneity and isotropy by studying the distribution of galaxies. But then the distribution that we are observing just now at different distances pertains to different epochs depending on the time that the light has taken to reach us. Thus even a spatially homogeneous distribution will appear nonhomogeneous in a nonstatic universe.

In more technical language, while the 3-spaces of homogeneity are orthogonal to the congruence of world-lines which are timelike lines, our observations are limited to a 3-surface spanned by the past directed null lines from our here and now. Thus the observed distribution will involve, besides the actual distribution, the Hubble parameter and the deceleration parameter which are associated with the temporal change of distribution. The things that we directly observe are the redshift z, the apparent magnitude m (which is essentially a measure of the apparent luminosity on a logarithmic scale: $m = \text{constant} - 2.5 \log l$, l being the apparent luminosity), the angular diameter θ of objects, and we can also count the number of source N. Theoretical relations between them (e.g., $z - m$, $N - z$, $\theta - z$) have been derived on the basis of the Friedmann metric but, unfortunately, as these involve H_0 and q_0, these relations cannot be used for the verification of the homogeneity postulate. Rather, they have been used in attempts to determine H_0 and q_0.

In passing, we may note that the theoretical $\theta - z$ relation shows that θ should have a minimum, but for the de Sitter metric (on which the steady state theory is based) should have no such minimum. It has been claimed that the observations do show a minimum of θ (Swarup, 1975).

A somewhat intriguing result has come from the study of a number N of radio sources above some specific flux density S (defined as the power received per unit area per unit frequency). For a uniform distribution of sources, the classical relation is $N \sim S^{-3/2}$ in the absence of any evolutionary change in the sources. Observations, however, give $N \sim S^{-\beta}$ where β is significantly greater than $\frac{3}{2}$. Apparently, this indicates a striking nonuniformity with a paucity of sources in our neighborhood. However, the prevalent trend is to explain it away as due to evolutionary changes. Thus homogeneity and isotropy have become almost an article of faith to be defended at all costs.

In any case, a test of homogeneity based on the observations of luminous matter alone is inadequate, as the universe seems to be dominated by nonluminous matter about which we give some information below.

The Dark Matter

The spiral galaxies show a rotation about a nuclear center where the luminosity falls off exponentially, the e-folding distance being a few kpc (1 pc $\approx$ 3 light years). Then the luminous matter is concentrated within a few kpc. However, the rotational velocity of objects within the galaxy remains practically constant up to all distances where there are luminous objects to show the rotation. Applying now the simple formula for equilibrium between centrifugal force and gravitational attraction, we obtain

$$\frac{GM_r}{r} = v_r^2 = \text{constant},$$

where M_r is the mass enclosed in the region of radius r. Hence the density ρ of matter comes out as

$$\rho = \frac{1}{4\pi r^2} \frac{dM_r}{dr} \propto \frac{1}{r^2}.$$

Thus the gravitating mass density falls off much slower than the luminuous density. This discrepancy between gravitating mass and luminous mass is attributed to dark matter, i.e., matter which does not radiate.

Again there are the so-called regular clusters of galaxies which have a smooth distribution with a pronounced central concentration. This indicates that they are in virial equilibrium. The virial theorem then gives

$$GM = \langle v^2 \rangle R,$$

where M is the mass of the cluster, R its linear dimension, and $\langle v^2 \rangle$ the velocity dispersion. Observation of $\langle v^2 \rangle$ leads to a value of M, an order of magnitude higher than the total luminous mass of the cluster. Similar conclusions regarding the preponderance of dark matter follows from a study of giant elliptical galaxies. However, we have no direct way to detect the dark matter that may be present in the so-called voids between galactic clusters, and hence the homogeneity of the distribution of matter cannot be verified by simple direct observations.

The observed redshifts should be isotropic if the universe is of the Friedmann type. A marked anisotropy in redshifts was first detected by Rubin and Ford in 1976. Subsequently, this anisotropy was interpreted as indicating an infall of our local group of galaxies towards the Virgo cluster. More recently, it has been claimed that, besides this, there is a further peculiar motion existing out to a range of 100 Mpc, and that this motion is consistent with a spherically symmetric mass concentration of a total amount of $\sim 3 \times 10^{16} \, M_{\odot}$. If confirmed, that will be a departure from homogeneity on a fairly large scale.

The most impressive evidence in favor of isotropy comes from the microwave background radiation. We shall give a somewhat detailed description of this radiation later on. But, before that, we shall consider the observational determination of H_0, q_0, etc.

The Hubble Parameter and the Age of the Universe

On the basis of the Friedmann metric, we can deduce a relation between the observable quantities m (the apparent magnitude) and z (the redshift). The relation is (see, e.g., Raychaudhuri (1979))

$$m = 5 \log_{10} z + 1.086(1 - q_0)z + M - 5 \log_{10} H_0 - 5. \quad (16.29)$$

In the above expression, terms involving higher powers of z have been neglected, M is the standard absolute magnitude of the sources (in this case, some selected galaxies) being observed (the absolute magnitude is defined in terms of the luminosity at a distance 10 parsecs from the source). We omit a formal derivation of (16.29) but note the salient points that are considered. As the radiation spreads out to a coordinate distance r, its intensity falls off inversely as the area of the spherical surface at r, which is $4\pi R^2 r^2 (1 + kr^2/4)^{-2}$. Thus the curvature and expansion come in. Again there is a degradation of energy received per unit time due first to the redshift and second to time dilation.

Besides the errors arising due to technical difficulties associated with observation, there is an intrinsic uncertainty associated with M, and thus from (16.29) we can determine H_0 only up to an uncertainty factor of 2. We shall take the value of H_0 as $(2 \cdot 10^{10} \, \text{yr})^{-1}$. In principle, we can also determine q_0 if the observations extend to sufficiently large values of z. However, large z means that the emission took place in the distant past and the question then arises that the sources might have a different value of M at those times due to evolutionary changes in the meantime.

Anyway, as $\ddot{R}/R$ is necessarily negative, the maximum possible age is H_0^{-1}, i.e., $\sim 2 \times 10^{10}$ yr. For the Einstein–de Sitter universe, we have from (16.21) the age $= \frac{2}{3}H_0^{-1} = 1.3 \times 10^{10}$ yr. The age of some astronomical objects such as the globular clustars is close to the value 1.3×10^{10} yr.

We can only fix rather wide bounds to the value of q_0. Thus it has been suggested that q_0 may well be above $\frac{1}{2}$ making the space closed ($k = +1$) or may be vanishingly small making $k = -1$. Even negative values of q_0 have sometimes been proposed although that is inconsistent with standard cosmology without the cosmological term.

A similar uncertainty hangs over the value of the present energy density ρ_0. Considering luminous matter alone, ρ_0 appears to be less than 10^{-30} g cm^3 or the density parameter less than unity, so that the space should be open. However, as we have already noted, there is considerable uncertainty about the amount of dark matter, so that it may well be that the density parameter Ω is ≥ 1 and the space is flat or closed. Observations thus fail to give a definitive answer to point (3) which we raised at the beginning of this section.

The Microwave Background Radiation—Evidence in Favor of Isotropy and the Big Bang

The salient features that have been observed about the microwave radiation background are:

(1) The distribution of intensity of radiation at different wavelengths agrees with Planck's formula for black body radiation at a temperature ~ 2.7 K. The peak intensity thus occurs at $\lambda \sim 1$ mm.

(2) The radiation is isotropic, i.e., comes uniformly from all directions. There is a small dipole anisotropy, i.e., if T_θ is the temperature of the radiation coming in a direction θ with respect to a particular axis about which there appears to be rotational symmetry, then

$$T_\theta = T_0(1 - \alpha \cos \theta),$$

T_0 being the temperature observed for $\theta = \pi/2$. Now it can be shown that black body radiation which appears isotropic in a particular coordinate frame will appear anisotropic in a relatively uniformly moving frame, although it will still show a Planckian distribution. The anisotropy is represented by the formula

$$T_\theta = \frac{T_0(1 - v/c \cos \theta)}{(1 - v^2/c^2)^{1/2}} \approx T_0\left(1 - \frac{v}{c \cos \theta}\right) \qquad \text{(for } v \ll c\text{)}.$$

Hence the observed anisotropy in the temperature is interpreted as indicating a peculiar velocity of ourselves relative to the frame in which the isotropy, assumed in cosmology, strictly holds good.

Barring this dipole anisotropy, the isotropy is so exact that we wonder why there is no imprint of the formation of structure in the universe (i.e., the galaxies, their clusters, and superclusters). The Cosmic Background Explorer (COBE) satellite has very recently confirmed the Planckian distribution and found the anisotropy (besides the dipole term) $\Delta T/T < 10^{-5}$ (Mather et al., 1990).*

(3) Since the discovery of radiation in 1965, up until now there has been no appreciable change in the radiation.

In view of this last observation, as well as of isotropy, it is clear that the radiation is not a local or transient phenomenon. Isotropy also demands that if the radiation is due to discrete sources, then these sources must be distributed very densely and uniformly. Indeed, an estimate of the requisite number density of discrete sources rules out this possibility. Again, the thermal nature of the radiation indicates that the radiation at some stage must have experienced a very effective interaction with matter. Such an interaction is absent now, otherwise, the quasars with large redshifts would not appear as sharp sources.

All the foregoing features find a very natural explanation in standard cosmology. Indeed, the existence of such radiation was anticipated by Gamow from the consideration of the expanding universe long before its discovery. If we trace the history of the expanding universe backward in time, we come across an ever-increasing density of matter and its temperature. Ultimately, there would be a mixture of radiation and a hot dense plasma interacting via the Thomson scattering. A thermal equilibrium will then prevail and the radiation will be thermalized. The presently observed background radiation is just a relic of this radiation where temperature has fallen enormously due to expansion. Reversing this line of reasoning, we may say that the thermalized MBR is evidence in favor of an earlier hot dense plasma state of the universe which was prevalent shortly after the big bang.

The strongest observational support in favor of the assumptions of homogeneity and isotropy is provided by the isotropy of the background radiation. The photons that reach us have apparently come from different distances after suffering a scattering by electrons. However, recall that in the kinetic theory of gases we may consider that each and every molecule describes the same distance (the mean free path) between two collisions. In the case of photons too, we may consider all of them to have suffered a scattering at an identical distance before reaching us. This gives rise to the idea of a last scattering surface as the source of the photons received by us, and the isotropy of the background radiation then indicates that there is an isotropy of the universe at least up to the last scattering surface about our locale.

To estimate the distance of the last scattering surface, we have to know the values of H_0 and q_0 and the number density of free electrons in the intervening space. As there are uncertainties in each of these the estimate can vary widely. However, the argument that we represent will be valid if the distance of the last scattering surface is anything greater than what can be considered a local region of the universe. Estimates of a lower bound to the distance of the last

---

* Recent analysis of data from the satellite COBE (Cosmic Background Explorer) apparently has detected anisotropy of $(\Delta T/T) = (1.1 \pm 0.2) \times 10^{-5}$ (Silk, 1992).

scattering surface correspond to a redshift of $z \approx 8$ which makes it a fairly large nonlocal region. (A realistic estimate of the last scattering surface is $z \approx 1000$.)

We may, of course, say that the observed isotropy merely indicates that the universe has a center of symmetry at which we are placed and nothing can be said about other locales. However, such an idea might have been considered acceptable in the pre-Copernican days; it seems repugnant to the spirit of our times. Thus it has been presumed that the isotropy that we observe is merely a part of the general isotropy of the universe at all locales. We can then show that isotropy at all locales leads to homogeneity as well. (This is a geometrical result which can be obtained without the help of any field equations.)

16.4. Summary

We have given an outline of the most basic observations in cosmology. We now enumerate some other results that have come from observational studies.

(a) The Angular Diameter Study

Suppose we have a family of objects all of a standard proper size, then their apparent angular diameter θ will depend on their position, as well as on the geometry of space. We can obtain relations between θ and the redshift z. These are on the basis of the Friedmann metric

$$\theta = \text{constant}(1 + z)^2 H_0 q_0^2 \{q_0 z + (q_0 - 1)[(1 + 2q_0 z)^{1/2} - 1]\}^{-1} \quad \text{for } q_0 > 0,$$

$$= \text{constant}(1 + z)^2 H_0 z^{-1}(1 + z/2)^{-1} \quad \text{for } q_0 = 0.$$

A negative value of q_0 is not consistent with the general theory of relativity and the vanishing cosmological constant. Hence the above two relations exhaust the possibilities for relativistic models. It is easy to see that with both the above forms θ has a minimum for some z. However, for the steady state theory which was at one time proposed as an alternative to relativistic models, $q_0 = -1$ (the general relativity equations were violated due to a continuous creation process) and the $\theta - z$ relation is

$$\theta = \text{constant } H_0(1 + z)z^{-1}.$$

In this case, θ shows a monotone decrease with z and has no minimum. Thus the existence of a minimum in the $\theta - z$ relations was considered a crucial test to decide between the relativistic and the steady state cosmology.

As a result of the analysis of angular diameter versus flux-density plots, Swarup (1975) and Kapahi (1975) claimed to have found a minimum, but they had to adopt an evolution formula for the radio source studies. The evolution assumed was questioned by Narlikar and Chitre (1977) who showed that alternative evolutions may remove the minimum.

(b) The Luminosity–Volume Test

Suppose that the survey of a sample of objects is complete in the sense that all objects, having a flux density above a certain limiting value at a particular frequency, have been detected. We may now compute for each source the limiting redshift Z_m at which it would be just included in our sample. We may also calculate the volume V_m corresponding to the redshift Z_m. Again there is the volume V up to the actual redshift z of the source. As $0 \leq z \leq Z_m$, we will have V/V_m lying between 0 and 1. For a uniform distribution and a complete sample, the average value of V/V_m would be 0.5. Considering the $3CR$ survey, Rowan–Robinson as well as Schmidt found that $(V/V_m)_{av}$ was appreciably higher (~ 0.64). It was argued that this indicated a strong evolution effect rather than a definite nonuniformity in the distribution.

Problems

1. The presence of dark matter is concluded by using Newton's laws of gravitation and motion. Alternatively, we may rule out the existence of dark matter by introducing modifications in Newton's laws. Suggest a suitable modification of the law of gravitation and critically examine it.

2. If L is the rate of emission of radiation by a point source at coordinate position r per unit solid angle and l is the rate of energy incident on the unit area at the origin, placed normal to the rays, then the luminosity distance D is defined as

$$l = \frac{L}{D^2}.$$

   Show that

$$D = \frac{R_0 r(1 + z)}{(1 + kr^2/4)},$$

   where R_0 is the value of R at the instant of reception.

3. Deduce the relation

$$T_\theta = \frac{T_0(1 - v/c \cos \theta)}{(1 - v^2/c^2)^{1/2}}.$$

4. Deduce the relation $N \sim S^{-3/2}$ for the classical case.

5. Suppose that observations are made in a spatially closed universe at an epoch very close to the stage of maximal expansion. What will be the form of the observed velocity–distance relation? Make an estimate of the age of the universe in such a case in terms of H_0 and q_0 (see Raychaudhuri and Mukherjee, 1984).

17. The Singularity Problem

17.1. Introduction

Before going over to a discussion of physical cosmology we introduce what is currently considered the most outstanding problem of theoretical cosmology. We have seen that the expanding Friedmann universe has a singularity (the big bang) corresponding to $R = 0$ at a finite time in the past. At first sight, we may think that the singularity is merely a consequence of our oversimplifying assumptions of isotropy and homogeneity and in more realistic models there would be no such singularity. This was indeed the idea put forward by Tolman and Eddington. However, later researches have shown this to be incorrect and we give a sketch of the developments that have led to the conclusion that a singularity in classical cosmology is almost inescapable.

For this purpose, we give a derivation of the Raychaudhuri equation (Raychaudhuri, 1955, 1979) which has played a crucial part in the derivation of singularity theorems.

17.2. The Raychaudhuri Equation

As we have already seen, the Riemann–Christoffel tensor may be defined by the commutator of the second covariant derivative of any vector; namely,

$$v^{\mu}_{;\alpha;\beta} - v^{\mu}_{;\beta;\alpha} = R^{\mu}_{\gamma\beta\alpha}v^{\gamma}. \tag{17.1}$$

Now contract the above equation with v^{α} and also over the indices μ and β to obtain

$$v^{\mu}_{;\alpha;\mu}v^{\alpha} - v^{\mu}_{;\mu;\alpha}v^{\alpha} = R_{\gamma\alpha}v^{\gamma}v^{\alpha}. \tag{17.2}$$

Suppose that v^{μ} is a unit vector and introduce the following definitions (the origin of these definitions will be given later):

(1) $v^{\mu}_{;\mu} \equiv \theta$ is the expansion scalar.
(2) $v^{\mu}_{;\alpha}v^{\alpha} \equiv \dot{v}^{\mu}$ (the departure of v^{μ} from geodesicity) is the acceleration vector. Because v^{μ} is a unit vector, $\dot{v}^{\mu}v_{\mu} = 0$.
(3) $\frac{1}{2}(v_{\alpha;\beta} - v_{\beta;\alpha}) - \frac{1}{2}(\dot{v}_{\alpha}v_{\beta} - \dot{v}_{\beta}v_{\alpha}) \equiv \omega_{\alpha\beta}$ is the vorticity tensor. Note that it is antisymmetric and orthogonal to the vector v^{α}, i.e., $\omega_{\alpha\beta}v^{\alpha} = \omega_{\alpha\beta}v^{\beta} = 0$.

We can also introduce a vorticity vector with the help of the Levi-Civita tensor

$$\omega^\mu = \tfrac{1}{2}\eta^{\mu\nu\rho\sigma}\omega_{\rho\sigma}v_\nu = \tfrac{1}{2}\eta^{\mu\nu\rho\sigma}v_{\rho,\sigma}v_\nu$$

(because of the antisymmetrization, the covariant derivative may be replaced by the ordinary derivative) also $\omega^\mu V_\mu = 0$.

(4) $\tfrac{1}{2}(v_{\mu;\nu} + v_{\nu;\mu}) - \tfrac{1}{2}(\dot{v}_\mu v_\nu + \dot{v}_\nu v_\mu) - \tfrac{1}{3}v^\alpha_{;\alpha}(g_{\mu\nu} - v_\mu v_\nu) \equiv \sigma_{\mu\nu}$ is called the shear tensor. It is symmetric, tracefree, and also

$$\sigma_{\mu\nu}v^\nu = \sigma_{\mu\nu}v^\mu = 0.$$

With these definitions, we get

$$v_{\mu;\nu} = \sigma_{\mu\nu} + \omega_{\mu\nu} + \tfrac{1}{3}\theta(g_{\mu\nu} - v_\mu v_\nu) + \dot{v}_\mu v_\nu. \tag{17.3}$$

Also, (17.2) may be written as

$$(v^\mu_{;\alpha}v^\alpha)_{;\mu} - v^\mu_{;\alpha}v^\alpha_{;\mu} - \theta_{,\alpha}v^\alpha = R_{\nu\alpha}v^\nu v^\alpha.$$

Using (17.3), the above equation becomes

$$v^\mu_{;\mu} - 2(\sigma^2 - \omega^2) - \tfrac{1}{3}\theta^2 - \theta_{,\alpha}v^\alpha = R_{\nu\alpha}v^\nu v^\alpha, \tag{17.4}$$

where σ and ω are defined by

$$\sigma^2 \equiv \tfrac{1}{2}\sigma_{\mu\nu}\sigma^{\mu\nu}, \tag{17.5}$$

$$\omega^2 \equiv \tfrac{1}{2}\omega_{\mu\nu}\omega^{\mu\nu}. \tag{17.6}$$

Note that if v^μ is timelike, σ^2 and ω^2 are positive definite, vanishing only if all the components of the shear tensor and the vorticity tensor vanish, respectively. Equation (17.4) is the desired relation and is known as the Raychaudhuri equation (Raychaudhuri, 1955, 1979).

17.3. The Meaning of Shear, Vorticity, and Expansion

To see how the names mentioned above arise, consider the case of elastic deformation in Newtonian physics. If $u(x, y, z)$ is the displacement of the point with coordinates (x, y, z) the increase in volume of a region bounded by a surface S is given by $\oint u\, ds$ where the integral is over the entire surface S. By Gauss's theorem

$$\oint u\, ds = \int_V \nabla u\, dv,$$

the integration volume V being bounded by the surface S. Hence the dilation defined by $\partial(dV)/(dV)$ is given by $\theta = \nabla \cdot u$.

As is well known, $\nabla \times u$ represents a rotation and we may split up the Cartesian tensor $\partial u_i/\partial x^k$ as follows:

$$\frac{\partial u_i}{\partial x^k} = \left[\frac{1}{2}\left(\frac{\partial u_i}{\partial x^k} + \frac{\partial u_k}{\partial x^i}\right) - \tfrac{1}{3}\theta\delta_{ik}\right] + \frac{1}{2}\left(\frac{\partial u_i}{\partial x^k} - \frac{\partial u_k}{\partial x^i}\right) + \tfrac{1}{3}\theta\delta_{ik}. \tag{17.7}$$

The three parts are physically significant, namely:

(a) The first part gives a change of shape without a change in volume because it is tracefree. This is the shear.
(b) The second part is simply a rotation.
(c) The third part represents an isotropic change of length leading to a volume expansion.

To take over these ideas to general relativity, we adopt the following procedure:

(i) The displacement vector is replaced by the velocity vector v^μ.
(ii) The shear, rotation, and expansion should all have tensorial form. In particular, expansion should be a scalar, while shear and rotation tensors should be in the 3-space orthogonal to the velocity vector v^μ.

Any vector (or tensor) may be projected into the 3-space orthogonal to v^μ by contracting with the "projection tensor" $g_{\mu\nu} - v_\mu v_\nu$. If we now project $v_{\mu;\nu}$ to the 3-space we get

$$v_{\mu;\nu}(g_\alpha^\nu - v^\nu v_\alpha) = v_{\mu;\alpha} - \dot{v}_\mu v_\alpha.$$

(Note that $\dot{v}^\mu = v^\mu_{;\alpha} v^\alpha = \partial v^\mu / \partial t$ in the local Lorentz frame in which the fluid is at rest, hence the name acceleration.) Replacing $u_{i,k}$ in the Newtonian expressions for shear and vorticity by $(v_{i;k} - \dot{v}_i v_k)$, we get the expressions we have identified with shear and vorticity in general relativity.

17.4. An Elementary Singularity Theorem

It is instructive to correlate the Raychaudhuri equation with the Friedmann equations for the isotropic universe. Identifying v^μ with the velocity vector of the cosmic fluid we see that for the Friedmann universe $\dot{v}^\mu = \omega^\mu = \sigma_{\mu\sigma} = 0$. Indeed, we may characterize the Friedmann universe as one in which there is an isotropic expansion without acceleration or rotation. In this case, $\theta = 3\dot{R}/R$ and the Raychaudhuri equation reduces to (16.17). We may next go over to more general cases and define a scale factor R by the relation $\theta = 3\dot{R}/R$. The Raychaudhuri equation is now

$$\frac{\ddot{R}}{R} = -\frac{4\pi G}{3}(\rho + 3p) - \tfrac{2}{3}\sigma^2 + \tfrac{2}{3}\omega^2 + \tfrac{1}{3}\dot{v}^\mu_{;\mu}.$$

The first term on the right-hand side is the gravitational interaction term and brings about a deceleration of the expansion. The signature of the terms show that any shear (or anisotropy of expansion) will merely augment the gravitational attraction and thus would only help in bringing about a collapse. The vorticity term, however, is effectively a repulsion (see the centrifugal force) and may presumably counterbalance the effect of gravitation and shear and bring about a bounce instead of a collapse. We shall return to this question a little later. The acceleration term may be of either sign and in the case of perfect fluids arises from the pressure gradient for

$$\dot{V}^\gamma = p_{,\sigma} \frac{(g^{\sigma\gamma} - v^\sigma v^\gamma)}{p + \rho}.$$

However, the work of Chandrasekhar on white dwarfs, and the subsequent investigations on the basis of different equations of state in connection with neutron star models, have shown that the pressure gradient cannot counterbalance the effect of gravitation beyond a certain degree of compactification. It is true that all these results have been obtained for steller bodies where the pressure has to vanish at the boundary and this condition is inapplicable in cosmology. Nevertheless, from the way in which the results have been obtained, it seems that the conclusion will remain valid if the pressure is to be nonnegative everywhere. In particular, we may enunciate the theorem: "In the absence of a pressure gradient and vorticity a singularity corresponding to a collapse of spatial volumes is unavoidable."

What if vorticity is present? Before answering this question, we will present a discussion of the Gödel universe which shows that vorticity may be associated with some awkward situations.

17.5. The Gödel Universe

Gödel (1949) gave the metric

$$ds^2 = a^2 \left[(dx^0 + \exp(x^1)\, dx^2)^2 - dx^{1^2} - \frac{\exp(2x^1)}{2} dx^{2^2} - dx^{3^2} \right].$$

The x^0 lines are the world-lines of matter so that the covariant components of the velocity vector are

$$v_0 = a, \qquad v_1 = v_3 = 0, \qquad v_2 = a \exp(x^1),$$

and hence the curl of the velocity vector has a nonvanishing component

$$v_{2,1} - v_{1,2} = ae^{x^1},$$

i.e., the vorticity ω^μ is nonvanishing. Gödel presented the metric as a solution of the field equations with a cosmological constant, and the matter in the form of dust, of density, ρ, where

$$8\pi\rho G = a^{-2} = -2\Lambda = 2\omega^2.$$

However, instead of having the cosmological constant Λ, we may consider the matter to have a pressure p equal in magnitude to the energy density ρ. The equation of state $p = \rho$ gives the velocity of compressional waves equal to that of light, and is usually referred to as Zel'dovich's limiting case. Gödel's universe is singularity free but has a peculiar feature as shown below.

The metric admits a transitive group of motions, namely:

(i) a translation along x^0 which indicates its stationary character;
(ii) a translation along x^2;
(iii) a translation along x^3; and
(iv) a translation along x^1 combined with a contraction along x^2.

Thus the space–time is completely homogeneous (see the de Sitter metric). More interesting results follow from a transformation of the Gödel metric to a cylindrically symmetric form:

$$dS^2 = 4a^2[dt^2 - dr^2 - dz^2 + (\sinh^4 r - \sinh^2 r)\, d\phi^2 + 2\sqrt{2}\, \sinh^2 r\, d\phi\, dt.$$

(We omit the transformation formulas which are given in Gödel's paper.) Apparently, the above metric has a singularity at $r = 0$ where $g_{\phi\phi}$, as well as the determinant of the metric tensor, vanish. However, if we make the transformation

$$\xi = r \cos \phi, \qquad \eta = r \sin \phi,$$

then for $r \to 0$, the r and ϕ 2-space (i.e., the space obtained by taking t and z constants) metric

$$-dr^2 + (\sinh^4 r - \sinh^2 r)\, d\phi^2 \to -dr^2 - r^2\, d\phi^2 = -d\xi^2 - d\eta^2.$$

This shows that $r = 0$ is a regular point and that ϕ is an angular coordinate with domain $0 \le \phi \le 2\pi$.

However, although ϕ is spacelike for small values of r, it is timelike for $r > \log(1 + \sqrt{2})$. Thus, beyond this value of r, the ϕ lines, although still closed, are timelike and, in principle, we can come back to the same space–time point after traveling all the while in the future oriented direction and can even manage to return to a point earlier than we started. Such behavior is repugnant to cause–effect relationships but is nevertheless not ruled out in the general theory of relativity.

17.6. General Singularity Theorems

We may now state some general singularity theorems. So far, we have considered singularities to be characterized by infinite values of physical variables. However, we may cut out the portions of space–time where physical variables blow up, and the truncated region may be presented as an everywhere regular space–time. Hence comes the question of the "completeness" of a space–time and space–time is now defined to be complete only if all timelike and null geodesics can be extended to arbitrary values of their affine parameter. Such a definition has an obvious physical appeal, a timelike or null geodesic is a possible world-line of a free particle (massive or massless); thus, if a space-–time is to qualify as regular, a freely moving particle must have its entire history within that space–time, it must not face a beginning or end of its career.

The new definition is much more powerful in the mathematical sense also, and quite a number of singularity theorems have been proposed. We may mention here a particular one due to Hawking and Penrose. Broadly speaking, it states that a singularity in the form of geodesic incompleteness is inevitable unless one of the following conditions is satisfied:

(a) The positive nature of energy is violated or, equivalently, the gravitational interaction becomes repulsive for positive energy density. Indeed, the continuous creation term introduced by Hoyle and Narlikar in their steady state theory can be considered as a negative energy term in Einstein's field equations. Similarly, the cosmological term, introduced ad hoc by Einstein, and the false vacuum energy–stress tensor considered in connection with inflationary models are associated with negative energy. In the case of the false vacuum, $T_\mu^\nu = \Lambda \delta_\mu^\nu$, thus the energy density ρ is positive but $(\rho + 3p)$, which occurs in the gravitational term in the Raychaudhuri equation, is negative.

(b) Closed timelike lines, as we have met with in the Gödel universe, occur. This would mean a breakdown of causality and an inability to make a unique past–future separation. There are some additional conditions; however, in the present discussion it is not possible to spell them out. In any case, they are not very significant in the problem of cosmology (see Hawking and Ellis, 1973).

However, two things should not be overlooked. First, the singularity theorem does not demand that abnormal conditions should occur everywhere, it may be that just one or a few geodesics are incomplete. Second, the singularity may not involve the blowing up of any physical variable. Such singularities, as distinct from the big bang singularity involving infinite density, have sometimes been called "whimper singularities." In view of our previous discussion, whimper singularities can only occur if there is vorticity in the fluid motion.

When there is vorticity, the world-lines of the fluid are not hypersurface orthogonal. Nevertheless, for homogeneous spaces, we can choose the time axis as orthogonal to the homogeneous 3-spaces. The metric then will be of the form

$$dS^2 = g_{00}\, dt^2 + g_{ik}\, dx^i\, dx^k.$$

A whimper singularity is then found to be associated with a sudden jump of g_{00} from $+1$ to -1. Of course, the homogeneous subspaces also change their character at this stage (Shepley, 1969; Ellis and King, 1974).

In view of the singularity theorems, the general attitude is to regard this as a pointer to a breakdown of the theory of general relativity if the field is sufficiently strong. While such a breakdown was anticipated by Einstein himself, the present outlook is somewhat different. Einstein speculated on a more perfect field theory where the sources of the field will not appear as artificial inputs; the present tendency is to attempt a quantization of the gravitational field where hopefully such singularities will not occur. However, neither of these lines of thought have so far met with success and the problem of singularity remains a challenging one.

A recent paper by Senovilla (1990) has aroused considerable interest. He presents a cosmological solution in which there is no blowing up of physical variables. The fluid motion is irrotational but is nongeodesic because of pressure gradient forces. The model is thus nonhomogeneous and from the point of view of the Raychaudhuri equation, the focusing of the world-lines of matter is prevented by the acceleration term. However, the authors could not decide whether his space–time is geodesically complete or not.

Problems

1. Newtonian cosmology is based on Euler's equations of hydrodynamics and Poisson's equation for the gravitational potential. Deduce the analogue of Raychaudhuri's equation in Newtonian cosmology.

2. Justify the result that in Newtonian cosmology homogeneity requires that the velocity field $v_i = a_{ik}r_k$ where the a_{ik}'s are functions of time alone, and v_i and r_i are the velocity and position vector components in Cartesian coordinates. What restriction is to be put on the a_{ik}'s if the universe is isotropic as well?

3. Obtain, for Newtonian cosmology, the equation corresponding to Friedmann's equations in relativistic cosmology.

4. In a homogeneous Newtonian universe, the matter is pressure-free dust and the motion is shear free. Show that there will be no big bang singularity if vorticity is present.

5. The result given in Problem 4 cannot be taken over to relativistic cosmology—why?

18. Thermal History of the Universe—
Cosmological Nucleosynthesis

18.1. The Thermal History

Whatever may be the doubts in accepting the singular origin of the universe (i.e., the big bang), the microwave background radiation forces us to consider seriously a hot dense state of the early universe. The first discussion of the physics of the early universe was done by Gamow and his school in the late 1940s, long before the discovery of the microwave background. His motivation was to explain the relative abundances of different nuclei, in this he was a little too ambitious. It is now generally held that only the lightest nuclei like helium and deuteron were manufactured in the early universe scenario, while the heavier nuclei owe their origin to processes occurring in stars. But before going into a discussion of cosmological nucleogenesis, we shall consider the thermal history of the universe.

To explore the thermal history of the universe, the idea of a temperature is introduced through the distribution formulas of Bose and Fermi

$$dn = g \frac{4\pi p^2 \, dp}{h^3} \frac{1}{e^{(\mu+E)/kT} \pm 1}, \tag{18.1}$$

where the upper positive and the lower negative signs hold for Fermi and Bose statistics, respectively. The formulas give the number density dn of particles having momenta lying between p and $p + dp$ and energy between E and $E + dE$ (the energy E includes the rest mass energy). The only other variable in the formulas is the temperature T, while g is a constant depending on the spin of the particle species considered, h is the Planck constant, and the constant μ, called the chemical potential, may be determined if the integrated number density over all momenta is known.

The validity of the formulas needs some consideration. First, the distributions obtain only if there is thermodynamic equilibrium and the particles of a species are noninteracting, except for elastic collisions which bring about the equilibrium. Equilibrium again is at two levels (an equilibrium amongst the particles of the same species) this ensures the applicability of formula (18.1) for the particular species, and an equilibrium between different species which will mean that the temperature T is identical for the different species. We shall proceed with the assumption that both these equilibria obtain and later examine the consistency of the results.

The chemical potential μ vanishes for photons and is conserved in reactions. Hence, from the creation–annihilation reaction of a pair of particles and antiparticles, i.e.,

$$\text{particle} + \text{antiparticle} \rightleftarrows \text{photons},$$

we get

$$\mu + \bar{\mu} = 0,$$

where the overbar indicates the antiparticle.

Now if μ (and consequently $\bar{\mu}$) are significantly different from zero, the numbers of particles and antiparticles, as given by (18.1), would differ significantly. However, the microwave background radiation shows that the number density of photons exceeds the number density of baryons by a large factor ($\sim 10^8 - 10^{10}$). We therefore put $\mu = \bar{\mu} = 0$, this corresponds to a universe in which the particle–antiparticle symmetry obtain exactly and the small number of baryons is neglected. This approximation $\mu = \bar{\mu} = 0$ is then extended over leptons as well.

The distribution formulas show that dn will be appreciable only as long as the exponential term $\exp(E/kT)$ in the denominator is not large. As $E^2 = (m_0^2 C^4 + p^2 c^2)$, this means that the number of particles for which $m_0 c^2 > kT$ will be extremely poor, and to evaluate the total energy density we need only consider particles with $m_0 c^2 \lesssim kT$ and put $E = pc$. Thus the total energy density at high temperatures may be written as

$$\rho = \int_0^\infty g \frac{4\pi p^2 \, dp}{h^3} \frac{pc}{e^{pc/kT} \pm 1}. \tag{18.2}$$

The infinite integrals for fermions and bosons are very simply related for

$$\int_0^\infty \frac{x^3 \, dx}{e^x - 1} - \int_0^\infty \frac{x^3 \, dx}{e^x + 1} = \int_0^\infty \frac{2x^3 \, dx}{e^{2x} - 1} = \frac{1}{8} \int_0^\infty \frac{x^3 \, dx}{e^x - 1},$$

or

$$\int_0^\infty \frac{p^3 \, dp}{e^{pc/kT} + 1} = \frac{7}{8} \int_0^\infty \frac{p^3 \, dp}{e^{pc/kT} - 1}.$$

For radiation, $g = 2$ and $\rho_\gamma = aT^4$ where a is the so-called radiation constant. Hence, for any boson, the energy density will be $\rho_b = (g/2)aT^4$ and for any fermion $\rho_f = \frac{7}{8}(g/2)aT^4$.

Thus, in this ultrarelativistic state, with an extremely high temperature of $\rho = bT^4$ where the constant b is determined by the number and nature of the ultrarelativistic particle species. In this stage, the pressure and density are also simply related.

$$p = \frac{\rho}{3}. \tag{18.3}$$

(We use the symbol p for both pressure and momentum, in what follows momentum will not appear, so there will not be any confusion.)

Plugging these relations into the conservation relation

$$d(\rho P^3) + p\,d(R^3) = 0, \tag{18.4}$$

we get

$$\rho R^4 = \text{constant}, \tag{18.5}$$

$$RT = \text{constant}. \tag{18.6}$$

In the Friedmann equation

$$\left(\frac{\dot{R}}{R}\right)^2 = \frac{8\pi\rho G}{3} - \frac{k}{R^2}, \tag{18.7}$$

the density term therefore behaves as R^{-4} and hence, in comparison, the curvature term can be neglected. Eliminating R from (18.7) by using (18.6) we get the simple relations

$$T = \left(\frac{3}{32\pi b}\right)^{1/4} t^{-1/2} \quad \text{and} \quad R \sim t^{1/2}, \tag{18.8}$$

where we have taken $t = 0$ at $T \to \infty$ (the big bang).

We can now justify the assumption of thermodynamic equilibrium at these stages. Integrating (18.1) over all momenta for $E = pc$ and $\mu = 0$, we get

$$n \propto T^3.$$

Hence the probability of a collision $\alpha n^2 \propto T^6$ and the characteristic time for attaining the equilibrium $\propto T^{-6}$, whereas due to expansion,

$$\frac{dT}{T} = -\frac{1}{2t} \propto T^{-2}$$

the temperature will change appreciably at a much slower rate. Thus there is no difficulty for the equilibrium to be obtained at these early stages.

Note, however, that the equilibrium will not exist between relativistic and nonrelativistic particles. For the nonrelativistic particles, the pressure is much less, e.g., for monatomic gas with $m_0 c^2 \gg kT$ the temperature will obey relation $TR^2 = \text{constant}$ instead of $TR = \text{constant}$. Thus, temperature differences will arise between the relativistic and nonrelativistic species.

The dominance of the relativistic species in the energy density is not valid at low temperatures. Thus, at present, the massless particles like the photon, and maybe the neutrino, contribute only about 10^{-34} g cm^{-3} to the energy density while the nonrelativistic baryons give a density of $\sim 10^{-30}$ g cm^{-3}. The energy density due to the relativistic particles is $\sim nkT$, whereas that due to the nonrelativistic particles is $\sim n'm_0 c^2$. In the present case, $n'/n \ll 1$, but $m_0 c^2$ is so great compared to kT that $n'm_0 c^2 \gg nkT$.

Consider the temperature range T given by $10^{11} > T > 10^9$. In this range the hadron masses are above kT and so their contribution to energy density can be neglected. The case of τ particles and muons is similar. Thus the contribution to energy will come from the photon, the electrons and positrons, and different species of neutrinos and antineutrinos. For the positrons and electrons ($s = \frac{1}{2}, g = 2$), the contribution and antineutrinos we get ($g = 1$ for each type because of helicity) a contribution

$$\rho_v = 4 \cdot \frac{7}{8}\frac{1}{2}aT^4 = \frac{7}{4}aT^4.$$

(We have not included the τ neutrino, the inclusion would increase the factor $\frac{7}{4}$ to $\frac{21}{8}$.) Thus the total energy density comes out as

$$\rho = \rho_v + \rho_e + \rho_v = (1 + \frac{7}{4} + \frac{7}{4})aT^4 = \frac{9}{2}aT^4.$$

Consequently, the constant b in (18.8) is to be replaced by $\frac{9}{2}a$ to give

$$T = \left(\frac{1}{48\pi a}\right)^{1/4} t^{-1/2}. \tag{18.9}$$

Equation (18.9) is somewhat defective in the sense that, in the still earlier stages of the universe, there would be many more species in the ultrarelativistic regime and thereby contribute effectively to the energy density. However, as in (18.9), the fourth root of the energy density is involved, any plausible increase in ρ would introduce a factor of the order of unity without affecting the form of dependence of T on time. Putting in numerical values we get

$$T \sim 10^{10}t^{-1/2}, \tag{18.10}$$

where T is in degrees Kelvin and t is in seconds. Thus, we estimate that a temperature of $\sim 10^{28}$ K was prevalent at $t \sim 10^{-36}$ s while the temperature $\sim 10^9$ K are at about 10^2 s. At the temperature $\sim 10^9$ K, kT falls below 1 MeV and, consequently, the annihilation of electron–positron pairs is no longer compensated for by the pair production from photons. The electrons thus quickly disappear and their energy and entropy goes to the radiation field. Thus the radiation temperature rises above the neutrino temperature. The magnitude of the rise can easily be calculated. The entropy of the radiation, electron–positron, and neutrino systems are, respectively,

$$S_\gamma = \frac{4}{3}\rho_\gamma \frac{V}{T} = \frac{4}{3}aT^3V,$$

$$S_e = \frac{4}{3}\rho_e \frac{V}{T} = \frac{7}{4}S_\gamma,$$

$$S_v = \frac{4}{3}\rho_v \frac{V}{T} = \frac{7}{4}S_\gamma.$$

So that when the electrons disappear, the entropy of radiation becomes $S_\gamma(1 + \frac{7}{4}) = 11S_\gamma/4$, whereas the neutrino entropy remains frozen. Hence, as the temperatures of both vary as the cube root of the entropy, the neutrino temperature T_v will be related to the radiation temperature T_γ as

$$\frac{T_\gamma}{T_v} = \left(\frac{11}{4}\right)^{1/3} \approx 1.40. \tag{18.11}$$

Even after this decoupling of the photons and neutrinos, both T_v and T_γ will follow the relation $RT =$ constant, assuming that the neutrino rest mass vanishes. Consequently, the ratio (T_γ/T_v) will be frozen at the value 1.4 and, if T at present is 2.7 K, T_v is about 1.9 K.

Below 10^9 K the energy density consists principally of photons and neutrinos. However, while these decrease as R^{-4}, the density of any matter decreases less fast. Hence, at a certain stage, the predominance of ultrarelativistic particles (i.e., photons and neutrinos) is lost, and the expansion is controlled progressively by the matter which consists principally of electrons and protons. The plasma, however, is strongly coupled with the radiation field through Thomson scattering processes and so the radiation and matter temperatures fall together. However, at about ~ 4000 K, the protons and electrons combine to form neutral hydrogen and then the matter temperature falls according to the relation $T_{mat}R^2 =$ constant (this corresponds to the adiabatic index $\gamma = \frac{5}{3}$ for a monatomic gas). The radiation temperature drops less fast, namely, $T_\gamma R =$ constant.

Numerical computation shows that, up to a temperature of 0.6×10^{10} K, the fall in temperature obeys the relation $RT =$ constant. Below that and up to 0.2×10^{10} K, the electrons are not ultrarelativistic, electronic pressure falls below one-third of the electronic energy density, and RT shows a progressive decrease. From 0.2–0.03×10^{10} K, the electron–positron annihilation goes on and RT_{rad} shows a progressive increase, finally, by a factor of 1.4 at $T \sim 3 \times 10^8$ K when the annihilation is complete. Presently, we shall see that the nucleosynthesis occurs rather sharply at a temperature of $\sim 10^9$ K which is reached after about 3 min from the beginning.

We can thus divide the history of the universe broadly into the following parts:

(a) the early universe dominated by radiation and ultrarelativistic particles when $p = \rho/3$;
(b) a later phase in which the energy density of radiation does dominate, but there is an appreciable amount of nonrelativistic matter which being completely ionized is in thermal equilibrium with the radiation;
(c) a still later stage in which the protons and electrons unite to form neutral hydrogen so that matter is decoupled from radiation, however, the expansion is still controlled predominantly by radiation; and
(d) the present phase, in which matter dominates the expansion and the pressure is almost negligible compared to the density.

18.2. Cosmological Nucleosynthesis

So long as the temperature is, say, higher than 10^9 K, the photons are energetic enough to disintegrate any deuteron that may be formed by the union of a proton and a neutron. Again, with the expansion reducing the number densities at a fast rate, the union between proton and neutron becomes rare when the temperature falls sensibly below 10^9 K. Thus, only in a narrow critical period, in the career of the universe the production of deuterons can be effective. The deuterons again produce helium nuclei by the following reactions:

$$^2H + {}^2H \rightarrow {}^3He + {}^1n,$$

$$^2H + {}^2H \rightarrow {}^3H + {}^1p,$$

$$^3H + {}^2H \rightarrow {}^4He + {}^1n.$$

All the above reactions have large cross sections and so an increase in baryon number density allows them to proceed rapidly leading to a depletion of ^2H nuclei. Thus there exists an upper bound to the baryon density if the observed deuteron abundance is to survive. Wagoner (1973) found that the relation between baryon density ρ_b and deuteron abundance obeys the inequality

$$(\rho_b)_0 X(^2H) \leq 2 \times 10^{-34} \text{ g cm}^{-3},$$

where the subscript zero refers to the present epoch and $X(^2H)$ is the ratio of the number of deuterons to that of protons. Hence putting the observed value $X(^2H) \sim 2.10^{-4}$ we have $(\rho_b)_0 < 10^{-30}$ g cm^{-3}. This value is consistent with the observed density of luminous matter but the density of dark matter violates this inequality. This is an argument for excluding baryons as the constituent of dark matter.

In actually computing the abundance of deuteron and helium produced in the early universe, Wagoner (1973) started at a temperature of $\sim 6 \times 10^{10}$ K. At this temperature there is thermodynamic equilibrium between the neutrons and protons brought about by the weak interaction reactions

$$p + \bar{\nu} \rightleftharpoons n + e^+, \qquad p + e^- \rightleftharpoons n + \nu, \qquad n \rightleftharpoons p + e^- + \bar{\nu}.$$

Hence

$$\frac{X_p}{X_n} = \exp\left[\frac{(m_n - m_p)c^2}{kT}\right],$$

where X_p and X_n are the number densities of protons and neutrons at temperature T. Substituting values we find that at that stage, i.e., $T \sim 6 \times 10^{10}$ K, neutrons constitute nearly 30% of the total nucleons. In the computation, thermonuclear reactions up to mass number 23 were considered. Schematically, the equations are of the following form:

Rate of change of the ith nuclear species	$=$	Rate of change due to universal expansion	$+$	Rate of change due to nuclei being used up in reactions (including spontaneous decay)

$$+ \quad \text{Rate of change due to production by reactions.}$$

The first factor on the right leads to a progressive decrease in the number of nucleons of each species, in such a manner that it does not lead to any change in the relative abundances. The second factor will involve the cross section of different nuclear reactions as functions of temperature and spontaneous decay constants. It will be of the form

$$-\sigma_{ik} n_i n_k - \lambda n_i,$$

where σ_{ik} is the cross section of a reaction between the ith and kth species, in which the ith nuclear species is used up and λ is the decay constant for the ith species. The third term is the only one which causes an increase in the species and will be of the form

$$\sum_{k,l} \sigma_{kl \to i} \, n_k n_l,$$

where $\sigma_{kl \to i}$ is the cross section of a reaction between the kth and lth species leading to the production of the ith species. Conceiveably, there may be terms of the form λn_k where the kth species spontaneously decays to produce an ith nucleus.

It is clear that the equation system considered is nonlinear, hence even the relative abundances will depend on the total baryon density and not simply on the initial proton–neutron ratio. Wagoner found that if the present baryon density is in the range of $(1-3) \times 10^{-31}$ g cm^{-3}, then there is good agreement between the observed and calculated abundances of both deuterium and helium. As we have already noted this is consistent with the observed luminous matter density.

The reader may wonder why steller processes, which can apparently account for the abundance of most elements, fail to do so for deuterium and helium. (On the basis of steller processes, helium atoms should constitute about 1% of the total number of atoms while observation shows an abundance of about 9%. Again, while the deuterium abundance observed is about 10^{-4}, in steller processes no appreciable deuterium should survive.) The reason is basically that if we allow the reaction between deuterons to continue for any appreciable time they are all used up. In the case of the early universe the deuterons–deuteron reactions are halted by the rapid fall in temperature and density due to the expansion. As, currently, there is no other satisfactory explanation for the deuteron–helium abundances, the theory which we have just outlined is considered to be a remarkable success of the standard cosmological model.

Note. The reader should not have the impression that everything is all right and the subject of cosmological nucleosynthesis is a closed chapter. It is true that the abundances of not only deuterium and helium, but also of lithium (of mass number 7) are presumably in agreement with the theory, but doubts have been voiced that this agreement will not survive if we take into account the chemical evolution in galaxies (Vidal-Madjar and Gry, 1984). Again the estimated baryonic mass density, while not inconsistent with luminous matter density, raises the vexatious problem of the nature of dark matter (a problem which has baffled any satisfactory solution so far).

There have been quite a number of investigations involving some novel inputs such as quark nuggets, unstable massive neutrinos, and hypothetical particles demanded by the considerations of supersymmetry. However, this is not the place to report on these highly speculative theories, but we would like to impress upon the reader that the subject is not quite closed.

Problems

1. Calculate the temperatures above which the following particles can be considered relativistic (use Boltzmann statistics):
   (a) electron; (b) proton; (c) muon; and (d) pion.

2. Estimate the age of the universe when the temperature was:
   (a) 10^{28} K; (b) 10^{12} K; and (c) 10^9 K.

3. Dimensions of length, time, and mass can be formed by combining the three universal constants c, h, and G and they are named after Planck. Calculate the Planck length, time, and mass. What was the temperature of the universe when the age was equal to the Planck time? (Order of magnitude only need be considered.)

4. If shear is present, effectively, gravitation is enhanced. What effect will it have on the abundances of deuteron and helium? (Only qualitative discussion is demanded.)

19. Structure Formation in the Universe

19.1. The Problem

The observed universe departs remarkably from the idealized Friedmann model in the form of galaxies, their clusters, and superclusters, and also large voids devoid of any luminous matter. True, the isotropy of the microwave background radiation does indicate a homogeneity and an isotropy on a large scale, but these departures from nonuniformity cannot be wished away. We may give some quantitative measure of these nonuniformities in a number of ways which we mention below.

1. *Galaxy correlations.* In the case of a perfectly random or uniform distribution the two-point correlation function $\xi(r)$, defined as the excess probability above random for a galaxy to be at a distance $\dot{r}$ from an arbitrary galaxy, should vanish. In reality, this function shows a peculiar behavior, it is positive and obeys a power law for distances up to ~ 10 Mpc, vanishes, and then becomes negative beyond some distance.
2. *The redshift for the isotropic universe should be strictly isotropic.* A departure from isotropy arises from what is called peculiar velocity, it may be identified with velocity relative to the frame in which the microwave background radiation is strictly isotropic. A coherent systematic departure may indicate the presence (or absence) of gravitating matter in the neighborhood. Thus a peculiar motion of a large number of galaxies including our local group has led to the idea of the so-called "great attractor," a large concentration of mass.
3. *There are large voids in the universe, devoid of any luminous matter.* It has been found that large, quasi-spherical, voids dominate space and the galaxies show a foamlike distribution.

The problem has been posed whether we can explain these peculiarities in the structure of the universe as a consequence of natural processes in an original out-and-out homogeneous and isotropic universe. The usual idea is that small perturbations (or perhaps not so small) have evolved to form these structure. We are then faced with a number of questions such as the following:

(a) What brought about the perturbation and what were the amplitude and spectrum of the perturbations?

(b) Do the perturbations evolve to form the structures as seen within a reasonable time span less than the age of the universe and consistent with the estimate of the age of galaxies and galactic clusters?

(c) How did the perturbation in their careers affect the microwave background and does observation of the microwave background reveal such effects?

19.2. The Linear Growth Formula

In olden days, i.e., before the impact of particle physics on cosmology, the perturbations were imagined as thermal fluctuations. Such perturbations, to start with, were small and a linearized treatment seemed adequate. We give an outline of the discussion for the fate of infinitesimal density perturbations in the background of the Friedmann universe.

We shall assume the cosmic material to have a perfect fluid energy–stress tensor

$$T^\mu_\nu = (p + \rho)v^\mu v_\nu - p\delta^\mu_\nu,$$

and the basic equations of our discussion will be the Raychaudhuri equation (17.4) and the divergence relations

$$\dot{\rho} = -(p + \rho)\theta, \tag{19.1}$$

$$\dot{v}^\nu = p_{,\mu}\frac{(g^{\mu\nu} - v\mu v^\nu)}{(p + \rho)}. \tag{19.2}$$

For the perturbed universe, the metric can be taken in the form

$$ds^2 = dt^2 + 2g_{0i}\,dx^i\,dt + g_{ik}\,dx^i\,dx^k, \tag{19.3}$$

where the Latin indices run from 1 to 3 (i.e., excludes the time coordinate) and the index 0 refers to the time coordinate. The smallness of the perturbation will mean that each of the following quantities (which vanish for the Friedmann universe) is small

$$v_i = g_{0i};\ p_{,i};\ \rho_{,i};\ g_{ik} + \frac{R_F}{(1 + kr^2/4)^2}\delta_{ik}.$$

The subscript F refers to the values in the Friedmann universe. The acceleration vector $\dot{v}^\mu$ is related to the time derivative of g_{0i} and (19.2) gives

$$\dot{g}_{0i} = \frac{(p_{,i} - \dot{p}g_{0i})}{p + \rho}, \tag{19.4}$$

whereas the vorticity tensor ω_{ik} is related to the space derivatives of g_{0i}

$$\omega_{ik} = \tfrac{1}{2}(g_{0i,k} - g_{0k,i}),$$

so that, using (19.4),

$$\dot{\omega}_{ik} = -\dot{p}\frac{\omega_{ik}}{p+\rho}.$$

In the case of vanishing pressure, ω_{ik} is thus constant and hence

$$\omega = (\tfrac{1}{2}g^{ik}g^{lm}\omega_{il}\omega_{km})^{1/2}$$

would satisfy

$$\omega = \omega_0 R_F^{-2},$$

which leads to the conservation of angular momentum. For the case of an ultrarelativistic gas $p = \rho/3$ and we obtain from the above

$$\omega = \omega_0 R_F^{-1}.$$

Thus, in either case, the vorticity decreases with expansion. In what follows we are primarily interested in the possibility of growth of perturbation and so shall neglect the effect of vorticity, i.e., spatial derivatives of the g_{0i}'s.

A straightforward calculation using (19.1) and (19.2), as well as the metric (19.3), yields

$$\dot{v}^\alpha_{;\alpha} = \frac{1}{p+\rho}[\nabla^2 p - \dot{p}(-g)^{-1/2}(g_{i0}g^{ik}\sqrt{-g})_{,k}], \qquad (19.5)$$

where ∇^2 is the Laplacian operator for the 3-space metric

$$d\sigma^2 = g_{ik}\,dx^i\,dx^k.$$

Introducing the condensation S, defined by

$$\rho = \rho_F(1 + S),$$

and the scale factor R in the perturbed universe defined by

$$\theta = \frac{3\dot{R}}{R},$$

we get from (19.1)

$$\frac{\dot{R}_F}{R_F} - \frac{\dot{S}}{3}\left(\frac{\rho_F}{\rho_F+\rho_F}\right) = \frac{\dot{R}}{R}\left[1 - \frac{p_F S - (p - p_F)}{(p_F + \rho_F)}\right]. \qquad (19.6)$$

As a simplifying assumption we assume $p = \alpha\rho$ where α is a constant; this is a justifiable assumption in the ultrarelativistic regime ($\alpha = \tfrac{1}{3}$) and also in the low temperature matter dominated phase ($\alpha = 0$). Equation (19.6) then gives

$$\frac{\dot{R}}{R} = \frac{\dot{R}_F}{R_F} - \frac{\dot{S}}{3(1+\alpha)} \qquad (19.7)$$

and, on differentiation,

$$\frac{\ddot{R}}{R} = \frac{\ddot{R}_F}{R_F} - \frac{\ddot{S}}{3(1+\alpha)} - \frac{2\dot{S}}{3(1+\alpha)}\frac{\dot{R}_F}{R_F}. \qquad (19.8)$$

Equation (19.4) may be formally integrated to give

$$g_{0i} = \frac{\alpha}{1 + \alpha} R_F^{-3\alpha} \int R_F^{3\alpha} S_{,i} \, dt + f_i R_F^{-3\alpha},$$

where the f_i's are functions of space coordinates only. The integral on the right is a gradient and hence can be removed by a coordinate transformation of the form $t = t + \phi(x^i)$. (A gradient g_{0i} does not involve vorticity.)

Further, if g_{0i} is continuous at the onset of the perturbation the f_i's must vanish. Hence the term involving g_{0i} in the right-hand side of (19.5) may be neglected. Substituting now from (19.5), (19.7), and (19.8) in the Raychaudhuri equation, we get

$$\ddot{S} + \frac{2\dot{R}_F}{R_F} \dot{S} + \alpha \nabla^2 S - 4\pi \rho_0 G(1 + 3\alpha)(1 + \alpha)S = 0. \qquad (19.9)$$

This is the well-known linear perturbation equation for the evolution of condensations.

Equation (19.9) may be handled by the method of separation of variables. Putting $S = \mu(t)\psi(r)$, (19.9) splits up into two equations

$$\ddot{\mu} + 2\dot{\mu}\frac{\dot{R}_F}{R_F} + \alpha \frac{A^2 \mu}{R_F^2} - 4\pi \rho_0 G(1 + 3\alpha)(1 + \alpha)\mu = 0, \qquad (19.10)$$

$$\frac{1}{q^{3/2}r^2}\frac{\partial}{\partial r}\left\{q^{1/2}r^2\frac{\partial\psi}{\partial r}\right\} + \frac{1}{r^2 \sin\theta}\frac{\partial}{\partial\theta}\left\{\sin\theta\frac{\partial\psi}{\partial\theta}\right\} + \frac{1}{r^2}\frac{\partial^2\psi}{\partial\phi^2} = -A^2\psi, \qquad (19.11)$$

where $q \equiv (1 + kr^2/4)^{-2}$ and A^2 is the separation constant.

We are interested in perturbations extending over only a small part of the universe, so we may take $|kr^2/4| \ll 1$, $q \approx 1$. In that case, the radial part of ψ which is regular at the origin is of the form $\sin Ar/r$. We can then take $\lambda = 2\pi A^{-1}R_F$ to define the linear dimension of the perturbation. As $C_s^2 = \partial p/\partial \rho = \alpha$, where C_s is the velocity of compressional waves, (19.10) may be rewritten in the form

$$\ddot{\mu} + 2\dot{\mu}\frac{\dot{R}_F}{R_F} + \left[\frac{C_s^2 A^2}{R_F^2} - 4\pi \rho_0 G(1 + 3\alpha)(1 + \alpha)\right]\mu = 0. \qquad (19.10)$$

Consider, for simplicity, the case of the static background ($\dot{R}_F = 0$) and put $\mu = \mu_0 e^{i\beta t}$. We get

$$\beta = \pm\left[\frac{C_s^2 A^2}{R_F^2} - 4\pi \rho_0 G(1 + 3\alpha)(1 + \alpha)\right]^{1/2}.$$

Obviously, the perturbation will have a possibility to grow with time if β is imaginary, i.e., if

$$4\pi \rho_0 G(1 + 3\alpha)(1 + \alpha) > \frac{C_s^2 A^2}{R_F^2}.$$

Thus the condition for the perturbation to grow (or the system to the unstable) is that the linear dimension λ must exceed a critical value λ_c given by

$$\lambda_c = \pi C_s [G\rho_0 (1 + 3\alpha)(1 + \alpha)]^{-1/2}. \tag{19.12}$$

This condition for instability for $\alpha = 0$, namely, $\lambda > \pi c_s (G\rho)^{-1/2}$ was obtained by Jeans on the basis of Newtonian physics and is known as the Jeans condition of instability. It is obviously not valid for the nonstatic cosmological situation. The expansion of the universe ($\dot{R}_F > 0$) is likely to hinder the growth of condensation and thus the Jeans condition may be necessary but not sufficient for the growth of condensation.

We can translate the constraint on linear dimensions to one on the mass of the condensation. Thus condensations can grow only if its mass exceeds the critical value M_c given by

$$M_c \approx \frac{4\pi}{3} \lambda_c^3 \rho_0 \approx \tfrac{4}{3}\pi^4 C_s^3 \rho_0^{-1/2} (1 + 3\alpha)^{-3/2} (1 + \alpha)^{-3/2}.$$

According to the above relation, in the radiation-dominated early universe M_c will increase rapidly as the temperature falls (because $C_s^2 = \tfrac{1}{3}$ and $\rho_0 \propto T^4$) attaining a peak value just prior to the recombination of protons and electrons to form neutral hydrogen. At recombination C_s drops sharply to a low value and M_c will also show a large drop. After recombination, M_c continues to decrease due to the decrease of C_s (with a fall in temperature) which dominates over the decrease of ρ_0.

Consider how the Jeans mass varies with epoch. Before the recombination of protons and electrons to form neutral hydrogen, we have a mixture of radiation with pressure $p_\gamma = \rho_\gamma/3$ and of matter with density ρ_m whose pressure is negligible compared to p_γ. However, due to the coupling between the ionized matter and the radiation via Thomson scattering the temperature of matter and radiation remain the same, although their interconversion is negligible. We have $\rho_\gamma R^4 = $ constant, $\rho_m R^3 = $ constant, and the sound velocity C_s comes out as

$$C_s^2 = \left(\frac{\partial p}{\partial \rho}\right)_s = \left(\frac{\partial p}{\partial \rho_\gamma}\right)\left(1 + \frac{\partial \rho_m}{\partial \rho_\gamma}\right)^{-1}$$

$$= \frac{1}{3}\left(1 + \frac{3}{4}\frac{\rho_m}{\rho_\gamma}\right)^{-1}.$$

Thus, at this stage, C_s does not change very much and the change of the Jeans mass occurs primarily due to the decrease in ρ due to expansion. Thus there is a steady increase of the Jeans mass reaching a maximum value just before recombination.

After recombination, C_s drops abruptly to only a few km s^{-1} (in units $C = 1$, $C_s \sim 10^{-5}$). Consequently, the Jeans mass has a sharp fall (note the occurrence of the cube of C_s in the expression for the Jeans mass). Numerical computation shows that the Jeans mass drops from $\sim 10^{18} M_\odot$ to $\sim 10^6 M_\odot$. As the universe cools still further, the sound velocity C_s varies as the square root of the temperature and the Jeans mass falls approximately as the $\frac{3}{2}$th power of the temperature.

However, if the condensation is in the radiation dominated era, there is an additional constraint to its growth. This arises from a diffusion of radiation out of the condensed region. In view of the scattering of photons, we may consider a mean free path for them $L_f = (\sigma n_e)^{-1}$, where σ is the Thomson scattering cross section for electrons and n_e their number density. Obviously, the fluctuations will be smoothed out if the mean free path is larger than (or comparable to) the wavelength of the fluctuation. This leads to a lower bound for the mass of condensations to grow, usually, called the Silk mass, whose value comes out to be 10^{13}–10^{14} $M_{\odot}$. Hence, finally, we are led to the following conclusions.

Condensations of mass $> 10^{14}$ $M_{\odot}$ (which is of the order of the galactic cluster masses) can grow in the radiation era for only a limited period prior to recombination, while in the post-recombination phase the bound is much less ($< 10^6$ $M_{\odot}$).

The interesting (or rather disturbing) point that emerges is that the characteristic galactic mass (10^{11}–10^{12} $M_{\odot}$) does not appear with any particular significance, and the indication is that in the above theory the galactic clusters were the primary products of fluctuations and the galaxies came into being later from a break-up of the clusters.

However, when we integrate (19.9) we find for the two limits $\alpha = \frac{1}{3}$ and $\alpha = 0$ (using the appropriate dependence of R_F and ρ_0 on time t)

$$S = at + bt^{-1} \qquad (\alpha = \tfrac{1}{3})$$
$$= at^{2/3} + bt^{-1} \qquad (\alpha = 0).$$

There are, in either case, growing modes but the growth is only to a very limited extent. During the later stage of the pre-recombination era there had been no growth, for the Jeans mass at this stage exceeded the mass of any condensation with $M < 10^{18}$ $M_{\odot}$. Hence not much error is committed by considering the growth to be only in the matter-dominated era ($\alpha = 0$), i.e., from $T < 4000$ K. With the $\frac{2}{3}$rd power relation this would give an amplification of S by a factor of $\sim 10^3$ (the exact value depends on the value of q_0). This is too small if the original perturbation is thermal. Anyway, we may back-calculate and say that the perturbation $\delta\rho/\rho$ (whatever its origin) must have been greater than 10^{-3} at the recombination epoch. However, such $\delta\rho/\rho$ should leave an imprint $\delta T/T$ of the same order in the microwave background on the small angle scale. But observations set an upper bound to $\delta T/T$ which is less, namely, $\delta T/T < 10^{-4}$.

All the foregoing considerations have been based on the idea that the matter is essentially baryonic. However, as we have already seen, by far the predominant constituent of matter is dark which is presumably not baryonic. Indeed, the constraints on baryonic density from the cosmological nucleogenesis calculation, as well as the present difficulty in explaining structure formation, are strong arguments in favor of the nonbaryonic nature of dark matter. However, before going into the discussion of galaxy formation with nonbaryonic matter, we shall have a look at the development of finite perturbations.

19.3. Finite Perturbation

The discussion of finite perturbation has been on the basis of some simplifying assumptions. We assume that the perturbation, to begin with, is infinitesimal in a background universe of zero spatial curvature. After a period of growth, a spherically symmetric region is cut off from the background universe and its subsequent evolution is calculated on the basis of Newtonian physics. Just after this cut-off, the spherical region (the perturbed density is assumed to be uniform in this region) still expands although the rate of expansion is somewhat less than in the background universe. However, as the perturbed density is somewhat greater than the critical density, the spherical region has a stage of maximum expansion and then contracts. Even the collapse is not allowed to proceed indefinitely (to a singularity) but within the collapsing sphere, smaller objects like stars form in which energy generation due to thermonuclear reaction sets in. It is then no longer proper to make the continuous gas approximation in the study of the perturbed region. We consider an assembly of particles moving under their mutual gravitational attraction. "N-body simulation" of dissipationless gravitational collapse has been investigated and the result shows that there is first a dense crunch and, finally, a centrally condensed distribution which is remarkably stable. This relaxation to the equilibrium state occurs mainly under the influence of the collective gravitational field and has a time scale of $\sim (GP)^{-1/2}$ which is much smaller than the typical time scale for relaxation due to collision. It is thus called "violent relaxation."

In the final equilibrium state, the virial theorem is satisfied, so that the total energy is equal in magnitude to half the potential energy in this stage. Again in the state of maximum expansion prior to the collapse, the total energy is the same as the potential energy. Combining these facts with the result that the gravitational potential energy varies inversely as the linear dimension of the system, we get the result that the equilibrium radius of the system is just half the maximum radius.

Theoretical considerations have shown that in the equilibrium state following violent relaxation, the distribution is Maxwellian in velocities, independent of the mass of the particles. Still the distribution is commonly referred to as isothermal.

The above picture of collapse suffers from several obvious defects. It is unlikely that perturbation would be strictly spherical and we would expect the perturbed density would show at the beginning a maximum at the center and decrease continuously to merge with the background density at the edge of the perturbed region. Then again the use of Newtonian physics leaves out possible general relativity effects. The N-body simulation assumes dissipationless collapse; while this may be true for dark matter (i.e., matter that does not interact with electromagnetic radiation), charged baryonic matter is likely to emit radiation. As a result we would expect the baryonic matter to lose its energy and sink towards the center while the outer region would become richer in dark matter.

19.4. Structure Formation with Dark Matter

Accepting the usual argument for the nonbaryonic nature of dark matter, we are faced with the question: What is it made of? Unfortunately, no clear or satisfactory answer to this question seems available; indeed, dark matter remains enigmatic. Still, we just mention the different ideas that have been advanced.

The possible constituents of dark matter have been classified into three types—hot, warm, and cold. They differ essentially so far as the streaming velocity is concerned, and this in turn may arise either due to their difference in masses or due to some peculiar circumstances such as Bose condensation.

The popular hot dark matter constituent is the light neutrino which is not massless and there is some observational support for the nonzero mass of some neutrinos. We can put an upper bound to the neutrino masses from the observed limit to the present energy density in the universe. In our study of the thermal history of the universe we have seen that following annililation of electron–positron pairs, the neutrino temperature becomes $(\frac{4}{11})^{1/3}$ times the radiation temperature. Assuming that the neutrinos were at that temperature ultrarelativistic, i.e., $m_\nu c^2 \ll 1$ MeV, we get, for the ratio of the number density of neutrinos of any species n_ν to the number density of photons, the value

$$\frac{n_\nu}{n_\gamma} = \frac{3}{4}\left(\frac{4}{11}\right).$$

The factor $\frac{3}{4}$ comes from the relation between the Fermi and Bose integrals

$$\int_0^\infty \frac{x^2\,dx}{e^x + 1} = \frac{3}{4}\int_0^\infty \frac{x^2\,dx}{e^x - 1},$$

and the factor $\frac{4}{11}$ from the T^3 dependence of the number density for relativistic particles. The ratio n_ν/n_j remains frozen to the above value during the subsequent expansion as the neutrino and photon numbers are both conserved. Hence using 2.7 K for the present temperature of the MBR we get $n_\nu \sim 10^2$. With the present cosmological density $\rho_0 \lesssim 10^{-29}$ g cm^{-3}, i.e., the possible upper bound, we get

$$\sum_{\substack{\text{sum over all} \\ \text{neutrino species}}} \lesssim 10^2 \text{ eV}.$$

After decoupling, the neutrinos stream freely with a velocity of $\sim c$ till the temperature falls to a level when $m_v c^2 > KT$. The age of the universe corresponding to $KT \sim 100$ eV is 10^{12} s, so that the distance covered by freely streaming neutrinos is $\sim 10^{22}$ cm. Hence at that epoch a fluctuation could survive only if its linear dimension was greater than $\sim 10^{22}$ cm. The corresponding mass, called the Jeans mass for free streaming comes out as $\sim 10^{15} M_\odot$ (the energy density in the universe at a temperature of ~ 100 eV is $\sim 10^{-18}$ g cm^{-3}).

Thus, we could expect that if the dark matter is composed of neutrinos, the primary structure would be of a mass characteristic of superclusters and that the galaxies came into being as a result of the fragmentation of the superclusters. We are thus led to conclude that galaxies are younger than superclusters and they were formed fairly recently ($z < 2$). However, these are not in conformity with observations and a number of galaxies having $z > 3$ have been observed, while quasar redshifts go to $z > 4$.

There are also other problems with a neutrino dark matter picture; we shall not go into them here. We shall just mention that the cases of warm and cold dark matter also face difficulties. The most important fact is that there is a natural inhibition in accepting the hypothetical particles like the photino, the axion, etc., or quark nuggets which no one has observed as yet and are dubious products of speculations in theoretical physics.

The warm dark matter candidates may have a mass of ~ 1 keV and consequently the Jeans mass for free streaming is $\sim 10^{12} M_\odot$. (Note that the temperature at which a particle becomes nonrelativistic, and so free-streaming becomes ineffective, varies directly as the mass and the corresponding age of the universe varies inversely as the square of the mass, hence the critical linear dimension L varies as the square of the mass. The Jeans mass which is proportional to $L^3 \cdot nm$ varies as m^{-2}.) For the warm particles the large galaxies are thus the primary condensations and the smaller ones, as the dwarf sphericals, result by fragmentation of these. But the dwarf galaxies cannot keep the warm particles confined and hence the amount of dark matter should be comparatively poorer in the dwarf sphericals than in the large galaxies. However, observations do not support this prediction.

With cold matter, the universe should be much more homogeneous than we actually find it. In particular, while we observe a coherent non-Hubble velocity of ~ 600 km s^{-1} for a region of ~ 60 Mpc, the cold dark matter prediction is that for regions of ~ 40 Mpc, there should not be any coherent peculiar velocity.

We shall consider the problem of the origin of fluctuations after our discussion of the inflationary universe scenario.

Problems

1. Picture a nonhomogeneous universe in the form of a spherical condensation separated from the Friedmann universe by a vacuous region. In the vacuum, consider a free particle falling radially. What will be its peculiar velocity?

2. Deduce the Jeans formula on the basis of Newtonian physics.

3. Estimate the Jeans mass at different stages in the universe.

20. Grand Unified Theory and Spontaneous Symmetry Breaking

20.1. Introduction

We proceed to give the reader an idea of the grand unified theory (referred to in abridged form as GUT) which has come to play an important role in cosmology. GUT is a mathematically highly sophisticated and complicated theory and we shall make no attempt to present a comprehensive report on its different aspects. Indeed, we shall merely try to make the reader familiar with the jargon that is commonly used in the application of GUT to cosmology. We begin with gauge fields.

20.2. Gauge Fields

Consider the Schrödinger equation for a particle. (We may as well begin with the Dirac equation; however, we think the average reader will find the discussion of the Schrödinger equation a little easier.) The equation being linear, the solution ψ is determined only up to an arbitrary constant multiplier. The normalization condition fixes the modulus of this multiplier but still leaves a phase factor $e^{i\theta}$ undetermined where θ is a real constant. Now suppose we make θ a function of the coordinates and time. The resulting ψ is still a solution of the Schrödinger equation only if the Hamiltonian includes an electromagnetic interaction term, in that case corresponding to the change in ψ, there is a change of the vector potential from $\mathbf{A}$ to $\mathbf{A} - \nabla \cdot \theta$, but such a change does not alter the electromagnetic field $F_{\mu\nu} \equiv A_{\mu;\nu} - A_{\nu;\mu}$. Thus, while in the absence of electromagnetic interaction, ψ merely admits a group of transformations with a constant parameter, the group involving a variable parameter is admitted if there is an electromagnetic interaction. This one-parameter group is referred to as U(1) (U coming from the word unitary which describes the matrices representing the group), and the process of changing from constant θ to variable θ is known as gauging the group. We shall hypothesize that this association of gauging a group with an interaction is nothing peculiar with the electromagnetic field but is a general property of all interactions. Indeed, we may utilize this to discover different interactions and explore their properties. In particular, we insist on a Lagrangian which has the proper gauge symmetry and then obtain both the equation for ψ and the field equation for the interaction by the appropriate variational procedure. We next take the case of weak interaction.

20.3. Weak Interaction

The weak interaction operates between leptons. There are three pairs of leptons; first, the charged particles, namely the electron, the muon and the tau and their neutral partners, the neutrinos, which are of much smaller mass (if not altogether massless). This threefold characteristic is described by saying that there are three flavors of leptons.

Consider one such pair of interacting leptons; to fix our ideas, let us say the electron and the electronic neutrino. Although the electron and the neutrino are manifestly different, the theory in its elementary form will consider them as two states of the same particle. The state function ψ will then consist of two components, one for the electronic state say U_1 and the other for the neutrino state say U_2. However, we can make a linear combination of these components with constant coefficients, i.e., introduce $\bar{U}_1$ and $\bar{U}_2$ such that

$$\bar{U}_i = a_{ik} U_k,$$

where i and k run over the values 1 and 2 and there is a summation over the repeated index k. The $\bar{U}_i$ also serve as a basis. The system thus admits a group of transformations represented by (2×2) matrices. As the elements of this matrix may be complex, there are apparently eight numbers involved in the transformation. However, in view of the relations

$$\bar{U}_i^* \bar{U}_i = U_i^* U_i,$$

we have $a_{ik} a_{il} = \delta_{kl}$. This gives four relations between the a_{ik} and, further, the determinant of the coefficients a_{ik} is of modulus unity. Thus by adjusting the phase factor (which we have already considered), we may have an additional constraint on a_{ik}, namely, that their determinant has the value $+1$. Hence, finally, the number of independent parameters is just 3 ($=8 - 5$). In a similar manner, if there are n component wave functions, the number of independent parameters is seen to be $n^2 - 1$. We now apply the procedure of gauging the group, i.e., replace the three constant numbers by three functions and thereby arrive at an interaction field. A quantization of the field heads to the idea of three bosons (electrically positive, negative, and neutral, called, respectively, the W^+, W^-, and Z bosons) which mediate the weak interaction, just as the quantization of the electromagnetic field leads to the photon.

This gauge field corresponds to the SU(2) group (S indicating that the group is simple), U(2) tells us that the symmetry transformations involve unitary (2×2) matrices). The next step consists an attempt to unify the weak with the electromagnetic interaction which is formally achieved by a combination of the SU(2) group (of weak interaction) with the U(1) group (of electromagnetic interaction) into the product group SU(2) × U(1). We are thus led to a picture in which three bosons of the weak interaction and the photon stand on somewhat identical footing. However, there is a conspicuous difference observationally, the weak interactions are of short range whereas the electromagnetic interaction is of long range. Again this means that while the photon is massless, the bosons, mediating the weak interaction, must have a nonvanishing mass. In short, we have a somewhat paradoxical situation, we have put the weak and electromagnetic interaction within the same framework although there is a manifest difference between them. The way out was pointed out by Higgs and the mechanism is known as spontaneous symmetry breaking.

To get an idea of spontaneous symmetry breaking, consider a scalar field ϕ for which the potential function $V(\phi)$ is of the form

$$V(\phi) = -a\phi^2 + b\phi^4,$$

where a and b are positive constants. As $V(\phi) = V(-\phi)$ we say that $\phi = 0$ is the symmetric state. However, the minimum of $V(\phi)$ occurs at $\phi = \pm(a/2b)^{1/2}$ whereas $\phi = 0$ gives a maximum of $V(\phi)$. The minimum energy states are known as true vacuum and there the field has the asymmetric value of either $+(a/2b)^{1/2}$ or $-(a/2b)^{1/2}$, while the symmetric state is one of unstable equilibrium, and is referred to as false vacuum.

Symmetry breaking in the case of an electroweak interaction is however more complicated. Here we have four fields corresponding to the four bosons. The minimum of energy occurs in a situation where, of the four fields, we have a nonzero value, the symmetry breaking consists of selecting this one. It is this breakdown of symmetry that leads to three of the bosons becoming massive while the photon remains massless. The masses of the W and Z bosons are nearly 80 GeV and 90 GeV, respectively.

20.4. Strong Interaction and Grand Unification

Until recently, i.e., before the idea of quarks gained general acceptance, strong interaction sources were thought to be hadrons and mediating particles the pions. However, today the quarks are looked upon as the primary sources and the mediating particles are called gluons. Unlike the case of photons which do not carry any charge, and so do not interact amongst themselves, the gluons possess the corresponding "charge" (called the color charge) and hence interact among themselves. Again, just as leptons occur in three flavors, quarks also have three flavors but unlike the weak interactions which take place between two leptons (e.g., the electron and the electronic neutrino), in the strong interaction the number of participants is three, their distinct character is indicated by saying that there are three color states—red, blue, and green. Thus while the weak interaction is associated with the gauging of the SU(2) group, the relevant group is now SU(3). Such a group will involve eight parameters and thus we are led to the existence of eight bosons (i.e., the just-mentioned gluons) mediating the interaction between quarks.

Grand unification consists of obtaining all three fields, the strong, the weak, and the electromagnetic, from gauging a single group. The simplest group which can realize this objective is SU(5) and it involves altogether 24 bosons, thus there should be 12 as yet undiscovered bosons besides the eight gluons, the Z and W^- bosons, and the photon. (Of course, unification may be achieved by taking groups of an even larger number of parameters but then the theory will increase in complexity.)

The 12 additional bosons which follow as a consequence of the SU(5) theory can mediate reactions which convert quarks into leptons and vice versa. We can thus understand the equality of charges of the leptons and, say, the proton. Again, the possibility of such reactions means a departure from baryon conservation. Thus neither leptons nor baryons are conserved but the difference between the baryon number and the lepton number turns out to be a conserved quantity. However, in order that observed constraints on the proton decay may not contradict the theory, we have to assume that these bosons have masses at least as high as 10^{15} GeV.

The three interactions which are sought to be unified in GUT are apparently of very different strengths. One way to quantify the strength of an interaction is to calculate the value of the nondimensional parameter g^2/hc where g is the relevant coupling constant. However, if we take into account the role of the vacuum as a seat of virtual particle–antiparticle pair, the coupling constants become functions of the energies of the interacting particles. Consider, for example, the interaction between two electrons. There will be an excess of virtual positrons in the neighborhood of each electron due to the attraction between dissimilar charges and hence the effective field at a large distance from an electron is considerably reduced, and we may say that the coupling, i.e., the effective charge of the electron, is much less than the intrinsic charge of the electron. However, as the energies of the electrons increase, they approach each other closely, penetrating more and more the surrounding positron cloud so that the effective charge of the electrons goes on increasing. We may thus say that so far as electromagnetic interaction is concerned the coupling constant will increase with an increase in energy of the interacting particles.

Quite a different picture emerges when we consider the strong interaction. Here like charges (i.e., colors) attract one another, so that the virtual cloud enhances the effect. Again as the approach becomes closer, the effect of the virtual cloud is reduced and, consequently, the coupling constant becomes weaker. Ultimately, in the limit of vanishing distance, the interaction disappears altogether and we have the so-called "asymptotic freedom" for quarks. Thus, as the energy increases, the different interactions behave differently and calculation leads us to expect that if the particle energies are as high as $\sim 10^{15}$ GeV (see the masses of the superheavy bosons of SU(5) theory) all three interactions will have identical coupling constants. Energies of this order are referred to as GUT energy where we expect a perfect unification of the three interactions.

But such energies are enormously high, far beyond those that we obtain from accelerators. The only situation where conceivably such energies may occur is the early universe, recall that in big bang models, as we approach the singularity $(R \to 0)$, $T \to \alpha$. Thus, we may speculate that at sufficiently high temperatures, when the typical particle energies KT are much higher than the GUT energy, there will be full SU(5) symmetry, but as the temperature goes below the GUT energy there will be a spontaneous breakdown of symmetry from SU(5) to the product group SU(3) $\times$ SU(2) $\times$ SU(1). To get an insight into this symmetry breaking process, it is necessary to have an idea of the dependence of the potential on temperature.

Consider a real scalar field ϕ with the Lagrangian density

$$L = \frac{1}{2}\left[\left(\frac{\partial \phi}{\partial t}\right)^2 - \left(\frac{\partial \phi}{\partial x}\right)^2 - \left(\frac{\partial \phi}{\partial y}\right)^2 - \left(\frac{\partial \phi}{\partial z}\right)^2 \right] - V,$$

where V, at $T = 0$, is given by

$$V(\phi) = -\tfrac{1}{2}\mu^2\phi^2 + \left(\frac{\lambda}{4}\right)\phi^4.$$

In this case $V(\phi)$ has a maximum at $\phi = 0$ and the minima occur at values of $\phi = \pm\mu/\sqrt{\lambda}$. Thus we expect that fluctuations would lead to a transition from the state $\phi = 0$ (the false vacuum) to the broken symmetric state $\phi = +\mu/\sqrt{\lambda}$ or $-\mu/\sqrt{\lambda}$ (the true vacuum). Things, however, become more interesting when we introduce the temperature dependence. Then if

$$V(\phi, T) = -\frac{\pi^2}{90}T^4 + \frac{1}{8}\left[\lambda\phi^2 - \frac{\mu^2}{3} \right]T^2 + \frac{1}{2}\left(\frac{\lambda}{2}\phi^2 - \mu^2 \right)\phi^2,$$

the extrema of V (i.e., $\partial V/\partial \phi = 0$) occur at

$$\phi = 0 \quad \text{and} \quad \phi = \pm\left[\frac{\mu^2}{\lambda} - \frac{T^2}{4} \right]^{1/2}.$$

Thus there is a critical temperature of $T_c = 2\mu/\sqrt{\lambda}$ above which there is the only extremum at $\phi = 0$ and there

$$\frac{\partial^2 V}{\partial \phi^2} = \tfrac{1}{4}\lambda T^2 - \mu^2 > 0.$$

Thus, above T_c, the field will have the symmetric state as the ground state (this is referred to as the restoration of symmetry at high temperature). Below T_0, however, $\phi = 0$ gives a maximum of $V(\phi)$ and the minimum of ϕ occurs at two values of ϕ and thus the stable states are the symmetry broken states.

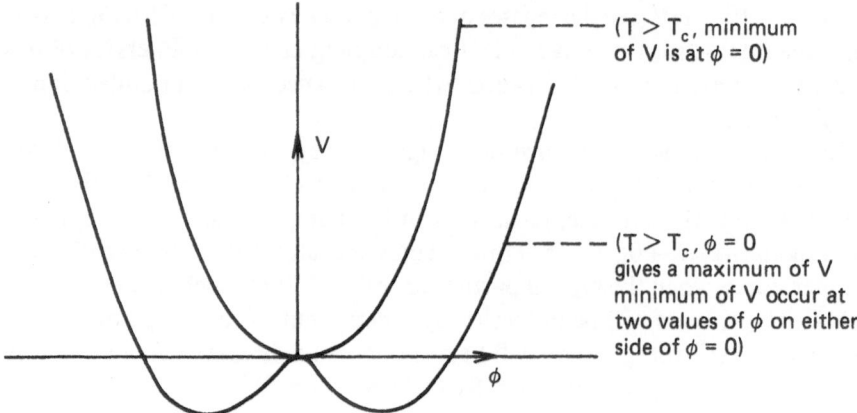

Figure 20.1. Zero of potential V has been fixed at $\phi = 0$ for both T.

In this case, there is no potential barrier between the metastable state and the stable state and the phase transition from the metastable state to the stable state is of second order.

However, a second-order phase transition occurs instantaneously and the phase transition originally considered in connection with inflationary models was of first order and takes place over a long time scale. In this case, there is, as in the previous case, complete symmetry above a temperature T_c. Below T_c, the global minimum of the potential occurs at a symmetry broken state $\phi \neq 0$; nevertheless, $\phi = 0$ still gives a minimum of V. Intervening the two minima is a maximum of V and thus classically the transition from the symmetric state to the broken symmetric state is forbidden. Quantum mechanically, however, a transition does take place by "tunneling" through the potential barrier but the time scale is large depending on the form of the barrier (see Fig. 20.2). The potential curves show a second critical temperature T_c' below which $V(\phi)$ has a maximum at $\phi = 0$; the potential barrier no longer exists and the transition again becomes very rapid (Sato, 1981).

The Coleman–Weinberg (1973) potential for symmetry breaking, SU(5) − SU(3) × SU(2) × U(1), is

$$V(\phi, T) = \left(\frac{18T^4}{\pi^2}\right)\int_0^\infty dx\, x^2 \ln\left\{1 - \exp\left[-\left(x^2 + \frac{25g^2\phi^2}{8T^2}\right)\right]\right\}$$
$$+ \left(\frac{5625}{512\pi^2}\right)g^4\left(\phi^4 \ln\left[\frac{\phi}{\phi_0}\right] - \frac{\phi^4}{4} + \frac{\phi_0^4}{4}\right),$$

where $\phi_0 \sim 10^{14}$–10^{15} GeV and $g^2 \sim \frac{1}{3}$ is the gauge coupling constant. For $T \gg \phi_0$, the only minimum of $V(\phi, T)$ occurs at $\phi = 0$, which means that at these high temperatures, the symmetry is restored. For lower temperatures, the global minimum of $V(\phi)$ is at $\phi \approx \phi_0$. However, $\phi = 0$ is always a local

[Zero of potential V has been fixed at $\phi = 0$ for both T]

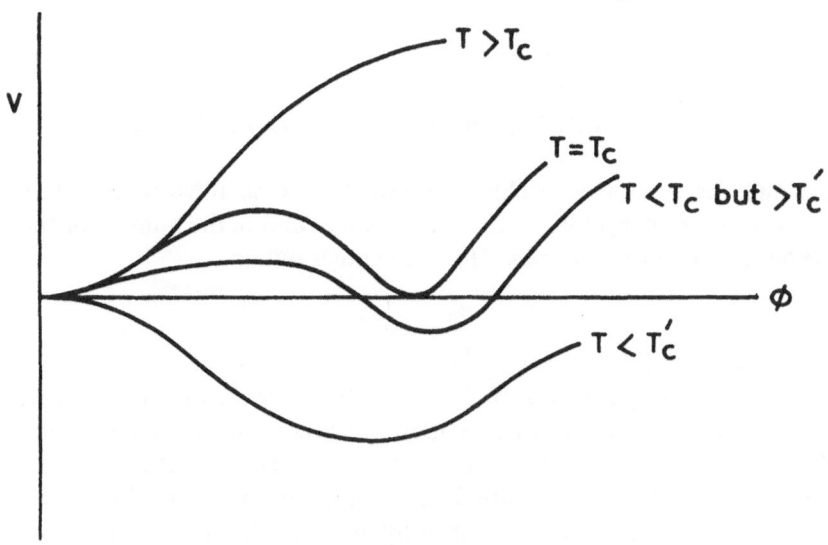

Figure 20.2

minimum of $V(\phi, T)$, as for small ϕ the potential expression is (for $T \neq 0$)

$$V(\phi, T) = \tfrac{75}{16} g^2 T^2 \phi^2 - \left(\frac{5625}{512\pi^2}\right) g^4 \phi^4 \ln\left(\frac{M_x}{T}\right) + \frac{9}{32\pi^2} M_x^4,$$

with

$$M_x^2 = \tfrac{25}{8} g^2 \phi_0^2.$$

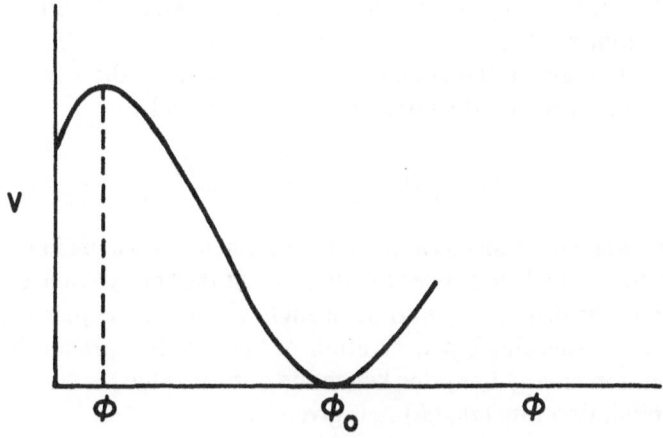

Figure 20.3

The potential has a small hump at, say, $\phi = \phi_1$ and is of the form shown in Fig. 20.3 for $T \ll \phi_0$. The phase transition takes place by tunneling through the barrier.

20.5. Baryon Asymmetry and the Baryon/Photon Ratio

Assuming the microwave background radiation to be universal and the dark matter to be nonbaryonic, we can estimate the ratio of the number density of baryons n_B to that of photons. The ratio comes out as

$$\frac{n_B}{n_\gamma} \sim 10^{-9},$$

with an uncertainty factor of 10. Linked with this is the apparent absence of antibaryons; so far as our own neighborhood is concerned, there is conclusive evidence against the presence of antimatter and there is hardly any reason to believe in the occurrence of antimatter in more distant parts of the universe. Yet, theoretically, it seems natural to hypothesize that in the early universe when the temperature and density were sufficiently high, baryon and anti-baryons could readily be produced and annihilated, and a thermodynamic equilibrium prevailed in which the net number of baryons vanished and, separately, the number of baryons (or antibaryons) was of the same order as the number of photons. (The difference in the baryon–photon number arose solely due to the different statistics and the weight factor g.) How are we then to understand the present state of affairs?

Let us try to work out the evolution of the baryon/photon ratio, assuming for the moment that the baryon conservation principle strictly holds good. As the universe cools down, the number density of baryons (and antibaryons) is reduced in step with photons maintaining the ratio near unity as both vary at T^3. This is the situation as long as the temperature T is high enough to allow neglect of the baryonic rest mass (i.e., $KT \gg M_B c^2$, the ultrarelativistic regime for baryons). With a further fall in temperature, when $M_B c^2 > kT$ (the nonrelativistic regime), the equilibrium distribution of the baryons assume the Boltzmann form and the number density is given by

$$n_B = \frac{g}{\hbar^3} \left(\frac{M_B kT}{2\pi} \right)^{3/2} e^{-M_B c^2 / KT}. \tag{20.1}$$

The exponential factor means a rapid decrease in n_B with a fall in temperature compared to n_γ which continues to fall at T^3. If thermodynamic equilibrium is to prevail, this drop in n_B is to be brought about by the pair annihilation of baryons. (At this stage, pair creation of baryons has practically stopped because photon energy has gone below the critical value.)

The annihilation rate $(dn_B/dt)_{\mathrm{ann}}$ is given by

$$\left(\frac{dn_B}{dt} \right)_{\mathrm{ann}} = -\sigma n_B^2 v, \tag{20.2}$$

where σ is the cross section for annihilation and v is a measure of the velocity of baryons which may be put equal to c without sensible error. On the other hand, the rate of change of n_B required for thermodynamic equilibrium is from (20.1), and the relation $T = \text{constant} \cdot t^{-1/2}$

$$\left(\frac{dn_B}{dt}\right)_{eq} = -\frac{M_B c^2}{2KT}\frac{n_B}{t}, \tag{20.3}$$

where we have considered only the exponential term which dominates the change in n_B. Equilibrium prevails as long as the annihilation rate, given by (20.2), is faster than that given by (20.3). However, as n_B falls, the annihilation rate (being proportional to n_B^2) will at one stage be less than the rate given by (20.3). The thermodynamic equilibrium will then be lost and the baryon/antibaryon numbers remain frozen. Physically, annihilation has become rare because of the low density and large separation and hence this freezing. From (20.2) and (20.3) this occurs for

$$\sigma n_B \approx \frac{M_B c}{2KT}\frac{1}{t}. \tag{20.4}$$

Plugging in the empirically obtained value of σ and using the proton mass for M_B, we find from (20.4) and (20.1) that $KT \sim 20$ MeV. With this value of KT in (20.1), and recalling that the photon number density is

$$n_\gamma \sim \left(\frac{KT}{c\hbar}\right)^3,$$

we get $n_B/n_\gamma \sim 10^{-18}$. This is, however, far less than the observed value of n_B/n_γ.

An obvious way out is to devise a mechanism for the separation of baryons and antibaryons at a much earlier epoch, so that there may be more surviving baryons. Unfortunately, such processes cannot give separated masses of the order of galactic masses due to the limitation set by the existence of horizons.

The presently interesting theories of the baryon–photon ratio envisages the possibility of baryon nonconserving interactions. As we have already seen, according to GUT, there are bosons which can violate the baryon-conservation principle but the stability of protons require them to be superheavy (i.e., $m > 10^{15}$ GeV). However, merely the existence of such an interaction is not enough to lead to a net baryon number differing from an original value zero. The other requirements are:

(i) C and CP violations—again this is admitted in the GUTs; and
(ii) a departure from thermodynamic equilibrium.

To see the necessity of these requirements, consider the scheme of baryon nonconserving reactions:

(i) $A \rightarrow B$ $(\Delta n_B > 0, \text{say})$,
(ii) $\bar{A} \rightarrow \bar{B}$ $(\Delta n_B < 0)$,
(iii) $B \rightarrow A$ $(\Delta n_B < 0)$,
(iv) $\bar{B} \rightarrow \bar{A}$ $(\Delta n_B > 0)$,

where overbars indicate antiparticles. If C and CP strictly hold good, reactions of types (i) and (ii) exactly counterbalance each other's effect (and so do (iii) and (iv)). If there is thermodynamic equilibrium then according to the principle of detailed balance (i) and (iii) (and (ii) and (iv)) neutralize each other. The need for thermodynamic equilibrium also follows from the statistical distribution formulas. The creation–annihilation reaction

$$\mathbf{B} + \overline{\mathbf{B}} \to \gamma + \gamma$$

shows that the chemical potential $\mu_{\mathbf{B}} = \mu_{\overline{\mathbf{B}}}$. Again a baryon nonconserving reaction like

$$\mathbf{B} + \overline{\mathbf{B}} \to \mathbf{B} + \overline{\mathbf{B}} + \mathbf{B}$$

shows that $\mu_{\mathbf{B}} = -\mu_{\overline{\mathbf{B}}} = 0$. Also, from CPT, $M_{\mathbf{B}} = M_{\overline{\mathbf{B}}}$. Hence, in equilibrium, the distributions will be identical, i.e., $n_{\mathbf{B}} = n_{\overline{\mathbf{B}}}$.

To examine the question of thermodynamic equilibrium, consider the decay of superheavy bosons. The rate of decay is $\Gamma_{\mathbf{D}} = \alpha g_* M_x$ where α is the coupling strength of the boson with fermions, and g_* is an effective sum of the statistical weight factors g:

$$g = \sum_{\text{bosons}} g + \tfrac{7}{8} \cdot \sum_{\text{fermions}} g,$$

where the sum extends over all particle species and M_x is the mass of the boson. We may neglect annihilation in the case of these bosons at the crucial temperature $T \sim M_x c^2/K$ as they have a much longer time scale. The other relevant rate concerns the fall in temperature

$$\frac{dT}{T} = H \sim \frac{1}{2t^{1/2}}.$$

There will be thermodynamic equilibrium at $T \sim M_x c^2/K$ if $\Gamma_{\mathbf{D}} \gg H$ and no baryon asymmetry will arise. If, however, $\Gamma_{\mathbf{D}} < H$ there will be a thermodynamic nonequilibrium in the form of an excess number of the bosons because the decay lags behind the fall in temperature. A baryon asymmetry may then arise. Writing out the values of $\Gamma_{\mathbf{D}}$ and H, the condition of nonequilibrium reduces to $M_x > g_*^{-1/2} \cdot \alpha \cdot 10^{19}$ GeV or $M_x > 10^{16}$ GeV (note that $g_* \sim 160$ and $\alpha = 1/45$ according to GUT models). This bound to the mass of X bosons is consistent with the bound we have arrived at from the stability of protons.

To get more specific ideas about baryon asymmetry, consider the decay of these bosons along two distinct channels. Let the subscripts 1 and 2 refer to the two channels; B denotes the baryon number of the products. Let the branching ratio be r and $1 - r$. Then the mean baryon number of the products per decay is

$$B_x = rB_1 + (1 - r)B_2.$$

Considering the $\overline{X}$ bosons we have, for their decay,

$$B_{\overline{x}} = -\bar{r}B_1 - (1 - \bar{r}_1)B_2.$$

Hence the decay of a pair will give rise to a net baryon number change

$$B_x + B_{\bar{x}} = (r - \bar{r})(B_1 - B_2).$$

A nonvanishing $(r - \bar{r})$ arises because of C and CP violation. (Note that both have to be violated.) This process occurs when baryon numbers are comparable to photon numbers, hence the resulting ratio of a baryon number to a photon is

$$\frac{n_B}{n_\gamma} = (r - \bar{r})(B_1 - B_2).$$

The precise calculation of n_B/n_γ thus require consideration of the various channels of reaction and the degree of C and CP violations. Different groups have performed the computation on different models, the whole thing is rather complicated and uncertainties remain. It is perhaps too much to claim that a theoretical computation has successfully explained the observed value of n_B/n_γ but we can legitimately say that, provided the basic ideas of the GUT are correct and $M_x > 10^{16}$ GeV, there is reason to believe that n_B/n_γ is not an arbitrarily fixed parameter.

Problems

1. The Lagrangian $L = \frac{1}{2}\phi^{,\mu}\phi_{,\mu} - (m^2/2)\phi^2$ leads to the Klein–Gordon equation

$$(\Box + m^2)\phi = 0,$$

where $\Box\phi = g^{ik}\phi_{,i;k}$. The mass of the particle is then identified with m. Consider the Lagrangian

$$L = \frac{1}{2}\phi^{,\mu}\phi_{,\mu} + \frac{\mu^2\phi^2}{2} - \frac{\lambda}{4}\phi^4$$

in which the potential has a minimum at $\phi_0 = \pm\mu\lambda^{-1/2}$. Expressing ϕ as a perturbation from ϕ_0, show that the effective mass is $\sqrt{2}\mu$.

2. Consider the Lagrangian of a vector field A_μ coupled with a complex scalar field x

$$L = -\frac{1}{4}(A_{\mu,\gamma} - A_{\gamma,\mu})^2 + \left|\left(\frac{\partial}{\partial x^\mu} + ieA_\mu\right)x^*\right|^2 + \mu^2|x|^2 - \lambda|x|^4$$

(for the moment we are considering flat space). Make the gauge transformation

$$A_\mu \rightarrow A_\mu - \frac{1}{e\phi_0}\frac{\partial\psi}{\partial x^\mu}$$

along with the substitution

$$x = \frac{1}{\sqrt{2}}\left[\phi + \phi_0 \exp\left(\frac{i\psi}{\phi_0}\right)\right].$$

Show that ψ does not appear in the resulting Lagrangian which now involves a term $(e^2/2)\phi_0^2 A_\mu A_\mu$, indicating that we have now a massive scalar field of mass $e\phi_0$.

21. The Inflationary Scenario

21.1. Introduction

We attempt to give an idea of the inflationary scenario which is thought to have prevailed at some stage in the early universe. The discussion may be divided into a number of parts:

(a) the problems of horizon and flatness and fine tuning and their implication in terms of entropy generation;
(b) the possibility of a de Sitter phase of exponential expansion (hence the name inflation) and the irreversible transformation of vacuum energy into radiation, accompanied by an enormous increase in entropy; and
(c) the different models for the inflation.

We have already mentioned the problems of horizon and flatness. We may recall that the horizon problem is related to the observed isotropy of the microwave background radiation. This radiation suffers attenuation due to interaction with matter as it proceeds and thus there is an exponential fall in intensity. The intensity $I = I_0 e^{-\tau}$ where τ, called the optical depth, depends both on the path length and the constitution of matter through which the radiation passes. It follows that the radiation we receive is effectively coming from regions with $\tau \lesssim 1$. To make estimates, we assume that all the radiation originating at $\tau = 1$ (called the last scattering surface). In the case of a flat universe with a hypothetical population of ionized hydrogen, the last scattering surface occurs at about $Z \approx 8$. However, present indications are that the universe consists primarily of dark matter which does not interact with radiation. The last scattering surface would then be at $Z \approx 1000$, which corresponds to the epoch of the recombination of protons and electrons to form neutral hydrogen when matter became decoupled from radiation.

So the isotropy of the Microwave Background Radiation (MBR) indicates that there was uniformity over the surface $z \approx 1000$. However, calculation shows that at $z \approx 1000$, almost any finite angular separation would make the points beyond their horizons. Thus there is conflict between causality and the observed isotropy of MBR. We may sum up:

isotropy of MBR $\rightarrow$ uniformity over the surface $z \approx 1000$,

standard model $\rightarrow$ no communication possible between points on the surface $z \approx 1000$,

hence uniformity is unattainable.

21.2. The Problems in Terms of Entropy

Following Guth (1981) we may quantify the flatness and horizon problems in terms of the entropy generation. Considering the volume $(cT)^3$, where T is the present age of the universe, the value of the entropy taking the MBR and three types of neutrino is $\approx 10^{87}$.

Consider a stage when the temperature was so high that the universe was dominated by ultrarelativistic particles. The energy density and the entropy density are given by the usual formulas

$$\rho = ag_* T^4; \qquad s = \tfrac{4}{3} ag_* T^3.$$

Consequently,

$$|\varepsilon| \equiv \frac{k}{R^2 T^2} = \left[\frac{4}{3} \frac{ag_*}{S} \right]^{2/3} \qquad (S = sR^3),$$

and

$$|1 - \Omega| = \left| \frac{\rho - \rho_c}{\rho} \right| \sim S^{-2/3} (ag_*)^{-1/3} \left(\frac{M_{pl}}{T} \right)^2,$$

where S is the total entropy $\approx 10^{87}$ and M_{pl} is the Planck mass. Considering the Planck temperature

$$|1 - \Omega|_{pl} \sim 10^{-58}.$$

It will be noted that this extremely low value of the ratio (or fine tuning) arises because of the high value of entropy at the Planck temperature. If we had the entropy $\sim O(1)$ at the Planck temperature and an irreversible process leading to an increase in entropy by a factor $Z^3 \sim (10^{29})^3$, then the flatness or fine tuning problem would be resolved. We shall now see that such an increase in entropy is obtained in the inflationary scenario by a rapid increase of R by a factor Z without the temperature undergoing any appreciable fall.

The increase of entropy also solves the horizon problem. The horizon distance at any time t is

$$l_h = R \int_0^t R^{-1} \, dt = 2t,$$

using $R \sim t^{1/2}$ for the early universe. Assuming the conservation of entropy, the presently observable region of the universe of linear dimension L_0 has evolved out of a region of dimension $L(t)$ at time t, where

$$L(t) = \left[\frac{S_0}{S(t)} \right]^{1/3} L_0.$$

This leads to a discrepancy between l_h and L at the Planck time of magnitude

$$\left(\frac{l_h}{L} \right)^3 \sim 10^{-87}.$$

Obviously, a generation of entropy of the order considered before will remove the horizon problem.

21.3. The Vacuum Energy–Stress Tensor and the de Sitter Phase

We have already seen in our discussion of the GUT that, at temperatures above a critical value, complete symmetry obtains and the ground state thus occurs for vanishing values of the Higgs' field variations; below the critical temperature the minimum of energy may correspond to symmetry broken states with nonvanishing values of the field variables. Considering, for simplicity, a single scalar field ϕ, the symmetric state $\phi = 0$ ceases to be the state of global minimum energy but, nevertheless, it may be metastable due to the potential having a local minimum at $\phi = 0$ or a maximum with a very flat form before falling to the minimum at $\phi \neq 0$. In either case, the universe stays for a long time in a more or less constant value of ϕ, say ϕ_0, so that we may take $\dot{\phi} \approx 0$ and $\nabla\phi = 0$ (homogeneity). We have, with the Lagrangian,

$$L = \tfrac{1}{2}\phi_{,\mu}\phi^{,\mu} - V(\phi),$$

and the energy–stress tensor will be of the form

$$T_\nu^\mu = V_0 \delta_\nu^\mu,$$

where $V_0 = V(\phi_0)$. The above energy–stress tensor can be written in perfect fluid form

$$T_\nu^\mu = (p + \rho)v^\mu v_\nu - p\delta_\nu^\mu,$$

with $\rho = -p = V_0$. The negative pressure, equal in magnitude to the energy density, has an important consequence. The divergence of the energy stress–tensor gives

$$\rho_{,\mu}v^\mu = -(p + \rho)v^\mu_{;\mu} = 0.$$

Thus in the expanding universe, the energy density is not reduced but remains constant, and hence the total energy in any comoving region continually increases. (Such a situation was hypothesized about four decades ago in the steady state theories of Bondi, Gold, and Hoyle, but there the constancy of the energy density was achieved by an assumption of continuous creation, violating the energy conservation principle. However, McCrea (1951) did suggest that a suitable negative pressure may lead to a simulation of continuous creation.)

The expansion leads to a progressive decrease in the matter density; in the early universe, the ultrarelativistic matter density falls off inversely as the fourth power of the scale factor R. Thus, if the situation envisaged continues for a sufficiently long time, the vacuum energy density $\rho \, (= V_0)$ will assume a dominating role over the matter density and the Einstein equations will assume the form

$$R_\nu^\mu - \tfrac{1}{2}R\delta_\nu^\mu = -8\pi V_0 G\delta_\nu^\mu.$$

We have already seen that in such a case, assuming the universe to be homogeneous, isotropic, and expanding, the solution is the de Sitter metric

$$dS^2 = dt - e^{Ht}(dr^2 + r^2\,d\theta^2 + r^2\sin^2\theta\,d\phi^2),$$

with

$$H^2 = \frac{8\pi V_0 G}{3} = \text{constant.}$$

(The reader may note that we are adopting a rather curious procedure, we started wondering how isotropy could be attained in the presence of horizons and we are now straightaway assuming isotropy, we shall return to this point later.)

The exponential expansion, i.e., the de Sitter metric, lasts so long as the vacuum energy dominates. The vacuum energy comes to dominate only after the universe cools to the critical temperature and as further expansion goes on there is supercooling, i.e., a fall in temperature much below the critical temperature. Finally, transition to the true vacuum occurs, the vacuum energy goes over to matter and, consequently, the matter temperature sharply rises, almost back to the critical temperature. The net result is an enormously expanded universe whose temperature has, however, not fallen in accordance with the relation $RT = \text{constant}$; it also has a much enhanced entropy and the energy of the matter has come almost entirely from the false vacuum energy. We have thus what has been called a free lunch.

We have seen that if the expansion is by a factor $Z \sim 10^{29}$, then the horizon and flatness problems would be solved and we now consider whether the expansion will actually be that large.

21.4. The Different Models of Inflation

(a) The Old Inflation (Guth, 1981)

In this original model, proposed by Guth, there was a potential barrier separating the local minimum at $\phi = 0$ from the global minimum at $\phi \neq 0$. Hence even when the universe cooled down to the critical temperature, the universe was trapped in the metastable state $\phi = 0$ (the false vacuum). The transition occurs by tunneling through the barrier and in the supercooled state, bubbles of true vacuum form and grow in size. However, the bubble nucleation probability was small and did not have time to expand and unite giving rise to a homogeneous universe. So what resulted was a number of isolated bubbles and a very nonhomogeneous universe. Again we could not consider a single bubble to have developed into the observable universe, for the entropy inside was of a much lower value compared to our presently observable universe. Thus this model was soon considered unsatisfactory.

(b) The New Inflation (Albrecht and Steinhardt, 1982; Linde, 1982)

The first model of new inflation was based on the Coleman–Weinberg potential that we considered in the previous chapter. As we have seen in this potential, at finite temperatures there is a small barrier near the origin, the critical temperature for the occurrence of this barrier being $\sim 10^{14}$ GeV and this barrier makes the symmetric state metastable. Only when the tempera-

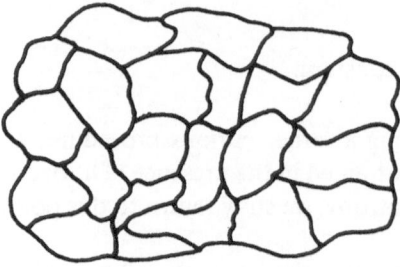

Figure 21.1. Spinodal or irregular fragmentation is new inflation.

ture falls considerably below 10^{14} GeV (say to about 10^9 GeV) the state $\phi = 0$ becomes unstable. The universe then breaks up into irregularly shaped regions (spinodal fragments) within each of which ϕ is approximately constant. To study the evolution of ϕ and the consequent inflation, we may write down the equation of motion of ϕ. It is

$$\ddot{\phi} + \frac{3\dot{R}}{R}\dot{\phi} + \Gamma\dot{\phi} + \frac{\partial V(\phi)}{\partial \phi} = 0, \tag{21.1}$$

where we have neglected any spatial variation of ϕ. The $\Gamma\dot{\phi}$ term arises due to particle creation and accounts for the conversion of the vacuum energy into the energy of ultrarelativistic particles. We may now consider two stages in the evolution of ϕ.

(i) The Slow Rolling Stage
For $T \to 0$ and $\phi \ll \phi_0$, the Coleman–Weinberg potential (introduced in the previous chapter) is of the $-\lambda\phi^4$ form, apart from some constants which normalize the potential to zero at $\phi = \phi_0$. Hence the potential is flat and the approximate solution of (21.1) is

$$\dot{\phi} \sim \frac{1}{3(\dot{R}/R)}\frac{\partial V}{\partial \phi},$$

as the $\ddot{\phi}$ and $\Gamma\dot{\phi}$ terms may be shown to be small. Also, as the derivatives of ϕ are all small, the energy–stress tensor of the ϕ field $T_\nu^\mu = V\delta_\nu^\mu$, and the metric

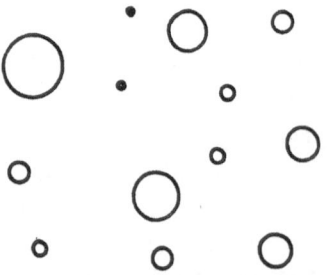

Figure 21.2. Bubble nucleation in old inflation of Guth.

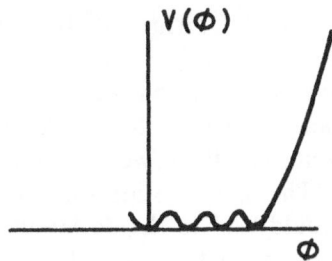

Figure 21.3. Potential in chaotic inflation of Linde.

assumes the de Sitter form with

$$H^2 = \frac{8\pi V G}{3M_{pl}^2},$$

where M_{pl} is the Planck mass $= (c\hbar/G)^{1/2}$ $(=G^{-1/2}$, in units, $c = \hbar = 1)$. It is thus about 2×10^{-5} g. Consequently, the scale factor R increases during this inflationary regime by $\exp(N)$ where

$$N = \int H\, dt \approx -3 \int H^2 \left(\frac{\partial V}{\partial \phi}\right)^{-1} d\phi$$

$$= -\frac{8\pi}{M_{pl}^2} \int V \left(\frac{\partial V}{\partial \phi}\right)^{-1} d\phi, \qquad (21.2)$$

the integral extending over the slow rolling phase. Thus we can have sufficient inflation ($N \sim 60$) with a suitable value of λ.

(ii) The Oscillating ϕ Phase
Near $\phi = \phi_0$, ϕ is in a potential well and we have a damped oscillation of ϕ, the damping coming from the terms $3(\dot{R}/R)\dot{\phi}$ and $\Gamma\dot{\phi}$. The ϕ field energy now goes over to the ultrarelativistic particles, ϕ settles down to the value ϕ_0, and the universe goes over to the Friedmann form with $R \sim t^{1/2}$.

(c) The Chaotic Inflation (Linde, 1983)
The chaotic inflation does not involve any peculiarity of the potential function in making the $\phi = 0$ state metastable. Instead, the potential function is simply of the form $V(\phi) = \frac{1}{4}\lambda\phi^4$. We assume that at a certain instant it is possible to have a region where $\phi_0 \gg M_{pl}$ but nevertheless $V(\phi) < M_{pl}^4$. Of course, this is possible only if λ is sufficiently small (weak coupling). The linear dimension of the region where this occurs is less than the horizon distance. The field ϕ rolls over to the state $\phi = 0$ and we have an inflation of magnitude $\exp(N)$ where (recall (21.2))

$$N = -\frac{8\pi}{M_{pl}^2} \int_{\phi_0}^{0} \frac{1}{4}\phi\, d\phi = +\frac{\pi\phi_0^2}{M_{pl}^2}.$$

Thus with say $\phi_0 \sim 5M_{pl}$ we can have $N \sim 80$. It is this local region which has grown to our observable universe.

21.5. A Critique of the Inflationary Models

We have mentioned only a few of the inflationary models, there are today a host of such models, each with its special appeal and problems. Instead of going into them, we shall give a criticism of the common feature of all of them.

The first question that arises is whether a vacuum energy–stress tensor would lead to the de Sitter form if the universe is originally in a general anisotropic form. Recall that we have started with a universe which is already isotropic and flat, and have thus practically assumed what we sought to explain. There have been claims that an inflationary situation will prevail even if these assumptions are dropped, however, in general, these claims cannot be sustained (see, e.g., Raychaudhuri and Modak (1988) and the papers cited therein). Second, there is a peculiar situation that, according to the inflationary models, the presently observed universe is merely an island in a much vaster background of an inhomogeneous universe, which we shall never see. Lastly, as yet, no model has been proposed which is completely satisfactory and does not involve any ad hoc assumption.

21.6. Fluctuations in the Inflationary Models

In the study of the evolution of perturbations in the inflationary background, the usual formalism has to be modified. To define a perturbation (of energy density, for example) we considered the difference between the energy density in the perturbed universe and that in the isotropic universe, from which the perturbed universe is imagined to have originated. Such a comparison requires an identification of space–time points in the perturbed and in the isotropic universe. However, the isotropic universe no longer exists and the identification becomes arbitrary. Put in another way, the identification means assigning the same coordinates x^α ($\alpha = 0$, 1, 2, 3) to two points, one in the perturbed universe and the other in the hypothetical isotropic universe. We are free to make a coordinate transformation in the perturbed universe alone, so that now the same coordinate x^α is assigned to a different point (technically speaking, we may change the mapping from one universe to the other). This change in the identification of points changes the magnitude of the perturbation. Such changes, which are extraneous to the physics of the situation, are referred to as gauge changes, and Bardeen (1980) developed a formalism to study perturbations in a gauge-invariant manner. Unfortunately, Bardeen's discussion, though not mathematically difficult, involves long and complicated calculations. Skipping over that we merely quote his results.

Bardeen finds that there are two linearly independent measures of gauge-invariant energy density perturbation. We may, of course, construct any number of gauge-invariant quantities by making combinations of these. It is important to note that the difference amongst these gauge-invariant quantities can be linked to a different choice of space sections, and the differences are significant only for perturbations whose characteristic wavelength is large compared to the horizon distance. Bardeen considers the 3-spaces orthogonal to the world-lines of matter to be the space sections, and deduces a second-order differential equation for the time evolution of the gauge-invariant density perturbation with this choice. But this equation is also too complicated to handle in its general form. To make it tractable, we shall assume that the spatial curvature of the isotropic universe vanishes (a situation consistent with the inflationary scenario) and we also neglect the shear stress terms. The equation then assumes the form

$$\ddot{Z} + (1 - r)H\dot{Z} - [(p + \rho) + \tfrac{1}{3}\gamma\rho]Z = -\frac{\delta p_k}{3}, \qquad (21.3)$$

where

$$Z \equiv \left(\frac{RH}{k}\right)^2 \frac{\rho_k}{\rho} = \frac{8\pi}{3}\left(\frac{R}{k}\right)^2 \rho_k.$$

The subscript k indicates the Fourier component of wave number k and

$$\gamma \equiv \frac{dp}{d\rho} = H^{-1}\frac{d}{dt}\ln(p + \rho).$$

Lyth (1985) argued that if $RH/k \gg 1$ (i.e., for perturbation extending well beyond the horizon), the δp term on the right may be neglected in comparison with the last term on the left, and then (21.3) can be integrated to give

$$Z + \frac{2}{3(1 + \omega)}\left(Z + \frac{\dot{Z}}{H}\right) = \text{constant}, \qquad (21.4)$$

where $\omega = p/\rho$. Lyth further found that the constant of integration in the above equation is equal to $\delta K/k^2$ where δK is the Fourier component of the curvature perturbation of wave number k.

The history of the universe contains three stages where ω can be taken as constant, e.g.,

$\omega = 0$ (for matter domination),

$\omega = \tfrac{1}{3}$ (for radiation domination),

$\omega = -1$ (for the de Sitter phase in the inflationary scenario).

We can readily integrate (21.4) for constant values of ω

$$Z - \frac{3\delta K}{k^2}\frac{(1 + \omega)}{(5 + 3\omega)} = A_e^{-Ht(5 + 3\omega)/2},$$

so that after a few Hubble times, Z settles down to a constant value

$$Z = \frac{3\delta K}{k^2}\frac{(1 + \omega)}{(5 + 3\omega)}. \qquad (21.5)$$

All the above considerations have been based on the assumption $RH/k > 1$

(i.e., for perturbations extending beyond the horizon). Consider that, initially, the perturbation is within the horizon, i.e., $RH/k < 1$. During the inflationary de Sitter regime, the Hubble parameter H remains constant while the scale factor R increases exponentially. Thus the inequality is reversed, i.e., the perturbation crosses the horizon during the inflationary epoch. Again, after the inflation stops and the universe goes over to a Friedmann regime, $R \sim t^{1/2}$ or $R \sim t^{2/3}$, according to whether radiation or matter is dominating and $H \sim t^{-1}$. Hence RH/k decreases, becomes less than unity, and the perturbation enters back into the horizon. At both these crossings, $RH/k = 1$ and hence $Z = \delta\rho_k/\rho$. Equation (21.5) then relates the values of $\delta\rho_k/\rho$ at the two crossings

$$\left(\frac{\delta\rho_k}{\rho}\right)_f = \frac{2}{5(1 + \omega_i)}\left(\frac{\delta\rho_k}{\rho}\right)_i. \tag{21.6}$$

The subscripts i and f indicate the initial and final crossings; we assume that the re-entry occurs in the matter-dominated phase. In case it occurs in the radiation-dominated regime, there is a small change in the numerical coefficient on the right, which is not of significance in this calculation and which aims at an order of magnitude estimate.

In the inflationary epoch $1 + \omega_i \approx 0$ and hence it is clear that there is an enormous amplification of the density perturbation in the intervening period between the two crossings. While this may appear to remove the difficulty of slow growth in the case of an ordinary Friedmann universe, trouble comes when we evaluate the right-hand side of (21.6).

The quantum fluctuation $\delta\phi$ in the field ϕ is given by

$$\delta\phi = \frac{H}{2\pi}, \tag{21.7}$$

and the energy density and pressure due to the ϕ field are

$$\rho = \tfrac{1}{2}\dot{\phi}^2 + V(\phi), \tag{21.8a}$$

$$p = \tfrac{1}{2}\dot{\phi}^2 - V(\phi). \tag{21.8b}$$

Note that we are considering the slow roll over regime in which $\dot{\phi}$, though small, does not vanish and $\nabla\phi$ may be neglected. Recall that during this regime $\dot{\phi} = -(3H)^{-1}(\partial V/\partial\phi)$, so using (21.7) and (21.8) we get

$$\left(\frac{\delta\rho/\rho}{1 + \omega}\right)_i = \frac{(\partial V/\partial\phi)\delta\phi}{\dot{\phi}^2} = \frac{9H^3}{2\pi}\left(\frac{\partial V}{\partial\phi}\right)^{-1}. \tag{21.9}$$

In the above calculation we have neglected the variation of H and $\dot{\phi}$. We now consider the Coleman–Weinberg potential which was introduced earlier.

In the slow rolling regime, $T \to 0$ and $\phi \ll \phi_0$, the Coleman–Weinberg potential may be written in the form

$$V(\phi) = \frac{B\phi_0^2}{4} - \lambda\frac{\phi^4}{4}, \tag{21.10}$$

where

$$B = \frac{5625}{512\pi^2} g^4 \approx 0.1,$$

and

$$\lambda = B\left[1 - \ln\left(\frac{\phi}{\phi_0}\right)^4\right].$$

Then from (21.6) we get, using (21.9) and (21.10),

$$\left|\left(\frac{\delta\rho}{\rho}\right)_f\right| = \frac{9H^3}{10\pi\lambda\phi^3}. \tag{21.11}$$

Again, the inflation enhances the scale factor by $\exp(N)$ during the slow roll over of ϕ where (see the second step in (21.2))

$$N \sim \frac{3H^2}{2\lambda\phi^2},$$

so that, finally,

$$\left|\left(\frac{\delta\rho}{\rho}\right)_f\right| \approx \frac{2}{5}\frac{\sqrt{6}}{\pi}\lambda^{1/2}N^{3/2}.$$

The first point to note about the above relation is that the dependence on the scale of perturbation comes through N. However, N varies logarithmically with the length or mass of the perturbation. Thus the perturbations are very nearly scale-independent and are consistent with the scale-invariant spectrum (Zel'dovich, 1972; Harrison, 1970). The scale-independence is necessary for the following reason: any scale-dependence may be represented by a relation of the form

$$\left(\frac{\delta\rho}{\rho}\right) \sim M^{-\alpha},$$

where M is the mass of the perturbation and α is different from zero. If α is positive, the fluctuations are on a small scale and would give rise to a large number of black holes, while with α negative large scale inhomogeneities would dominate. A vanishing α (scale-independence) saves the situation from either extreme. However, the recently observed large scale coherent velocity fields are indications against a scale-free spectrum.

However, the magnitude of $(\delta\rho/\rho)$ comes out too high. For inflation to be effective in removing the horizon and fine tuning problems, $N \to 60$, and with the usual GUT value of $\lambda \sim 4$, $(\delta\rho/\rho) \sim 10^2$, whereas the observed lack of patchiness in the MBR temperature requires $\delta\rho/\rho \lesssim 10^{-4}$. This uncomfortable difficulty can be resolved by choosing a much smaller value of λ but then the whole thing will be very artificial.

Problems

1. Show that, in general, the Klein–Gordon equation for a scalar ϕ is

$$\ddot{\phi} + \dot{\phi}\frac{\partial}{\partial t}(\ln|\sqrt{g}|) - g^{ij}\phi_{,i,j} - [g^{ij}_{,i} + g^{ij}(\ln\sqrt{|g|})_{,i}]\phi_{,j} = -\frac{\partial V}{\partial\phi}$$

if $ds^2 = -dt^2 + g_{ik}\, dx^i\, dx^k$. Investigate the influence of the nonhomogeneity terms on the rollover time of ϕ in case the metric departs only slightly from the de Sitter metric and the space derivatives of ϕ are also small.

2. Show that for a perfect fluid where the velocity field is hypersurface orthogonal, the following equations hold good:

$$\dot{\theta} + \tfrac{1}{2}\theta^2 = \Lambda - 4\pi(\rho + 3p) - 2\sigma^2 + a^\mu_{;\mu},$$

$$\tfrac{1}{2}(R^* - 2\sigma^2) + \tfrac{1}{3}\theta^2 - \Lambda = 8\pi\rho,$$

where θ, σ, and a^μ are the expansion scalar, shear scalar, and acceleration vector, respectively, and R^* is the scalar curvature for the local 3-space orthogonal to the velocity vector, Λ being the cosmological constant. Show that the metric will tend to the de Sitter form if R^* is nonpositive and $a^\mu = 0$.

3. Verify that the metric

$$dS^2 = dt^2 - X^2\, dr^2 - Y^2(d\theta^2 + \sin^2\theta\, d\phi^2),$$

with

$$Y = \Lambda^{-1/2}, \qquad X = X_0 \exp(\Lambda^{1/2}t),$$

satisfies the empty space field equations with the cosmological constant Λ. Show that although the expansion of the space is exponential, the shear and the spatial curvature remain constant in time and so the metric does not tend to the de Sitter form.

22. Concluding Remarks

In these concluding remarks, we propose to indicate some of the so-called anomalous observations which apparently throw some doubt on the validity of the standard picture of cosmology that we have built up. However, we shall also try to give the readers the reasons which led the majority of cosmologists to accept the standard model. Nevertheless, it is well to bear in mind that the next decade is likely to bring in considerable new information, and the situation that will emerge at the turn of the century may well be quite different.

The so-called standard model is today over 65 years old if we identify its birth with the publication of Friedmann's paper, giving nonstatic solutions for the Einstein equations exhibiting spatial homogeneity and isotropy. True there have been amplifications and extensions during this period but the fact that the model has survived over this long period is an indication of its intrinsic strength.

Yet it seems based on absurdly simple assumptions. The universe is assumed to be spatially homogeneous and isotropic while at first sight it is remarkably nonhomogeneous and anisotropic. True, we talk of a "large scale" in this connection but that large scale remains beautifully vague and undefined.

In judging the acceptability of a model, it is appropriate to raise the following questions:

(i)   How far are the assumptions consistent with observations?
(ii) Does the model introduce any otherwise unverified modification of the prevalent theories?
(iii) Are there observational data contradicting the model? If so, can they be explained?
(iv) How does the model fare as compared with other suggested models?

The assumptions of the standard model are the validity of the general theory of relativity and the cosmological principle. Right now there exists no observational evidence contradicting general relativity. Indeed, the observations on the Hulse–Taylor binary pulsar have brought in added confidence in the theory. True, there are doubts about the validity of the general theory of relativity when spatial separations are of the order of Planck length, but that would be relevant only at very early stages of the standard model and, hopefully, quantum gravity may remove the singular beginning and thus strengthen rather than weaken the standard model.

The cosmological principle stands on much weaker ground. While the isotropy of the microwave background radiation is put forward as evidence in its favor, the associated problem of horizon makes the whole argument suspect. The existence of structure in the universe, in the form of clusters and superclusters of galaxies as well as large voids, appears to be directly contradicting the cosmological principle, at least in its naive form.

The answer to point (ii) is in the negative. Indeed, the great merit of the standard model is the absence of any ad hoc modification of the prevalent theoretical ideas. Regarding (iii), the observational data which apparently contradict the standard model are, first, the so-called anomalous redshifts, objects which are presumably close together show significantly different redshifts. There is also the curious "quantization" of redshifts with Δz, multiples of Δz_0 corresponding to a velocity of 72 km s^{-1} (Arp (1987), and references therein). Both these cast doubt on the idea of a redshift being due to universal expansion. However, the large amount of available data showing the redshift apparent magnitude correlation cannot be brushed aside lightly and we suspect that while the standard cosmological explanation of the redshift is basically correct, there is another source of spectral shift whose nature is as yet not understood.

The other notable contradiction with observations is the existence of structure in the universe. So far, attempts to explain them as originating from perturbations have not been successful. Hopes have been raised that new inputs such as the inflationary scenario and cosmic strings may solve the problem but as yet these hopes are unrealized.

The observations on radio sources like the N–S relation, the angular size–flux density relations, and their interpretations remain controversial to the extent that it is not possible to say whether they support or contradict the standard model.

Lastly, a comparison with other models. Here we find that, excepting the steady state models which were carefully studied in the 1950s and 1960s, hardly any other model has been discussed seriously. The steady state idea has fallen into disrepute after the discovery of the microwave background radiation. The speculation that the redshift may be due to an interaction between photons (the "tired light" hypothesis) suffers from the defect of being an ad hoc idea which, nevertheless, does not explain the anomalous redshifts and the quantal effect. At one time, Ellis (1978) proposed a nonhomogeneous static model in which the redshifts arose from the differences in gravitational potential but then the model was found wanting when are considered the $N - m$ relation.

The linear velocity distance relation was questioned by Nicoll and Segel (1975) and an altogether novel cosmological model was put forward by them. But their conclusion about the quadratic relation between velocity and distance has not been accepted by the cosmologist community and even such a quadratic relation may be accommodated in the standard model (Raychaudhuri and Mukherjee, 1984).

What are the successes of the standard model? We usually cite the explanations of the microwave background radiation and the deuterium and helium abundances. But then there are vexatious problems as to why there is no imprint of the structure formation in the microwave background radiation. Again doubts regarding the exact primordial abundance of helium persist. Altogether, we must not feel confident about these so-called successes.

We have not mentioned the age problem and the consequent constraints on the values of the Hubble constant H, the energy density ρ, and the deceleration parameter q. Standard cosmology puts forward the equation

$$\frac{4\pi G\rho}{3} = qH^2.$$

Indeed, if and when the precise values of ρ_0, q_0, and H_0^2 are available, the decision about standard cosmology will be easy to reach.

To conclude, the standard model holds the field; if not because of its successes, at least by default, because a better alternative is simply not there.

Appendix. Differential Forms

A.1. Introductory Ideas and Definitions

The use of differential forms enables us to make some calculations much more quickly than is possible by conventional tensor calculus. We shall make no attempt to develop this subject in a manner that is done in some mathematical texts, we merely provide an introduction to some ideas and formulas so that the reader may follow the literature on general relativity where these forms are now widely used, and also to do some computations on his own.

At the beginning we shall emphasize two characteristics of differential forms:

(1) they are all scalars; and
(2) they are associated with tensors but only antisymmetric parts of the tensors appear.

The simplest case is a scalar f; the corresponding form is also f and is called a 0-form (zero form).

Next, if we have a vector A_μ, the corresponding form $\phi \equiv A_\mu \, dx^\mu$ is called a 1-form. Note that, in general, dx^μ may be any vector whatsoever, but for the present we identify it with the vector of coordinate differentiatials. For a tensor of rank 2, $F_{\mu\nu}$, the corresponding 2-form is

$$2! \, F_{\mu\nu} \, dx^{[\mu} \, dy^{\nu]}.$$

The factor 2! is merely to neutralize the $\frac{1}{2}$ that comes due to anti-symmetrization

$$dx^{[\mu} \, dy^{\nu]} \equiv \tfrac{1}{2}[dx^\mu \, dy^\nu - dx^\nu \, dy^\mu].$$

We can build up higher forms but for the present we leave that exercise to consider the algebra of differential forms.

The processes of addition (or subtraction) of forms consist of adding (or subtracting) the corresponding tensors and then building up a form by contracting with the vector dx^μ and antisymmetrizing. Obviously, as addition or subtraction of tensors is defined only when the ranks of the tensors are the same, these processes for forms are also defined subject to that restriction.

The product (called the exterior or wedge product) of a p-form ϕ with a q-form ψ is formed in the following manner. Multiply each component of the tensor in ϕ by each component of the tensor in ψ so as to get a tensor of rank

$p + q$. From this build up the corresponding form, e.g., if f is a 0-form and ϕ a p-form,

$$f \wedge \phi = f \wedge (A_{\alpha_1 \alpha_2 \dots \alpha_p} \, dx^{[\alpha_1} \, dy^{\alpha_2}, \dots, d\xi^{\alpha_p]})$$

$$= f A_{\alpha_1 \alpha_2 \dots \alpha_p} \, dx^{[\alpha_1} \, dy^{\alpha_2}, \dots, d\xi^{\alpha_p]}$$

$$= f\phi,$$

while if ϕ_1 and ϕ_2 are both 1-forms, say,

$$\phi_1 = A_\alpha \, dx^\alpha, \qquad \phi_2 = B_\alpha \, dy^\alpha,$$

then

$$\phi_1 \wedge \phi_2 = 2! \, A_\alpha B_\beta \, dx^{[\alpha} \, dy^{\beta]},$$

but

$$\phi_2 \wedge \phi_1 = 2! \, B_\alpha A_\beta \, dy^{[\alpha} \, dx^{\beta]}.$$

Obviously,

$$\phi_1 \wedge \phi_2 = -\phi_2 \wedge \phi_1.$$

Thus, the communication rule does not hold good and, in particular, if ϕ_1 and ϕ_2 are identical and both equal to ϕ (say)

$$\phi \wedge \phi = 0.$$

In general, if ϕ_p is a p-form and ψ_q is a q-form,

$$\phi_p \wedge \psi_q = (-1)^{pq} \psi_q \wedge \phi_p.$$

The reader may check the above relation by writing it out in full.

In tensor calculus, we have a formalism based on the choice of n linearly independent vectors $e^\alpha_{a_0}$ where the Greek letters α and β correspond to vector indices, while the Latin letters a, b, etc., label the vectors. In the four dimensions of general relativity we require four vectors e^α_1, e^α_2, e^α_3, and e^α_4, and the formalism is known as tetrad formalism. Here in differential forms also we will introduce a tetrad. Then writing g_{ab} for the scalar product of the ath and bth vectors

$$g_{ab} = g_{\alpha\beta} e^\alpha_a e^\beta_b. \tag{A.1}$$

As the vectors e^α_a are independent and $|g_{\alpha\beta}|$ is nonsingular we have a reciprocal matrix g^{ab} such that

$$g^{ab} g_{bc} = \delta^a_c.$$

We now construct a tetrad of reciprocal vectors $e^{(a)\alpha}$

$$e^{(a)\alpha} = g^{ab} e^{(\alpha)}_b.$$

Then $e^{(a)\alpha} e_{b\alpha} = g^{ac} e^\alpha_c e_{b\alpha} = g^{ac} g_{bc} = \delta^a_b$, and

$$e^{(a)}_\alpha e^\beta_a = \delta^\beta_\alpha.$$

Related to the basic tetrad, we build up the basic 1-forms

$$\theta^a = e^{(a)}_\alpha \, dx^\alpha, \tag{A.2}$$

which may be inverted to give

$$dx^\alpha = e_a^\alpha \theta^a. \tag{A.3}$$

We next proceed to define exterior differentiation. To form the exterior differential of say, a p-form, we differentiate the p-rank tensor covariantly to obtain a $(p + 1)$-rank tensor, and then use this new tensor to build up a $(p + 1)$-form which is called the exterior differential of the original p-form, e.g., for a 0-form, f, differentiation of the scalar f gives the vector $f_{,\alpha}$ and the form is $f_{,\alpha}\, dx^\alpha$. We write

$$df = f_{,\alpha}\, dx^\alpha.$$

Next, consider a 1-form $\phi = A_\alpha\, dx^\alpha$, differentiation of A_α gives $A_{\alpha;\beta}$ and the 2-form is $2!\, A_{\alpha;\beta}\, dy^{[\beta}\, dx^{\alpha]}$ (note the order of the indices). Thus

$$d\phi = 2!\, A_{\alpha;\beta}\, dy^{[\beta}\, dx^{\alpha]}.$$

Two things are to be noted:

(1) Due to antisymmetrization, $A_{\alpha;\beta}$ may be replaced by $A_{\alpha,\beta}$ if we are considering Riemannian geometry, as the Christoffel symbol $\Gamma_{\alpha\beta}^\sigma$ is symmetric in α and β.
(2) If we perform the exterior differentiation operation twice, then the tensor involving $P_{\alpha\beta,\mu,\nu}$ would be symmetric in μ and ν and so vanish on contraction with $dy^{[\mu}\, dx^{\nu]}$. Hence $d^2\phi = 0$, where ϕ is any form. Regarding the differential of a wedge product the following result follows directly:

$$d(\phi \wedge \psi) = d\phi \wedge \psi + (-1)^p \phi \wedge d\psi,$$

where p is the degree of ϕ.

A.2. Connection 1-Forms and Ricci Rotation Coefficients

Consider the vector $e_{a;\beta}^\alpha\, dx^\beta$. As any vector is expressible as a linear combination of basic vectors e_a^α we must have

$$e_{a;\beta}^\alpha\, dx^\beta = \omega_a^b e_b^\alpha, \tag{A.4}$$

or

$$\omega_a^b = e_{a;\beta}^\alpha e_\alpha^b\, dx^\beta. \tag{A.5}$$

Thus ω_a^b is a 1-form corresponding to the vector $e_{a;\beta}^\alpha e_\alpha^b$. But

$$dx^\beta = e_c^\beta \theta^c, \qquad \text{so that} \qquad \omega_a^b = (e_{a;\beta}^\alpha e_\alpha^b e_c^\beta)\theta^c. \tag{A.6}$$

They are called connection 1-forms. The above equation presents ω_a^b as a linear combination of the 1-form θ^a, the coefficient of θ^c is known as the Ricci rotation coefficient and are scalars (written γ_{ac}^b)

$$\omega_a^b = \gamma_{ac}^b \theta^c. \tag{A.7}$$

Let us calculate the exterior differential $d\theta^a$. We have, since $e_{\alpha;\beta}^a = \gamma_{bc}^a e_\alpha^b e_\beta^c$, from the definition of γ^a_{bc}

$$
\begin{aligned}
d\theta^a &= 2! \, e^a_{\alpha;\beta} \, dy^{[\beta} \, dx^{\alpha]} \\
&= -2! \, \gamma^a_{bc} e^c_\beta e^b_\alpha \, dy^{[\beta} \, dx^{\alpha]} \\
&= -\gamma^a_{bc} \theta^c \wedge \theta^b \\
&= -\omega^a_b \wedge \theta^b.
\end{aligned}
\tag{A.8}
$$

Differentiating (A.1) we get

$$
d(g_{ab}) = (e^\alpha_{a;\beta} e^\alpha_b + e^\alpha_a e_{b\alpha;\beta}) \, dx^\beta.
$$

Substituting from (A.4), the above equation becomes

$$
\begin{aligned}
d(g_{ab}) &= \omega^c_a e^\alpha_c e_{b\alpha} + \omega^c_b e_{c\alpha} \alpha e^\alpha_a \\
&= \omega^c_a g_{bc} + \omega^c_b g_{ac}.
\end{aligned}
$$

Introducing the notation $g_{ab}\omega^b_c = \omega_{ac}$, we find

$$
dg_{ab} = \omega_{ba} + \omega_{ab}.
\tag{A.9}
$$

Very often we choose the basis vectors in such a way that the g_{ab}'s are constants, so that (A.9) gives an antisymmetry for the ω_{ab}'s.

A.3. Cartan's Equations of Structure

We are now in a position to express the Riemann–Christoffel tensor in the language of differential forms. As the Riemann–Christoffel tensor is defined as the commutator of a second covariant derivative, we begin by calculating the exterior differential of ω^a_b

$$
\begin{aligned}
d\omega^a_b &= 2! \, (e^\alpha_{b;\beta} e^a_\alpha)_{;\gamma} \, dx^{[\gamma} \, dy^{\beta]} \\
&= 2! \, (e^\alpha_{b;\beta;\gamma} e^a_\alpha \, dx^{[\gamma} \, dy^{\beta]}) + 2! \, (e^\alpha_{b;\beta} e^a_{\alpha;\gamma} \, dx^{[\gamma} \, dy^{\beta]}).
\end{aligned}
\tag{A.10}
$$

The first term on the right gives $R^\alpha_{\delta\gamma\beta} e^\delta_b e^a_\alpha \, dx^{[\gamma} \, dy^{\beta]}$ which we write as Ω^a_b, i.e.,

$$
\begin{aligned}
\Omega^a_b &= \tfrac{1}{2} R^\alpha_{\delta\gamma\beta} e^a_\alpha e^\delta_b e^\gamma_c e^\beta_d \theta^c \wedge \theta^d \\
&= \tfrac{1}{2} R^a_{bcd} \theta^c \wedge \theta^d,
\end{aligned}
$$

where we have used (A.2) or (A.3). The Ω^a_b's are called curvature 2-forms. Regarding the second term on the right of (A.10), consider the wedge product $\omega^a_b \wedge \omega^b_c$

$$
\begin{aligned}
\omega^a_b \wedge \omega^b_c &= 2! \, e^\alpha_{b;\beta} e^a_\alpha e^\sigma_{c;\gamma} e^b_\sigma \, dx^{[\beta} \, dy^{\sigma]} \\
&= -2! \, e^\alpha_b e^a_\alpha e^\sigma_{c;\gamma} e^b_{\sigma;\beta} \, dx^{[\beta} \, dy^{\sigma]} \\
&= -2! \, e^a_{\sigma;\beta} e^\sigma_{c;\gamma} \, dx^{[\beta} \, dy^{\sigma]},
\end{aligned}
$$

so that equation (A.10) becomes

$$
d\omega^a_b = \Omega^a_b - \omega^a_c \wedge \omega^c_b,
$$

or

$$\Omega_b^a = d\omega_b^a + \omega_c^a \wedge \omega_b^c. \tag{A.11}$$

Equations (A.8) and (A.11) are known as Cartan's first and second equation of structure, respectively. These, along with (A.9), enable an easy calculation of the Riemann–Christoffel tensor, as we shall see a little later.

A.4. Bianchi Identities and Symmetry Properties of the Riemann–Christoffel Tensor

Taking the exterior differential of (A.11) to get

$$
\begin{aligned}
d\Omega_b^a &= d\omega_c^a \wedge \omega_b^c - \omega_c^a \wedge d\omega_b^c \\
&= (\Omega_c^a - \omega_d^a \wedge \omega_c^d) \wedge \omega_b^c - \omega_c^a \wedge (\Omega_b^c - \omega_d^c \wedge \omega_b^d) \\
&= \Omega_c^a \wedge \omega_b^c - \omega_c^a \wedge \Omega_b^c), \tag{A.12}
\end{aligned}
$$

equation (A.12) is just the Bianchi identities $R^\alpha_{\beta[\gamma\delta;\epsilon]} = 0$. Also from the first equation of structure

$$d\theta^a = -\omega_b^a \wedge \theta^b,$$

we get

$$\Omega_b^a \wedge \theta^b = 0,$$

which gives $R_{a[bcd]} = 0$.

A.5. An Example of the Calculation of the Riemann–Christoffel Tensor

To calculate the ω_{ab}'s we utilize the two sets of relations (A.8) and (A.9)

$$dg_{ab} = \omega_{ab} + \omega_{ba},$$

and

$$d\theta^a = -\omega_b^a \wedge \theta^b.$$

Bringing in the Ricci rotation coefficients, we can write the above relations in the form

$$dg_{ab} = 2\gamma_{(ab)c}\theta^c,$$

$$g_{ab}d\theta^b = -2\gamma_{a[bc]}\theta^c\theta^b.$$

Thus we can compute both $\gamma_{(ab)c}$ and $\gamma_{a[bc]}$ and finally find ω_{ab} in the form

$$
\begin{aligned}
\omega_{ab} &= \gamma_{abc}\theta^c \\
&= [\gamma_{(ab)c} + \gamma_{(ac)b} - \gamma_{(bc)a} + \gamma_{a[bc]} + \gamma_{b[ca]} - \gamma_{c[ab]}]\theta^c. \tag{A.13}
\end{aligned}
$$

In practice, matters are much simpler, we choose the basic 1-forms θ^a in such a way that the ω_{ab}'s are antisymmetric, so that the first three terms on the right-hand side of (A.13) fall off. It is often not even necessary to go all the way to (A.13) but read off the antisymmetric ω_{ab}'s from (A.8).

As an illustration we take the Gödel metric

$$dS^2 = (dt + e^x \, dy)^2 - dx^2 - \frac{e^{2x}}{2} dy^2 - dz^2.$$

Let us choose the following basic 1-forms

$$\theta^0 = dt + e^x \, dy,$$

$$\theta^1 = dx,$$

$$\theta^2 = e^x / \sqrt{2} \, dy,$$

$$\theta^3 = dz,$$

so that, when the metric $dS^2 = (\theta^0)^2 - (\theta^1)^2 - (\theta^2)^2 - (\theta^3)^2$, we get

$$d\theta^0 = e^x \, dx \wedge dy = \sqrt{2}(\theta^1 \wedge \theta^2),$$

$$d\theta^1 = d\theta^3 = 0,$$

$$d\theta^2 = e^x \sqrt{2} \, dx \wedge dy = (\theta^1 \wedge \theta^2).$$

Without going into (A.5), we may easily see that

$$\omega_{01} = \theta^2 / \sqrt{2} = \omega^0_1 = \omega^1_0,$$

$$\omega_{02} = -\theta^1 / \sqrt{2} = \omega^0_2 = \omega^2_0,$$

$$\omega^1_2 = -\theta^2 + \theta^0 / \sqrt{2} = -\omega^2_1,$$

so that

$$\Omega^0_1 = d\omega^0_1 + \omega^0_2 \wedge \omega^2_1 = \tfrac{1}{2}\theta^1 \wedge \theta^0,$$

$$\Omega^0_2 = \tfrac{1}{2}\theta^2 \wedge \theta^0,$$

$$\Omega^1_2 = \tfrac{1}{2}\theta^1 \wedge \theta^2.$$

Using the relation $\Omega^0_b = \tfrac{1}{2} R^a_{bcd} \theta^c \wedge \theta^d$ yields the following frame components of the Riemann–Christoffel tensor

$$R^0_{110} = \tfrac{1}{2} = -R^1_{001}, \qquad R^1_{212} = \tfrac{1}{2} = -R^2_{112} = -R^1_{221},$$

$$R^0_{222} = \tfrac{1}{2} = -R^2_{002},$$

Hence

$$R_{00} = R^1_{001} + R^2_{002} = -1,$$

$$R_{11} = R^0_{110} + R^2_{112} = 0,$$

$$R_{22} = 0,$$

$$R_{12} = 0.$$

The actual Ricci tensor components are

$$R_{\alpha\beta} = R_{ab} e_\alpha^{(a)} e_\beta^{(b)},$$

which in the present case becomes

$$R_{\alpha\beta} = -e_\alpha^{(0)} e_\beta^{(0)},$$

i.e., $R_{00} = -1$, $R_{11} = -e^2$, and $R_{02} = -e^x$, which obviously satisfies the Einstein equation

$$R_{\mu\nu} = -8\pi(T_{\mu\nu} - \tfrac{1}{2} g_{\mu\nu} T),$$

with

$$T_{\mu\nu} = (p + \rho)v_\mu v_\nu - p g_{\mu\nu},$$

where

$$8\pi p = 8\pi\rho = \tfrac{1}{2},$$

$$v_0 = 1, \qquad v_2 = e^x.$$

(Note the apparent difference with the Gödel values because we have taken Gödel's $a = 1$ and instead of a dust with a cosmological term, we have considered a fluid with $p = \rho$.)

Problem

The Gödel metric can also be written as

$$dS^2 = \left(\frac{e^x}{\sqrt{2}} dy + \sqrt{2}\, dt\right)^2 - dx^2 - dt^2 - dz^2.$$

Choose $\theta^0 = e^x/\sqrt{2}\, dy + \sqrt{2}\, dt$, $\theta^1 = dx$, $\theta^2 = dt$, $\theta^3 = dz$, and make a calculation of the Riemann–Christoffel tensor.

References

Ables, J.G., McConnel, D., Jacka, C.E., McCulloch, P.M., Hall, P.J., and Hamilton, P.A. (1989), *Nature* **342**, 158.

Albrecht, A. and Steinhardt, P.J. (1982), *Phys. Rev. Lett.* **48**, 1220.

Anand, S.P.S. (1968), *Astrophys. J.* **153**, 135.

Arnett, W.D. and Bowers, R.L. (1977), *Astrophys. J. Suppl.* **33**, 415.

Arp, H. (1987), *J. Astrophys. Astronom.* **8**, 241.

Bahcall, J.N. (1979), *Ann. Rev. Astronom. Astrophys.* **16**, 241.

Bailyn, C. (1989), *Nature* **334**, 298.

Banerjee, A. (1966), *Phys. Rev.* **150**, 1086.

Bardeen, J.M. (1980), *Phys. Rev.* **D22**, 1882.

Bardeen, J.M. and Wagonar, R. (1971), *Astrophys. J.* **167**, 359.

Baym, G., Bethe, H.A., and Pethick, C.J. (1971), *Nucl. Phys.* **A175**, 225.

Baym, G. and Pethick, C. (1975), *Ann. Rev. Nucl. Sci.* **25**, 27.

Baym, G. and Pethick, C. (1979), *Ann. Rev. Astron. Astrophys.* **27**, 415.

Baym, G., Pethick, C., and Sutherland, P. (1971), *Astrophys. J.* **170**, 299.

Bekenstein, J. (1972a), *Phys. Rev.* **D5**, 1239.

Bekenstein, J. (1972b), *Phys. Rev.* **D5**, 2403.

Bekenstein, J. (1973), *Phys. Rev.* **D7**, 2333.

Blandford, R., Applegate, J.H., and Harnquist, L. (1983), *Monthly Notices Roy. Astronom. Soc.* **204**, 1025.

Blandford, R., Narayan, R., and Romani, R.M. (1984), *J. Astrophys. Astron.* **5**, 369.

Bondi, H. (1952), *Monthly Notices Roy. Astronom. Soc.* **112**, 195.

Bondi, H. and Gold, T. (1948) *Monthly Notices Roy. Astronom. Soc.* **108**, 252.

Boriakoff, W., Ferguson, D.C., Haugan, M.P., and Terzian, Y. (1982), *Astrophys. J.* **261**, L97.

Boyer, R.M. and Lindquist, R.W. (1967), *J. Math. Phys.* **8**, 265.

Braginskii, V.B. and Panov, V.I. (1972), *Soviet Phys. JETP* **34**, 463.

Brinkman, W. (1980), *Astron. Astrophys.* **85**, 146.

Buchdahl, W.A. (1959), *Phys. Rev.* **116**, 1027.

Burke, W.L. (1981), *Astrophys. J.* **244**, L1.

Butterworth, E.M. and Ipser, J.R. (1975), *Astrophys. J.* **200**, L103.

Butterworth, E.M. and Ipser, J.R. (1976), *Astrophys. J.* **204**, 200.

Canuto, V. (1974), *Ann. Rev. Astron. Astrophys.* **12**, 167.

Canuto, V. (1975), *Ann. Rev. Astron. Astrophys.* **13**, 335.

Carmelie, M. (1982), *Classical Theory of Fields, General Relativity and Gauge Theory* (New York: Wiley).

Carter, B. (1979), in *General Relativity*: An Einstein Centenary Survey, edited by S.W. Hawking and W. Israel (Cambridge University Press, Cambridge, England).

Chandrasekhar, S. (1939), *An Introduction to the Study of Stellar Structure* (Chicago: University of Chicago Press).

Chandrasekhar, S. (1964), *Astrophys. J.* **140**, 417.

Chitre, S.M. (1988), in *Highlights in Gravitation and Cosmology*, edited by B.R. Iyer, A. Kembhavi, J.V. Narlikar, and C.V. Vishveshwara (Cambridge: Cambridge University Press).

Coleman, S. and Weinberg, E. (1973), *Phys. Rev.* **D7**, 1888.

Cowie, L.L. and Hu, E.H. (1987), *Astrophys. J.* **318**, L33.
De, U.K. and Raychaudhuri, A.K. (1968), *Proc. Roy. Soc. London Ser A*, **303**, 97.
de Sitter, W. (1917), *Proc. Akad. Sci. (Amsterdam)* **19**, 1217; **20**, 229.
Dicke, R.H., Roll, P.G., and Krotkov, R. (1964), *Ann. Physics* **28**, 442.
Eddington, A.S. (1935), *Observatory* **58**, 37.
Einstein, A. (1917), *Sitzungsber Akad. Wiss. Preuss.* **142**.
Einstein, A. (1936), *Science* **84**, 306.
Einstein, A. (1939), *Ann. of Math.* **40**, 922.
Eisenhart, L.P. (1925), *Riemannian Geometry* (Princeton: Princeton University Press).
Eisenhart, L.P. (1933), *Continuous Groups of Transformation* (Princeton: Princeton University Press).
Ellis, G.F.R. (1978), *Gen. Relativity Gravitation* **9**, 87.
Ellis, G.F.R. and King, A.R. (1974), *Commun. Math. Phys.* **38**, 119.
Eötvös, R.V., Pekar, D., and Fekete, E. (1922), *Ann. Physik* **68**, 11.
Ernst, F.J. (1968), *Phys. Rev.* **167**, 1175.
Fischback, D., Sudarsky, D., Szafer, A., Talmadge, C., and Anderson, S.H. (1986), *Phys. Rev. Lett.* **56**, 3.
Fomalont, E.B. and Sramek, R.A. (1975), *Astrophys. J.* **199**, 749.
Fomalont, E.B. and Sramek, R.A. (1976), *Phys. Lett.* **36**, 1475.
Forman, W., Jones, C., Cominsky, I., Julien, P., Murry, S., Peters, G., Tananbaum, H., and Gioccani, R. (1978), *Astrophys. J. Suppl.* **38**, 357.
Friedman, A. (1922), *Ann. Phys.* **10**, 377.
Friedman, J.L., Ipser, J.R., and Parker, L. (1984) *Nature* **312**, 255.
Friman, B.L. and Maxwell, O.V. (1979), *Astrophys. J.* **232**, 541.
Ghosh, P. (1984), *J. Astrophys. Astron.* **5**, 307.
Ginzburg, V.L. and Ozemoi, L.M. (1965), *Soviet Phys. JETP* **20**, 689.
Godel, K. (1949), *Rev. Mod. Phys.* **21**, 447.
Goldreich, P. and Julian, W.H. (1969), *Astrophys. J.* **157**, 869.
Gott, J.R. (1985), *Astrophys. J.* **288**, 422.
Greenstein, J.L., Boksenberg, A., Carswell, R., and Shortbridge, K. (1977), *Astrophys. J.* **212**, 186.
Grossman, S.A. and Narayan, R. (1989), *Astrophys. J.* **344**, 637.
Gullahorn, G.E.R. and Rankin, J.M. (1982), *Astrophys. J.* **250**, 520.
Guinan, E.F. and Maloney, F.P. (1985), *Astron. J.* **90**, 1519.
Guth, A. (1981), *Phys. Rev.* **D23**, 347.
Harnden, F.R. Jr. and Seward, F.D. (1984), *Astrophys. J.* **283**, 279.
Harrison, B.K., Thorne, K.S., Wakano, M., and Wheeler, J.A. (1965), *Gravitation Theory and Gravitational Collapse* (Chicago: University of Chicago Press).
Harrison, E. (1970), *Phys. Rev.* **D1**, 2726.
Hartle, J.B. (1971), *Phys. Rev.* **D3**, 2938.
Hartle, J.B. (1972), in *Magic without Magic*, edited by J. Klauder (San Francisco: W.H. Freeman).
Hawking, S.W. (1971), *Monthly Notices Roy. Astronom. Soc.* **152**, 75.
Hawking, S.W. and Ellis, G.F.R. (1973), *The Large Scale Structure of Space Time* (Cambridge: Cambridge University Press).
Hellings, R., Adams, P.J., Anderson, J.D., Keesey, M.S., Lau, E.L., Standish, E.M., Canuto, V.M., and Goldman, I. (1983), *Phys. Rev. Lett.* **51**, 1609.
Hillebrandt, W. (1982), in *Supernovae: A survey of current research*, Proc. NATO Adv. Study Inst. (Dordrecht; Reidel).
Holding, S.C., Stacey, F.D., and Tuck, G.J. (1986), *Phys. Rev.* **D33**, 3487.
Hoyle, F. (1947), *Monthly Notices Roy. Astronom. Soc.* **107**, 231.
Hoyle, F. (1948), *Monthly Notices Roy. Astronom. Soc.* **108**, 372.
Hoyle, F., Narlikar, J.V., and Wheeler, J.A. (1964), *Nature* **203**, 914.
Iwamoto, N. (1980), *Phys. Rev. Lett.* **44**, 1637.
Jackson, J.D. (1975), *Classical Electrodynamics* (New York: Wiley).
Kapahi, V.K. (1975), *Monthly Notices Roy. Astronom. Soc.* **172**, 513.

Kasner, E. (1927), *Amer. J. Math.* **43**, 217.

Kluzniak, W., Ruderman, M., Shaham, J., and Tavani, M. (1988), *Nature* **334**, 225.

Krisher, T.P., Anderson, J.D., and Campbell, J.K. (1990), *Phys. Rev. Lett.* **64**, 1322.

Kuroda, K. and Mio, N. (1990), *Phys. Rev.* **D42**, 3903.

Lamb, D.Q. and Van Horn, H.H. (1975), *Astrophys. J.* **200**, 306.

Landau, L.D. (1932), *Phys. Z. Soviet Union* **1**, 285.

Landau, L.D. and Lifshitz, E.M. (1982), *Quantum Electrodynamics* (New York: Pergamon Press).

Laplace, P.S. (1809), *The System of the World* (London: W. Flint).

Liebes, S. (1964), *Phys. Rev.* **B133**, 835.

Lightman, A.P., Press, W.H., Price, R.M. and Tevkolsky, S.A. (1975), *Problem Book in Relativity and Gravitation* (Princeton: Princeton University Press).

Linde, A. (1982), *Phys. Lett.* **108B**, 389.

Linde, A. (1983), *Phys. Lett.* **129B**, 177.

Lyne, A.G. (1984), *Nature* **310**, 300.

Lyth, D.H. (1985), *Phys. Rev.* **D31**, 1792.

Majumdar, S.D. (1947), *Phys. Rev.* **72**, 390.

Manchester, R.N. and Taylor, J.H. (1977), *Pulsars* (San Francisco: W.H. Freeman).

Mather, J.C. et al. (1990), *Astrophys. J. Lett.* **354**, L37.

Maxon, S. (1972), *Phys. Rev.* **A5**, 1630.

Maxwell, O.V. (1979), *Astrophys. J.* **231**, 201.

Maxwell, O.V., Brown, G.E., Campbell, D.K., Dashen, R.F., and Manassah, J.T. (1977), *Astrophys. J.* **216**, 77.

Mazur, P.O. (1982), *J. Phys.* **A15**, 3173.

McCrea, W.H. (1951), *Proc. Roy. Soc. London* **A206**, 562.

Misner, C.W. and Wheeler, J.A. (1957), *Ann. Phys.* **2**, 525.

Misner, C.W. and Zapolsky, H.S. (1964), *Phys. Rev. Lett.* **12**, 635.

Narasimha, D. and Chitre, S.M. (1988), *Astrophys. J.* **332**, 75.

Narlikar, J.V. and Chitre, S.M. (1977), *Monthly Notices Roy. Astronom. Soc.* **180**, 525.

Nelson, P.G., Graham, D.M., and Newman, R.D. (1990), *Phys. Rev.* **D42**, 963.

Nicoll, J.F. and Segel, I.E. (1975), *Proc. Nat. Acad. Sci. U.S.A.* **72**, 4691.

Nobili, A.M. and Will, C.M. (1986), *Nature* **320**, 39.

Nomoto, K. and Tsuruta, S. (1981), *Astrophys. J.* **250**, L19.

Oppenheimer, J.R. and Snyder, H. (1939), *Phys. Rev.* **56**, 455.

Oppenheimer, J.R. and Volkoff, G.M. (1939), *Phys. Rev.* **55**, 374.

Ostriker, J.P. and Bodenheimer, P. (1968), *Astrophys. J.* **151** 1089.

Ostriker, J.P. and Bodenheimer, P., and Lynden-Bell, D. (1966), *Phys. Rev.* **17**, 816.

Paczynski, B.P. (1987), *Nature* **325**, 572.

Pandharipande, V.R. and Smith, R.A. (1975), *Phys. Lett.* **B59**, 15.

Penrose, R. (1969), *Nuovo Cimento* **1**, 252.

Petrov, A.Z. (1969), *Einstein Spaces* (New York: Pergamon Press).

Pines, D. (1980), *J. Phys. Colloq.* **41**, C2/111.

Podurets, M.A. (1965), *Soviet Astron. A. J.* **8**, 868.

Pound, R.V. and Rebka, G.A. (Jr.) (1960), *Phys. Rev. Lett.* **4**, 337.

Price, R.H. (1972a), *Phys. Rev.* **D5**, 2419.

Price, R.H. (1972b), *Phys. Rev.* **D5**, 2439.

Rainich, G.Y. (1925), *Trans. Ann. Math. Soc.* **27**, 106.

Raychaudhuri, A.K. (1953), *Phys. Rev.* **89**, 417.

Raychaudhuri, A.K. (1955), *Phys. Rev.* **98**, 1123.

Raychaudhuri, A.K. (1979), *Theoretical Cosmology* (Oxford: Clarendon Press).

Raychaudhuri, A.K. and Modak, B. (1988), *Classical Quantum Gravity* **5**, 225.

Raychaudhuri, A.K. and Mukherjee, G. (1984), *Monthly Notices Roy. Astronom. Soc.* **209**, 353.

Robertson, D.S. and Carter, W.E. (1984), *Nature* **310**, 572.

Rubin, V.C., Ford, W.K., and Rubin, J.S. (1976), *Astrophys. J.* **183**, L111.

Salpeter, E.E. (1961), *Astrophys. J.* **134**, 669.

Saslaw, W.C., Narasimha, D., and Chitre, S.M. (1985), *Astron. Astrophys.* **172**, L14.

Sato, K. (1981), *Phys. Lett.* **99B**, 66.

Senovilla, J.M.M. (1990), *Phys. Rev. Lett.* **64**, 2219.

Shepley, L.C. (1969), *Phys. Lett.* **A28**, 695.

Sieber, W. and Wielebinski, R. (1981), *Pulsars—IAU Symposium 95* (Dordrecht: Reidel), Specially, p. 423.

Silk, J. (1992), *Nature,* **356**, 741.

Soucail, G., Fort, B., Mellie, Y., and Picat, J.P. (1987), *Astronom. and Astrophys.* **172**, L14.

Stacey, F.D. and Tuck, G.J. (1981), *Nature,* **292**, 230.

Stacey, F.D., Tuck, G.J., Moore, G.I., Holding, S.C., Goodwin, B.D., and Zhou, R. (1987), *Rev. Mod. Phys.* **59**, 157.

Swarup, G. (1975), *Monthly Notices Roy. Astronom. Soc.* **172**, 501.

Sweeney, M.A. (1976), *Astronom. and Astrophys.* **49**, 375.

Synge, J.L. (1960), *Relativity—the General Theory* (Amsterdam: North-Holland).

Taylor, J.H. and Weisberg, J.M. (1982), *Astrophys. J.* **253**, 908.

Taylor, J.H. and Weisberg, J.M. (1989), *Astrophys. J.* **345**, 434.

Teitelboim, C. (1972), *Nuovo. Cimento Lett.* (11) No. 3, 326.

Tomimatsu, A. and Sato, H. (1972), *Phys. Rev. Lett.* **29**, 1344.

Trümper, J.W., Pietch, W., Reppin, C., Voges, W., Staubert, R., and Kendziorra, E. (1978), *Astrophys. J.* **219**, L105.

Vaidya, P.C. (1951), *Proc. Indian Acad. Sci.* **A33**, 264.

Van den Heuvel, E.P.J. (1984), *J. Astrophys. Astron.* **5**, 209.

Vidal-Madjar, A. and Gry, C. (1984), *Astron. Astrophys.* **280**, 629.

Vosset, R.F.C., Levine, M.W., Mattison, E.M., Blomberg, E.L., Hoffman, T.F., Nystrom, G.V., Farrel, B.F., Decher, P.B., Eby, P.B., Bangher, C.R., Watts, J.W., Teuber, D.L., and Wills, F.D. (1980), *Phys. Rev. Lett.* **45**, 2081.

Wagh, S.M. and Dadhich, N. *Phys. Lett. (Review Section)* (1989), **183**, 137.

Wagoner, R. (1973), *Astrophys. J.* **179**, 343.

Walsh, D., Carswell, R.F., and Weymann, R.J. (1979), *Nature,* **279**, 381.

Wang, Q., Chen, K., Hamilton, T.T., Ruderman, M., and Shaham, J. (1989), *Nature* **338**, 319.

Weber, J. (1961), *Phys. Rev.* **117**, 306.

Will, C.M. (1981), *Theory and Experiment in Gravitational Physics* (Cambridge: Cambridge University Press).

Wolszczan, A., Kulkarni, S.R., Middleditch, J., Backet, D.C., Fruchter, A.S., and Deway, A.J. (1989), *Nature* **337**, 53.

Woodward, J.F. (1984), *Astrophys. J.* **279**, 803.

Zel'dovich, Ya B. (1964), *Soviet Phys. Dokl.* **9**, 195.

Zel'dovich, Ya B. (1972), *Monthly Notices Roy. Astronom. Soc.* **160**, 1.

Bibliography

Accetta, F.S. and Krauss, L.M. (1988), *Cosmic Strings: The Current Status* (Singapore: World Scientific).

Adler, R, Bazin, M., and Schiffer, M. (1975), *Introduction to General Relativity* (New York: McGraw-Hill).

Alvarez, R.D., Tenreiro, J.M., Cabanell, I., and Quirós, M. (1987), *Cosmology and Particle Physics* (Singapore: World Scientific).

Anderson, J.L. (1967), *Principles of Relativity Physics* (New York: Academic Press).

Audouze, J. and Mathieu, N. (1986), *Nucleosynthesis and its Implications* (Dordrecht: Reidel).

Barnes, C.A., Clayton, D.D., and Schramm, D.N. (1982), *Essays in Nuclear Astrophysics* (Cambridge: Cambridge University Press).

Bertotti, B. (ed.) (1973), *Experimental Gravitation—Enrico Fermi School Course 56* (New York: Academic Press).

Bishop, R.L. and Crittenden, R.J. (1964), *Geometry of Manifolds* (New York: Academic Press).

Blades, J.C., Turnshek, D., and Norman, C.A. (1988), *QSO Absorption Lines Probing the Universe* (Cambridge: Cambridge University Press).

Bonnor, W.B., Islam, J.N., and MacCallum, M.A.H. (1984), *Classical General Relativity* (Cambridge: Cambridge University Press).

Boothby, W.M. (1975), *An Introduction to Differentiable Manifolds and Riemannian Geometry* (New York: Academic Press).

Braginsky, V.B. and Manukin, A.B. (1977), *Measurement of Weak Forces in Physics Experiments* (Chicago: University of Chicago Press).

Burbidge, G.R. and Burbidge, E.M. (1967), *Quasistellar Objects* (San Francisco: Freeman).

Carmelie, M. (1982), *Classical Fields—General Relativity and Gauge Theory* (New York: Wiley Interscience).

Chandrasekhar, S. (1967), *An Introduction to the Study of Stellar structure* (New York: Dover).

Cohen, N. (1988), *Gravity's Lens—Views of the New Cosmology* (New York: Wiley).

Demianski, M. (1985), *Relativistic Astrophysics* (New York: Pergamon Press).

De Witt, C. and DeWitt, B. (1964), *Relativity, Groups and Topology* (New York: Gordon & Breach).

Dicke, R.H. (1964), *Theoretical Significance of Experimental Relativity* (New York: Gordon & Breach).

Eddington, A.S. (1963), *The Mathematical Theory of Relativity* (Cambridge: Cambridge University Press).

Eisenhart, L.P. (1960), *Riemannian Geometry* (Princeton: Princeton University Press).

Fang, L.E. and Ruffini, R. (eds.) (1984), *Cosmology of the Early Universe* (Singapore: World Scientific).

Fang, L.E. and Ruffini, R. (eds.). (1985), *Galaxies, Quasars and Cosmology* (Singapore: World Scientific).

Field, G.B., Arp, H., and Bahcall, N. (1973) *The Redshift Controversy* (New York: W.A. Benjamin).

Flanders, H. (1963), *Differential Forms with Applications to the Physical Sciences* (New York: Academic Press).

Fock, V. (1959). *The Theory of Space–Time and Gravitation* (New York: Pergamon Press).

Frank, J., King, A.R., and Raine, D.J. (1985), *Accretion Power in Astrophysics* (Cambridge: Cambridge University Press).

Galeotti, P. and Schramm, D.N. (ed.) (1988), *Gauge Theory and the Early Universe* (Boston: Kluwer Academic).

Galor, B. (1981), *Cosmology, Physics and Philosophy* (New York: Springer-Verlag).

Giacconi, R. (ed.) (1981), *X-ray Astronomy with Einstein Satellite* (Dordrecht: Reidel).

Giacconi, R. and Ruffini, R. (1978), *Physics and Astrophysics of Neutron Stars and Black Holes— Enrico Fermi Course, 65* (Amsterdam: North Holland).

Gibbons, G., Hawking, S.W., and Siklos, S.T.C. (eds.) (1983), *Very Early Universe* (Cambridge: Cambridge University Press).

Göckeler, M. and Schücker, T. (1987), *Differential Geometry, Gauge Theories and Gravity* (Cambridge: Cambridge University Press).

Grünbaum, A. (1978) *Philosophical Problems of Space and Time* (Dordrecht: Reidel).

Gursky, H. and Ruffini, R. (1975), *Neutron Stars, Black Holes and Binary X-ray Sources* (Dordrecht: Reidel).

Halmos, P.R. (1958), *Finite Dimensional Vector Spaces* (New York: Van Nostrand).

Harrison, B.K., Thorne, K.S., Wakano, M., and Wheeler, J.A. (1965), *Gravitation Theory and Gravitational Collapse* (Chicago: University Chicago Press).

Harrison, E.R. (1981), *Cosmology* (Cambridge: Cambridge University Press).

Hawking, S.W. and Ellis, G.F.R. (1973), *The Large Scale Structure of Space Time* (Cambridge: Cambridge University Press).

Hawking, S.W. and Israel, W. (eds.) (1979), *General Relativity: An Einstein Centinary Survey* (Cambridge: Cambridge University Press).

Hawking, S.W. and Israel, W. (eds.) (1987), *Three Hundred Years of Gravitation* (Cambridge: Cambridge University Press).

Held, A. (ed.) (1980), *General Relativity and Gravitation* Vols. 1 and Vol. 2 (New York: Plenum).

Helgason, S. (1978), *Differential Geometry, Lie Groups and Symmetric Space* (New York: Academic Press).

Hewitt, A., Burbidge, G.R., and Fang, L.Z. (1987), *Observational Cosmology* (Dordrecht: Reidel).

Hicks, N.J. (1955), *Notes on Differential Geometry* (New York: Van Nostrand).

Hillebrandt, W., Kuhfuss, R., Miller, E., and Truran, J.W. (eds.) (1987), *Nuclear Astrophysics* (New York: Springer-Verlag).

Höhler, G. (ed.) (1973), *Astrophysics.* Springer Tracts in Modern Physics, No. 69 (New York: Springer-Verlag).

Hoyle, F. (1963), *Frontiers of Astronomy* (Heineman, ELBS).

Infeld, L. and Plebanski, J. (1960), *Motion and Relativity* (New York: Pergamon).

Irvine, J.M. (1978), *Neutron Stars* (Oxford: Oxford University Press).

Israel, W. (ed.) (1970), *Differential Forms in General Relativity* (Comm. Dublin Institute for Advanced Studies, Series A, No. 19).

Israel, W. (ed.) (1973), *Relativity, Astrophysics and Cosmology* (Dordrecht: Reidel).

Iyer, B.R., Kembhavi, A., Narlikar, J.V., and Vishveshwara, C.V. (1985), *Highlights in Gravitation and Cosmology* (Cambridge: Cambridge University Press).

Iyer, B.R., Kembhavi, A., Narlikar, J.V., Mukunda, N., and Vishveshwara, C.V. (1989), *Gravitation, Gauge Theories and the Early Universe* (Netherlands: Kluwar).

Kafatos, M. (ed.) (1988), *Supermassive Black Holes* (Cambridge: Cambridge University Press).

Kafatos, M. and Michalitsianos, A. (1988), *Supernova 1987 A in the Large Magellanic Gloud* (Cambridge: Cambridge University Press).

Kobayashi, S. and Nomizu, K. (1963), *Foundations of Differential Geometry*, Vols. 1 and 2 (New York: Interscience).

Kolb, E.W. and Turner, M.S. (1990), *The Early Universe* (Reading, MA: Addison-Wesley).

Kramer, D., Stephant, H., MacCallium, M., and Herlt, E. (1980), *Exact Solutions in Einstein's Field Equations* (Cambridge: Cambridge University Press).

Kundt, W. (1987), *Astrophysical Jets and their Engines* (Dordrecht: Reidel).

Kundt, W. (1990), *Neutron Stars and their Birth Events* (Kluwer).

Landau, L.D. and Lifshitz, E.M. (1975), *The Classical Theory of Fields* (Oxford: Pergamon).
Levi-Civita, T. (1954), *The Absolute Differential Calculus* (Glasgow: Blackie).
Lewin, W.H.G. and Van den Heuvel, E.P. (1983), *Accretion Driven Stellar X-Ray Sources* (Cambridge: Cambridge University Press).
Lightman, A.P., Press, W.H., Price, R.H., and Teukolsky, S.A. (1975), *Problem Book in Relativity and Gravitation* (Princeton: Princeton University Press).
Linde, A. (1990), *Particle Physics and Inflationary Cosmology* (Harwood Academic, Switzerland).
Lingenfelter, R.E., Hudson, H.S., and Worrall, D.M. (eds.) (1982), *Gamma-ray Transients and Related Astrophysical Phenomena* (American Institute of Physics).
Longair, M.S. (ed.) (1974), *Confrontation of Cosmological Theories with Observational Data* (Dordrecht: Reidel).
Longair, M.S. and Einasto, J. (eds.) (1978), *The Large Scale Structure of the Universe*, IAU Symposium, No. 79 (Dordrecht: Reidel).
Lord, E.A. (1976), *Tensors, Relativity and Cosmology* (New York: Tata, McGraw-Hill).
Manchester, R.N. and Taylor, J.H. (1977), *Pulsars* (San Francisco: Freeman).
Matsushima, Y. (1972), *Differentiable Manifolds* (New York: Marcel Dekker).
McVittie, G.C. (1965), *General Relativity and Cosmology* (London: Chapman and Hall).
Mehra, J. (1975), *Einstein, Hilbert and the Theory of Gravitation* (Dordrecht: Reidel).
Menzel, D.H., Whipple, F.L., and DeVancouteurs, G. (1970), *Survey of the Universe* (Englewood Cliffs, NJ: Prentice-Hall).
Misner, C.W., Thorne, K.S., and Wheeler, J.A. (1973), *Gravitation* (San Francisco: Freeman).
Møller, C. (1960), *The Theory of Relativity* (Oxford: Clarendon Press).
Møller, C. (ed.) (1962), *Evidence for Gravitational Theories—Proc. School Enrico Fermi, Course 20* (New York: Academic Press).
Narlikar, J.V. (1978), *General Relativity and Cosmology* (New York: Macmillan).
Narlikar, J.V. (1983), *Introduction to Cosmology* (Jones and Bartlett).
Narlikar, J.V. and Padmanabhan, T. (1986), *Gravity, Gauge Theory and Cosmology* (Dordrecht: Reidel).
North, J.D. (1965), *The Measure of the Universe* (Oxford: Oxford University Press).
Novilov, I.D. and Frolov, V.P. (1989), *Physics of Black Holes* (The Netherlands: Kluwer).
Pauli, W. (1958), *The Theory of Relativity* (Oxford: Pergamon).
Peebles, P.J.E. (1971), *Physical Cosmology* (Princeton University Press).
Peebles, P.J.E. (1980), *The Large Scale Structure of the Universe* (Princeton: Princeton University Press).
Petrov, A.Z. (1969), *Einstein Spaces* (Oxford: Pergamon).
Raychaudhuri, A.K. (1979), *Theoretical Cosmology* (Oxford: Oxford University Press).
Reichenback, H. (1988), *Philosophy of space and Time* (New York: Dover).
Robertson, H.P. and Noonan, T.W. (1968), *Relativity and Cosmology* (New York: Saunders).
Ryan, M.P. and Shepley, L.C. (1975), *Homogeneous Relativistic Cosmology* (Princeton: Princeton University Press).
Sachs, R.K. (ed.) (1971), *General Relativity and Cosmology—Proc. School Enrico Fermi, Course 47* (New York: Academic Press).
Schouten, J.A. (1954), *Ricci Calculus* (New York: Springer-Verlag).
Schutz, B.F. (1980), *Cosmological Methods of Mathematical Physics* (Cambridge: Cambridge University Press).
Schutz, B.F. (1985), *A First Course in General Relativity* (Cambridge: Cambridge University Press).
Schwarzschild, M. (1958), *Structure and Evolution of the Stars* (New York: Dover).
Sciama, D.W. (1971), *Modern Cosmology* (Cambridge: Cambridge University Press).
Sexl, R. and Sexl, H. (1979), *White Dwarfs–Black Holes: An Introduction to Relativistic Astrophysics* (New York: Academic Press).
Shapiro, S.L. and Teukolsky, S.A. (1983), *Black Holes, White Dwarfs and Neutron Stars* (New York: Wiley).
Sieber, W. and Wielebinoky, W.R. (eds.) (1981), *Pulsars* (Dordrecht: Reidel).

Stephani, H. (1982), *General Relativity: An Introduction to the Theory of Gravitational Field* (Cambridge: Cambridge University Press).

Straumann, N. (1984), *General Relativity and Relativistic Astrophysics* (New York: Springer-Verlag).

Synge, J.L. (1960), *Relativity: The General Theory* (Amsterdam: North-Holland).

Thompson, W.S., Carney, B.W., and Karwowski, H.J. (1990), *Primordial Nucleosynthesis* (Singapore: World Scientific).

Thorne, K.S., Price, R.H., and Macdonald, D.A. (1986), *Black Hole: the Membrane Paradigm* (New Haven, CT: Yale University Press).

Tolman, R.C. (1958), *Relativity, Thermodynamics and Cosmology* (Oxford: Oxford University Press).

Unsöld, A. (1983), *The New Cosmos* (New York: Springer-Verlag).

Wald, R.M. (1984), *General Relativity* (Chicago: University of Chicago Press).

Weedman, D.W. (1986), *Quasar Astronomy* (Cambridge: Cambridge University Press).

Wegner, G. (ed.) (1989), *White Dwarfs* (New York: Springer-Verlag).

Weinberg, S. (1972), *Gravitation and Cosmology* (New York: Wiley).

Weinberg, S. (1977), *The First Three Minutes* (New York: Basic Books).

West, R.M., Davidson, K., and Netzer, H. (eds.) (1983), *Highlights of Astronomy*, Vol. 6 (Dordrecht: Reidel).

Weymann, R.J., Carswell, R.F., and Smith, M.G. (1981), *Ann. Rev. Astron. Astrophys.* **19**, 41.

Will, C. (1981), *Theory and Experiment in Gravitational Physics* (Cambridge: Cambridge University Press).

Zel'dovich, Ya.B. and Nooikov, I.D. (1971, 1983), *Relativistic Astrophysics*, Vols. 1 and 2 (Chicago: University of Chicago Press).

Index

Dispersion measure   145–146
Doppler shift   73, 114, 180–181
Dual tensor   79
Duality rotation   84–85

Eddington, A.S.   131
Eigenvalue   25, 172
Eigenvector   25
Einstein's
    elevator   6
    field equations   35
    tensor   29–30
Electrovac solution   82
Elementary flatness   94
Ellis, G.F.R. and King, A.R.   233
Energy–momentum pseudotensor
    105–108
Energy–momentum tensor   45–47
    of electromagnetic field   79, 84
Entropy   178, 180
Eötvös experiment   5, 8
Equation of state   125–128, 134,
    161–163, 165
Equivalence principle   4–6
Ergosphere   102–103, 179
Ernst, F.J.   96
Euler–Lagrange equation   24
Evolution of stars   138–143, 167
Expansion scalar   229
Exterior product   278

Fermi
    distribution   126
    energy   162
    momentum   127
    surface   154
Field of a charged particle   79
Field equations, Einstein's   35
Fifth force   8
Fischbach, D.   8
Flatness problem   265
Fluctuations, in inflationary model
    270
Frequency shift, of light   71–73

Galaxy correlations   243
Galilean transformation   3
Gamow   235
Gauge
    condition   39
    coupling constant   258

fields   253
    symmetry   253
Geodesic
    coordinates   22
    deviation   32–34, 40–41, 71
    differential equation of   23–24
    null   24, 59, 102, 177, 185
Geodetic precession   66
Glitches   144
Gödel
    metric   99
    universe   231
Grand Unified Theory (GUT)   253
Gravitational radiation   108–111, 114
Gravitational waves   38, 108–111
    detection of   40
    plane   39
Guth, A.   267
Gyroscope   67

Hamiltonian   24
Hamilton–Jacobi method   103
Hamilton's variational principle   23
Hawking, S.W. and Ellis, G.F.R.   233
Hawking, S.W. and Penrose, R.   233
Helicity   40
Helium flash   141
Hewish   147
Higg's mechanism   255
Hilbert   35
Horizon   100–102, 175, 177–178, 196,
    197
Hubble parameter   213
Hulse–Taylor pulsar   114–119, 144,
    153, 158

Index
    lowering of   14, 15
    raising of   14, 15
Inflationary scenario   264
    chaotic   269
    model   269
    old, new   267
Isometry of space   89

Jacobian of transformation   15
Jeans mass   247

Kerr   96–97, 100, 102, 175–176,
    179–180

 # ASTRONOMY AND
ASTROPHYSICS LIBRARY

Continued from page ii

Modern Astrometry
By J. Kovalevsky

Astrophysical Formulae 3rd Edition (2 volumes)
Volume 1: Radiation, Gas Processes, and
  High Energy Astrophysics
Volume 2: Space, Time, Matter, and Cosmology
By K.R. Lang

Observational Astrophysics 2nd Edition
By P. Lena, F. Lebrun, and F. Mignard

Astrophysics of Neutron Stars
Editors: V.M. Lipunov and G. Börner

Galaxy Formation
By M.S. Longair

Supernovae
Editor: A.G. Petschek

General Relativity, Astrophysics, and Cosmology
By A.K. Raychaudhuri, S. Banerji, and A. Banerjee

Tools of Radio Astronomy 3rd Edition
By K. Rohlfs and T.L. Wilson

Atoms in Strong Magnetic Fields
Quantum Mechanical Treatment and Applications in Astrophysics and Quantum Chaos
By H. Ruder, G. Wunner, H. Herold, and F. Geyer

The Stars
By E.L. Schatzman and F. Praderie

Physics of the Galaxy and Interstellar Matter
By H. Scheffler and H. Elsässer

Gravitational Lenses
By P. Schneider, J. Ehlers, and E.E. Falco

Relativity in Astrometry, Celestial Mechanics, and Geodesy
By M.H. Soffel

The Sun
An Introduction
By M. Stix

Galactic and Extragalactic Radio Astronomy 2nd Edition
Editors: G.L. Verschuur and K.I. Kellermann

Reflecting Telescope Optics (2 volumes)
Volume I: Basic Design Theory and Its Historical Development
Volume II: Manufacture, Testing, Alignment, Modern Techniques
By R.N. Wilson

Tools of Radio Astronomy
Problems and Solutions
By T.L. Wilson and S. Hüttemeister